Hedgehog Signaling Protocols

METHODS IN MOLECULAR BIOLOGY™

John M. Walker, SERIES EDITOR

METHODS IN MOLECULAR BIOLOGY™

Hedgehog Signaling Protocols

Edited by

Jamila I. Horabin

Associate Professor
Department of Biomedical Sciences
College of Medicine
Florida State University
Tallahassee, Florida, USA

HUMANA PRESS ✳ TOTOWA, NEW JERSEY

This publication is printed on acid-free paper. ∞

ANSI Z39.48-1984 (American Standards Institute)

Permanence of Paper for Printed Library Materials.

Cover illustration: Background and bottom left inset: Provided by Dr. Sudipto Roy: One day old transgenic zebrafish embryos expressing a GFP reporter in the midline (notochord, floor plate and ventral brain tissues) under the control of the regulatory elements of the *sonic hedgehog* gene. Transgenic strain generated by F. Muller and image captured by S. Roy.

Top left inset: *Shh* expression in the nervous system, limb bud and branchial arches in Stage 21 chick embryo. *See* Chapter 2 by Eva Tiecke and Cheryll Tickle.

Central image of inset: 10 day chick wing with a digit pattern of 432234 after implanting a Shh-soaked bead. *See* Chapter 2 by Eva Tiecke and Cheryll Tickle.

Central upper image of inset: Stage 48 wild type Xenopus embryo. *See* Chapter 3 by Thomas Hollemann, Emmanuel Tadjuidje, Katja Koebernick & Tomas Pieler.

Top right inset: Dorsal head view of rat telencephalon marked for neural explant assay. *See* Chapter 4 by Rina Mady and Jhumku Khotz.

Bottom right inset: is an unpublished one of mine: Drosophila third instar wing imaginal disc stained for Wingless (green) and Cubitus interruptus (red). Provided by Jamila I. Horabin.

Production Editor: Rhukea J. Hussain
Cover design by Karen Schulz

For additional copies, pricing for bulk purchases, and/or information about other Humana titles, contact Humana at the above address or at any of the following numbers: Tel.: 973-256-1699; Fax: 973-256-8341; E-mail: orders@humanapr.com; or visit our Website: www.humanapress.com

Photocopy Authorization Policy:

Authorization to photocopy items for internal or personal use, or the internal or personal use of specific clients, is granted by Humana Press Inc., provided that the base fee of US $30.00 per copy is paid directly to the Copyright Clearance Center at 222 Rosewood Drive, Danvers, MA 01923. For those organizations that have been granted a photocopy license from the CCC, a separate system of payment has been arranged and is acceptable to Humana Press Inc. The fee code for users of the Transactional Reporting Service is: [978-1-58829-692-4/07 $30.00].

Printed in the United States of America. 10 9 8 7 6 5 4 3 2 1

eISBN: 978-1-59745-516-9

Library of Congress Control Number: 2007932488

Preface

When the *Hedgehog* gene was first described by Nusslein-Volhard and Wieschaus in their seminal paper of 1980 (*Nature* **287,** 795), it was one of many, identified because it affected patterning of the *Drosophila* embryo in a very specific way. Almost three decades and many experiments later, we have a good grasp of what their mutants were revealing. The story is not complete, but our understanding of developmental molecules and their mode of action is far beyond that of early embryologists who, with their insightful experiments told of morphogens (substances that regulate cell fates in relation to their concentration), but were without the technology and reagents to reveal them. From the gene collections of Nusslein-Volhard and Wieschaus as well as others who followed in their footsteps, we now know enough to appreciate the complexities of the Hedgehog (Hh)-signaling pathway and its importance in determining appendage and tissue types. Remarkably, the pathway is relatively well conserved across the animal kingdom, and utilized in very synonymous ways. We have also come to recognize that when functioning abnormally, the pathway can lead to various diseases and cancers, highlighting the importance of analyzing and understanding developmental signaling cascades.

This book is intended for Molecular Biologists, Geneticists, and Biochemists interested in manipulating and analyzing the Hh-signaling pathway. In the first half, it covers manipulating the Hh system in vertebrates, followed by a series of chapters describing various molecular and genetic tools available to the *Drosophila* experimentalist. The book winds down with chapters describing some biochemical approaches, done with *Drosophila* cells but the methods should be applicable to other cell types. Last, but not least, a chapter describing how to use sequence analyses to study the evolutionary history and determine functional conservation of Hh expression is included.

We begin with the chapter by Baker, Taylor, and Pepinsky, who describe how to purify the human and rat N-terminal signaling fragments of Sonic Hh (ShhN) from bacterial and insect cells. The ShhN protein is particularly sensitive to metal ion-induced oxidation, and the methods are devised to minimize this oxidation. As Hh is naturally modified in vivo, the authors also describe how to prepare *E. coli*-expressed human ShhN which has been modified at the N-terminus with various fatty acyl moieties.

Chapter 2 by Tiecke and Tickle discusses how to use purified Hh and apply it to chick wing buds to examine the resulting developmental effects. Varying the developmental timing and position of application uncovers different effects of Hh on tissues. To manipulate *Xenopus* development, Hollemann, Tadjuidje, Koebernick, and Pieler describe in Chapter 3 microinjection techniques for mRNA as well as inhibitors of Hh signaling. In Chapter 4, Mady and Kohtz describe a more accurate and reproducible use of rat neural explants to score the effects of Hh, a method which allowed them to demonstrate the induction and differentiation of unspecified neuronal progenitors.

These vertebrate systems provide good proof for the functional conservation of Hh and its homologs. With the Zebrafish genetic analyses can also be performed, and in Chapter 5 Roy describes a rapid and convenient assay that can be used to distinguish effects of loss of function or gain of function genes that affect Hh pathway activation during embryogenesis. To more rapidly analyze the effects of various alterations of the Hh system, cell culture experiments provide one of the best opportunities and in Chapter 6 Kasper, Regl, Eichberger, Frischauf, and Aberger describe how to use retroviral systems to introduce desired changes of Hh components in both dividing and quiescent mammalian cells. Detmer and Garner add to this with their use of flow cytometry in Chapter 7, which when combined with relevant signaling and differentiation markers can be powerful for isolating the population of Hh affected cells one wishes to analyze.

Chapters 8 to 15 describe some of the techniques used by those working with *Drosophila*. Callejo, Quijada, and Guerrero use green fluorescent protein tagged Hh in Chapter 8 to demonstrate in vivo immunocytochemistry techniques which analyze the extracellular distribution and intracellular trafficking of Hh. Chapter 9 by Gallet and Thérond gives an account of using the Confocal microscope to analyze fluorescent protein as well as *in situ* hybridization signals.

Although genetic analysis is the stronghold of *Drosophila*, the power of RNAi as a tool for regulating gene expression cannot be overlooked. Marois and Eaton in Chapter 10 describe a vector for temporally and spatially controlling expression of RNAi substrates in *Drosophila*. The mainstay in manipulating *Drosophila,* however, still relies on genetic analysis and with genes that are lethal, clonal analysis is indispensable. The generation of mutant clones to identify maternally acting genes in Hh signaling is described by Selva and Stronach in Chapter 11, while in Chapter 12 Bankers and Hooper walk you through the considerations for making somatic clones, how to induce them, and how to prepare and analyze the tissues, clones, and phenotypes. Chapter 13 by Busson and Pret is a tour de force compilation of the available GAL4/UAS reagents for the targeted expression of Hh pathway components and their variants. These allow tests of signaling in different cell types with varied developmental timing, and continue to provide insights into mechanism and function.

The next two chapters go into the realm of Biochemistry, with Stegman and Robbins in Chapter 14 describing biochemical fractionations and how to begin characterizing the proteins in the resulting fractions. Tong and Jiang in Chapter 15 give a detailed account of how to perform immunoprecipitations from cultured cells, imaginal discs, and embryos. Both chapters also introduce culture and transfection procedures for two commonly used *Drosophila* cell lines.

Chapter 16, the closing chapter, by Müller and Borycki reminds us of evolutionary context. They describe the use of sequence alignment to build and analyze phylogenetic trees, and search tools for phylogenetic footprinting and transcription factor-binding sites to characterize *cis*-regulatory elements of developmental genes and *hh*.

Each of the chapters give fairly detailed descriptions, including internet resources where relevant, and pointers for success. For one inexperienced in manipulating the Hh system, they should offer a valuable resource. Additionally, because the techniques are generally utilized, we hope this book will serve the broader function of explaining techniques that, with small modifications, are applicable to other signaling systems.

Jamila I. Horabin

Contents

Contributors

FRITZ ABERGER • *Department of Molecular Biology, University of Salzburg, Hellbrunnerstrasse, Salzburg, Austria*

DARREN P. BAKER • *Biogen Idec, Inc., 14 Cambridge Center, Cambridge, MA, USA*

CHRISTINE M. BANKERS • *Program in Molecular Biology, University of Colorado Health Sciences Center, Aurora, CO, USA*

ANNE-GAELLE BORYCKI • *University of Sheffield, Centre for Developmental Genetics, Department of Biomedical Science, Sheffield, UK*

DENISE BUSSON • *Institut Jacques Monod (UMR7592-CNRS/Universités Paris VI et VII), Tour 42 5éme étage 2-4, Place Jussieu Paris Cedex, France*

AINHOA CALLEJO • *Centro de Biología Molecular "Severo Ochoa" (C.S.I.C.-UAM), Universidad Autónoma de Madrid, Cantoblanco, Madrid, Spain*

KRISTINA DETMER • *Division of Basic Medical Sciences, Mercer University School of Medicine, Macon, USA*

SUZANNE EATON • *MPI for Molecular Cell Biology and Genetics, Pfotenhauerstrasse, Dresden, Germany*

THOMAS EICHBERGER • *Department of Molecular Biology, University of Salzburg, Hellbrunnerstrasse, Salzburg, Austria*

ANNA-MARIA FRISCHAUF • *Department of Molecular Biology, University of Salzburg, Hellbrunnerstrasse, Salzburg, Austria*

ARMEL GALLET • *EMBO YIP and CNRS ATIPE, ISDBC, Centre de Biochimie, Université de Nice Sophia Antipolis, Parc Valrose, France*

RONALD E. GARNER • *Division of Basic Medical Sciences, Mercer University School of Medicine, St., Macon, USA*

ISABEL GUERRERO • *Centro de Biología Molecular "Severo Ochoa" (C.S.I.C.-UAM), Universidad Autónoma de Madrid, Cantoblanco, Madrid, Spain*

THOMAS HOLLEMANN • *Institut für Physiologische Chemie, Hollystr. Halle, Germany*

JOAN E. HOOPER • *Program in Molecular Biology and Program in Cell and Developmental Biology, University of Colorado Health Sciences Center, Aurora, CO, USA*

JIN JIANG • *Center for Developmental Biology, Department of Pharmacology, University of Texas Southwestern Medical Center, Dallas, USA*

MARIA KASPER • *Department of Molecular Biology, University of Salzburg, Hellbrunnerstrasse, Salzburg, Austria*

KATJA KOEBERNICK • *Zentrum Biochemie und Molekulare Zellbiologie, Justus-von-Liebig-Weg, Göttingen, Germany*

xi

JHUMKU D. KOHTZ • *Program in Neurobiology and Department of Pediatrics, Children's Memorial Hospital and Feinberg School of Medicine, Northwestern University, N. Halsted, Chicago, USA*

RINA MADY • *Program in Neurobiology and Department of Pediatrics, Children's Memorial Hospital and Feinberg School of Medicine, Northwestern University, N. Halsted, Chicago, IL, USA*

ERIC MAROIS • *Institute of Molecular and Cellular Biology, Strasbourg Cedex, France*

FERENC MÜLLER • *Institute of Toxicology and Genetics, Forschungszentrum, Karlsruhe. Postfach. Karlsruhe, Germany*

R. BLAKE PEPINSKY • *Biogen Idec, Inc., Cambridge, MA, USA*

TOMAS PIELER • *Zentrum Biochemie und Molekulare Zellbiologie, Justus-von-Liebig-Weg, Göttingen, Germany*

ANNE-MARIE PRET • *Centre de Génétique Moléculaire (UPR2167-CNRS associé à l'Université Paris VI), de la Terrasse Gif sur Yvette, France*

LUIS QUIJADA • *Centro de Biología Molecular "Severo Ochoa" (C.S.I.C.-UAM) Universidad Autónoma de Madrid, Cantoblanco, Madrid, Spain*

GERHARD REGL • *Department of Molecular Biology, University of Salzburg, Hellbrunnerstrasse, Salzburg, Austria*

DAVID ROBBINS • *Department of Pharmacology and Toxicology, Dartmouth Medical School, Hanover, USA*

SUDIPTO ROY • *Institute of Molecular and Cell Biology, Proteos, Singapore*

ERICA M. SELVA • *Department of Biological Sciences, University of Delaware, Newark, USA*

MELANIE STEGMAN • *Department of Microbiology and Immunology, Cornell Weill Medical College, New York, USA*

BETH E. STRONACH • *Department of Biological Sciences, University of Pittsburgh, Pittsburgh, PA, USA*

EMMANUEL TADJUIDJE • *Division of Developmental Biology, CCRF, Cincinnati, OH, USA*

FREDERICK R. TAYLOR • *Biogen Idec, Inc., 14 Cambridge Center, Cambridge, MA, USA*

PASCAL P. THÉROND • *EMBO YIP and CNRS ATIPE, ISDBC, Centre de Biochimie, Université de Nice Sophia Antipolis, Parc Valrose, France*

CHERYLL TICKLE • *Division of Cell and Developmental Biology, School of Life Sciences, University of Dundee, Dundee DD1, UK*

EVA TIECKE • *Division of Cell and Developmental Biology, School of Life Sciences, University of Dundee, Dundee DD1, UK*

CHAO TONG • *Center for Developmental Biology, University of Texas Southwestern Medical Center, Dallas, TX, USA*

Overview

Overview of Hedgehog Signaling

Hedgehog (Hh) is a secreted protein that patterns and specifies cell fate in several different tissues during the development of both vertebrate and invertebrate animals. It generally acts as a morphogen, patterning in a concentration dependent manner (reviewed in *1–3*). Defects or misregulation of Hh signaling can lead to cancer, diseases and congenital defects such as basal cell carcinomas, holoprosencephaly, cyclopia and skeletal malformations to name a few (for a more complete description *see 4,5*).

The active Hh protein is synthesized as a precursor and undergoes auto-catalytic cleavage. It is additionally modified at both its amino and carboxy termini by palmitoyl and cholesterol adducts, respectively (**Fig. 1**). These modifications not only alter the activity of Hh, but affect its properties, influencing signal strength and range of effect.

Many of the initial aspects of Hh signaling came to light from work on *Drosophila*, and analyses on vertebrate systems have shown that the pathway is relatively well conserved. Recent discoveries have also uncovered differences. For simplicity, we shall first discuss the system in *Drosophila* and use this framework to highlight some of the divergences in the pathway.

The Hh Pathway in *Drosophila*

In *Drosophila*, response to Hh is mediated by Cubitus interruptus (Ci), a zinc finger transcription factor with both activator and repressor activities. Depending on the presence or absence of the Hh ligand, Ci is processed into either an activator or repressor. These fates of Ci are controlled by two membrane proteins, Patched (Ptc) a twelve-pass transmembrane protein, and Smoothened (Smo), a seven-pass transmembrane protein (*see* reviews *2,3* and references therein). In the absence of Hh, Ptc suppresses Smo and this triggers the events that lead to the proteolysis of Ci to its repressor form, a 75 kDa isoform. Hh relieves the Ptc-mediated suppression of Smo preventing the proteolysis of Ci to result in the activator form, the full length 155 kDa isoform.

The processing of Ci is achieved through a complex of Ci with the cytoplasmic components of the pathway, known members of which are Costal-2 (Cos2), Fused (Fu) and Suppressor of Fused [Su(fu)]. Cos2 has sequence similarity to the motor domain of kinesin, Fu appears to be a serine threonine kinase, while Su(fu) shows no homology to any known protein. The Hh cytoplamsic complex is tethered to the Smo cytoplasmic tail by Cos2 (*6*).

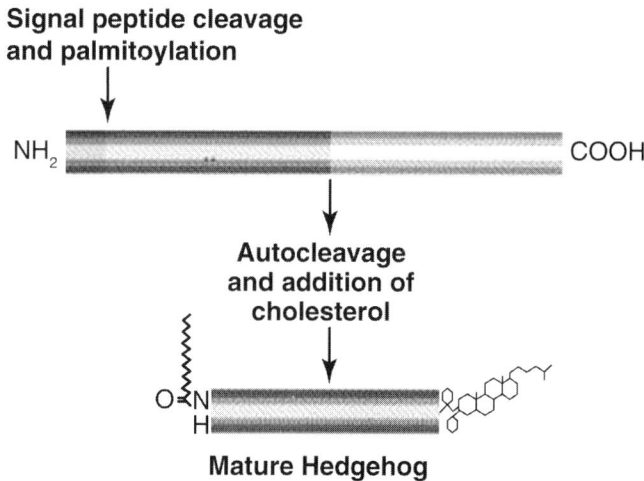

Fig. 1. Processing and modification of the Hh ligand. SP at the amino end of the protein represents the Signal Peptide which is removed.

Integral to regulating the processing of Ci is its phosphorylation by Protein kinase A (PKA), as well as Glycogen synthase kinase-3β and Casein kinase I (CKI). These kinases appear to use Cos2 as a scaffold (7) and promote the processing of Ci to the 75 kDa repressor, with the activity of the F-box protein, Slimb, a component of the SCF ubiquitin ligase complex. Hh reduces the phosphorylation of Ci by PKA which prevents its proteolysis, releases and activates it to result in full length Ci in the nucleus. Hh also promotes phosphorylation of Smo by PKA and CKI, activating it and increasing its levels at the plasma membrane **(Fig. 2)**.

Depending on the level of Ci activation, different downstream Hh targets are turned on. These include *wingless, decapentaplegic* and *ptc,* potent molecules which themselves direct cell fates and developmental processes. The upregulation of Ptc is one of many intriguing aspects of the effects of Hh, because Ptc binds to and limits the spreading of Hh. Elevating Ptc levels results in Hh shaping its own gradient and activity.

Other proteins also modulate the stability and spreading of Hh. Dispatched, a protein with homology to Ptc, appears to be dedicated to the release of Hh from secreting cells (8). The heparin sulfate proteoglycans (HSPGs), molecules which cover the cell surface, affect the spreading of Hh from producing cells as well as its receptor binding in receiving cells (reviewed in 9). The Wnt inhibitory factor-1 protein (WIF-1) also appears to regulate the specificity of Hh binding (10).

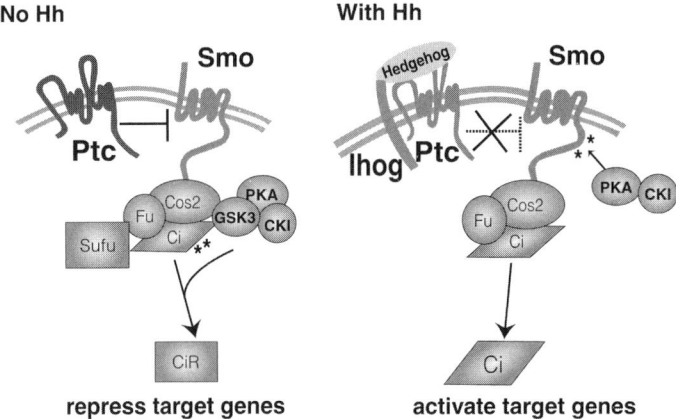

No Hh **With Hh**

repress target genes activate target genes

Fig. 2. Simplified depiction of Hh signaling. Without Hh, Ptc represses Smo and the Hh cytoplasmic complex in conjunction with the kinases, PKA, CKI and GSK3 which phosphorylate Ci (asterisks), lead to formation of the repressor form (CiR). Hh binds to Ptc and its co-receptor Ihog (and Boi, not shown) to relieve the inhibition of Smo by Ptc. This prevents the proteolysis of Ci which translocates to the nucleus and, with the transcriptional co-activator CREB, activates transcription. Smo is phosphorylated (asterisks) in the process. See text for additional details and abbreviations.

Completing the complex network of proteins that support both generation of the Hh ligand and its reception in receiving cells, are the recently described Ptc co-receptors, Ihog (interference hedgehog) and Boi (brother of Ihog). These transmembrane proteins bind to Hh and influence the ability of receiving cells to transduce the signal *(11)*.

Conservation and Divergences of the Hedgehog Pathway

Vertebrate orthologs of almost all the *Drosophila* Hh signaling proteins have been identified, including those that mature the ligand. In many cases they function similarly, but there frequently are multiple forms with each contributing differentially in their respective roles e.g. mammals have three Hh forms: Sonic (Shh), Indian and Desert, but Shh has the greatest scope of activity. The Ci counterpart is represented by three forms of Gli protein, Gli1, Gli2 and Gli3. The latter two are the primary targets of Hh signaling. In the absence of Hh, Gli3 is processed and appears to provide most of the repressor activity, the counterpart of Ci75 in *Drosophila*. The presence of Hh prevents the processing of Gli3 to the repressor form and activates Gli2 to provide the transcription activator function reminiscent of *Drosophila* Ci155 (*see* reviews *2,12,13* and references therein).

Molecules such as Dispatched, the Ptc co-receptors, Ihog and Boi (Boc and Cdo; *14,15*), and WIF-1 (Shifted) also have their vertebrate counterparts suggesting that many of the steps up to the binding of Hh to its receptor, including the effects of the HSPGs, are conserved. Mammals also have additional Hh binding proteins, such as Hip1 (Hh interacting protein), that bind to secreted Hh and shape the gradient, for which a *Drosophila* counterpart is not known.

The major difference, however, appears to involve Smo and the events downstream of the membrane components, with respect to the role of Cos2, the Hh cytoplasmic complex and the activation of Gli. The closest mammalian orthologs to *Drosophila* Cos2, Kif27 and Kif7 (kinesin family), do not affect Shh signaling *(16)*. Rather, it is Su(fu) that plays a role more akin to Cos2, unlike in *Drosophila* where Su(fu) is essentially dispensable *(17)*. Consistent with this functional difference, the tail of *Drosophila* Smo, where Cos2 binds, is much longer and only the residues near the last membrane spanning region are conserved between the vertebrate and fly counterparts. Note that in the Zebrafish, Kif7 a Cos2 like protein, is functionally similar to the *Drosophila* protein *(18)*.

This difference in Smo and the utilization of Cos2 may be due to the reliance on cilia and their intraflagellar transport (IFT) proteins for mammalian Hh signaling (*19* and references cited). Recent data suggest that in the presence of Hh, Smo is transported to the tip of the cilium where it activates Gli2 and prevents the processing of Gli3 to its repressor form. Gli2 is then subsequently transported down the cilium and then to the nucleus where it activates Hh targets. In the absence of Hh, Ptc appears to inhibit Smo from entering the cilium so inhibiting the activation process; an effect that is in principle similar to *Drosophila* but mechanistically different. Many vertebrate cells that respond to Hh signaling have cilia, suggesting this may be the normal setting.

Mammals also use the G-protein-coupled receptor kinase 2 to phosphorylate the Smo tail on activation of the pathway and β-arrestin 2, which binds to phosphorylated Smo, to activate the Hh signal. Intriguingly, Zebrafish with a Cos2-like counterpart and reliance on a β-arrestin 2, would appear to have a Hh signaling system that is intermediate between flies and mammals. There are no data indicating whether or not Zebrafish require IFT proteins and cilia for Hh signaling.

The conservation and differences in Hh signaling highlight the need for further analyses. We hope the following chapters facilitate the process and bring out more of the fascinating twists and turns, still to be appreciated.

References

1. McMahon, A. P., Ingham, P. W., and Tabin, C. J. (2003) Developmental roles and clinical significance of hedgehog signaling. *Curr. Topics Dev. Biol.* **53**, 1–114.
2. Hooper, J. E. and Scott, M. P. (2005) Communicating with Hedgehogs. *Nature Rev. Mol. Cell Biol.* **6**, 306–317.

3. Torroja, C., Gorfinkiel, N., and Guerrero, I. (2005) Mechanisms of Hedgehog gradient formation and interpretation. *J. Neurobiol.* **64,** 334–356.

4. Ingham, P. W. and McMahon, A. P. (2001) Hedgehog signaling in animal development: paradigms and principles. *Genes & Dev.* **15,** 3059–3087.

5. Pasca di Magliano, M., and Hebrok, M. (2003) Hedgehog signalling in cancer formation and maintenance. *Nature Rev. Cancer* **3,** 903–911.

6. Ogden, S. K., Ascano, M. Jr., Stegman, M. A., Suber, L. M., Hooper, J. E., and Robbins, D. J. (2003) Identification of a functional interaction between the transmembrane protein Smoothened and the kinesin-related protein Costal2. *Curr. Biol.* **13,** 1998–2003.

7. Zhang, W., Zhao, Y., Tong, C., et al. (2005) Hedgehog-regulated Costal2-kinase complexes control phosphorylation and proteolytic processing of Cubitus interruptus. *Dev. Cell* **8,** 267–278.

8. Burke, R., Nellen, D., Bellotto, M., et al. (1999) Dispatched, a novel sterol-sensing domain protein dedicated to the release of cholesterol-modified hedgehog from signaling cells. *Cell* **99,** 803–815.

9. Hacker, U., Nybakken, K., and Perrimon, N. (2005) Heparan sulphate proteoglycans: the sweet side of development. *Nature Rev. Mol. Cell Biol.* **6,** 530–541.

10. Gorfinkiel, N., Sierra, J., Callejo, A., Ibanez, C., and Guerrero, I. (2005) The Drosophila ortholog of the human Wnt inhibitor factor Shifted controls the diffusion of lipid-modified Hedgehog. *Dev. Cell* **8,** 241–253.

11. Yao, S. Lum, L., and Beachy, P. A. (2006) The Ihog cell-surface proteins bind Hedgehog and mediate pathway activation. *Cell* **125,** 343–357.

12. Huangfu, D. and Anderson, K. V. (2006) Signaling from Smo to Ci/Gli: conservation and divergence of Hedgehog pathways from Drosophila to vertebrates. *Development* **133,** 3–14.

13. Osterlund, T. and Kogerman, P. (2006) Hedgehog signalling: how to get from Smo to Ci and Gli. *Trends Cell Biol.* **16,** 176–180.

14. Zhang, W., Kang, J. S., Cole, F., Yi, M. J., and Krauss, R. S. (2006) Cdo functions at multiple points in the Sonic Hedgehog pathway, and Cdo-deficient mice accurately model human holoprosencephaly. *Dev. Cell* **10,** 657–665.

15. Tenzen, T., Allen, B. L., Cole, F., Kang, J. S., Krauss, R. S. and McMahon, A. P. (2006) The cell surface membrane proteins Cdo and Boc are components and targets of the Hedgehog signaling pathway and feedback network in mice. *Dev. Cell* **10,** 647–656.

16. Varjosalo, M., Li, S. P., and Taipale, J. (2006) Divergence of hedgehog signal transduction mechanism between Drosophila and mammals. *Dev. Cell* **10,** 177–186.

17. Svard, J., Henricson, K. H., Persson-Lek, M., et al. (2006) Genetic elimination of Suppressor of fused reveals an essential repressor function in the mammalian Hedgehog signaling pathway. *Dev. Cell* **10,** 187–197.

18. Tay, S. Y., Ingham, P. W., and Roy, S. (2005) A homologue of the Drosophila kinesin-like protein Costal2 regulates Hedgehog signal transduction in the vertebrate embryo. *Development* **132,** 625–634.

19. Scholey, J. M. and Anderson, K. V. (2006) Intraflagellar transport and cilium-based signaling. *Cell* **125,** 439–442.

1

Purifying the Hedgehog Protein and its Variants

Darren P. Baker, Frederick R. Taylor, and R. Blake Pepinsky

Abstract

The purification of recombinant versions of the N-terminal signaling fragment of Sonic hedgehog (ShhN) from *E. coli*, Hi-5™ insect cells, yeast, and mammalian cell sources reveals diverse post-translational modifications that affect the potency of the purified protein. Modifications to the N-terminal cysteine with fatty acyl groups results in significant increases in potency, up to 100-fold, when compared with the unmodified protein. Proteolytic clipping at sites near the N-terminus results in inactivation of signaling activity. The ShhN protein is particularly sensitive to metal ion-induced oxidation, and the methods described here were developed to minimize this oxidation. The purification methods developed for ShhN were applicable to human Indian and Desert hedgehog N-terminal signaling proteins, and therefore should be useful for hedgehog proteins from other species.

Key Words: Hedgehog; Sonic hedgehog; ShhN; cholesterol-modified; fatty-acylated.

1. Introduction

Hedgehog proteins constitute a family of extracellular signaling molecules that are involved in the regulation of invertebrate and vertebrate embryo development. Vertebrate organisms express multiple forms of hedgehog, and in mammals three homologs, Sonic hedgehog (Shh), Indian hedgehog (Ihh), and Desert hedgehog (Dhh), have been identified *(1,2)*. Shh is synthesized as a 45 kDa precursor protein that is cleaved autocatalytically to yield a 20 kDa N-terminal fragment (ShhN, amino acid residues 24–197 in the human gene sequence) with a palmitoyl group attached to the α-amine of the N-terminal cysteine (Cys-24) and a cholesterol molecule attached to the C-terminal glycine (Gly-197), and a 25 kDa C-terminal fragment that is responsible for peptide bond cleavage and for catalyzing the addition of the cholesterol *(3–7)*. ShhN is responsible for all known Shh-dependent signaling activities. Ihh *(8)*, and Dhh are processed similarly although less is known about the post-translational modifications of these proteins.

From: *Methods in Molecular Biology: Hedgehog Signaling Protocols*
Edited by: J. Horabin © Humana Press Inc., Totowa, NJ

In this chapter, we describe methods for the purification and characterization of human and rat ShhN, expressed from either a gene construct that encodes the N-terminal fragment alone (human ShhN expressed in *E. coli* and rat ShhN expressed in Hi-5™ insect cells), or from the full-length gene construct that encodes both the N- and C-terminal fragments (human Shh expressed in Hi-5 cells). We also describe methods for the preparation and purification of *E. coli*-expressed human ShhN modified at the N-terminus with various fatty acyl moieties. Modification of the N-terminal fragment of human ShhN with these hydrophobic groups has been shown to significantly improve the in vitro and ex vivo potency when compared with the unmodified protein *(9,10)*.

The purified N-terminal fragments of human and rat Shh serve as useful tools for understanding the mechanism(s) by which hedgehog proteins function, and have been used in both in vitro and in vivo studies *(10–14)*. However, in vivo, the unmodified protein has weak activity and a short half-life, and the fatty acid-modified forms have limited solubility as well as a short half-life. Engineering longer half-life forms with improved in vivo efficacy by PEGylation *(15)*, or by fusion to IgG Fc domains *(16)* has been accomplished. In addition, highly soluble forms with enhanced potency have been prepared in which hydrophobic amino acids replace the N-terminal cysteine residue *(9)*.

2. Materials

2.1. Enzyme

Calf-intestine enterokinase (Biozyme Laboratories International Ltd, San Diego, CA).

2.2. Chromatography Resins

1. SP-Sepharose fast flow resin (GE Healthcare, Piscataway, NJ).
2. Phenyl Sepharose (high sub) fast flow resin (GE Healthcare).
3. NTA-Ni^{2+} agarose (Qiagen, Valencia, CA).
4. Cyanogen bromide (CNBr)-activated Sepharose 4B resin (Sigma, St Louis, MO).
5. Pre-packed Bio-Scale S10 column (Bio-Rad, Hercules, CA).

2.3. DNA Constructs, Cell Lines, and Media

1. cDNA for full-length human Shh, as a 1.6 kb *Eco*RI fragment subcloned into pBluescript SK$^+$ *(17)* (provided by Curis, Inc., Cambridge, MA).
2. Purified anti-ShhN monoclonal antibody 5E1: from conditioned culture medium *(17)* (cell line provided by Curis, Inc.) using protein A sepharose, followed by dialysis against 20 m*M* sodium phosphate (pH 7.2), 150 m*M* NaCl (buffer NN below). Aliquot and store at –70°C.
3. Mouse embryonic fibroblast cell line C3H10T1/2 (American type culture collection).
4. Hi-5 insect cells (Invitrogen, Carlsbad, CA).
5. Sf-900 II serum-free medium (Invitrogen).

6. TB-MGB medium: 1% (w/v) tryptone, 42.3 mM dibasic sodium phosphate, 22 mM monobasic potassium phosphate, 18.7 mM ammonium chloride, 94.2 mM sodium chloride, 1 mM magnesium sulfate, 0.4% (w/v) glucose. Autoclave 10g tryptone and 5g Nacl in 890 mL of water. Add after autoclaving: 100 mL 10× M9 salts, 1 mL of 1 M magnesium sulfate, 10 mL of 40% glucose. 10× M9 salts: 423 mM sodium phosphate dibasic, 220 mM potassium phosphate monobasic, 187 mM ammonium chloride, and 86 mM sodium chloride.

2.4. Buffers and Non-Buffering Solutions (see Note 1)

All buffers and non-buffering solutions described in this chapter are given below. In a few cases, a buffer is used for the purification of more than one form of ShhN. In these cases, the composition of the buffer is given only in the section where it is first described, and cited in subsequent sections using the alphabetical nomenclature, e.g., buffers A, B, C, AA, BB, CC, etc. Therefore, the reader is advised to read the appropriate section prior to beginning a purification to ensure that all required buffers and solutions are prepared.

2.4.1. Buffers and Solutions for the Purification of Human ShhN Expressed in E. coli from a Construct Encoding the N-terminal Fragment

1. Buffer A: 25 mM sodium phosphate (pH 8.0), 150 mM NaCl, 1 mM EDTA, 0.5 mM dithiothreitol (DTT), and 1 mM phenylmethylsulfonylfluoride (PMSF) (Sigma, St Louis, MO).
2. Buffer B: 25 mM sodium phosphate (pH 5.5), 150 mM NaCl, and 0.5 mM DTT.
3. Buffer C: 0.5 M Mes (pH 5.0).
4. Buffer D: 25 mM sodium phosphate (pH 5.5), 300 mM NaCl, and 0.5 mM DTT.
5. Buffer E: 25 mM sodium phosphate (pH 5.5), 400 mM NaCl, and 0.5 mM DTT.
6. Buffer F: 25 mM sodium phosphate (pH 5.5), 800 mM NaCl, and 0.5 mM DTT.
7. Buffer G: 25 mM sodium phosphate (pH 8.0), 1 M NaCl, 20 mM imidazole, and 0.5 mM DTT.
8. Buffer H: 1 M imidazole (pH 7.0).
9. Buffer I: 1 M sodium phosphate (pH 8.0).
10. Buffer J: 25 mM sodium phosphate (pH 8.0), 1 M NaCl, 200 mM imidazole, and 0.5 mM DTT.
11. Buffer K: 25 mM sodium phosphate (pH 8.0), 400 mM NaCl, 1.25 M sodium sulfate, and 0.5 mM DTT.
12. Buffer L: 25 mM sodium phosphate (pH 8.0), 400 mM NaCl, and 0.5 mM DTT.
13. Buffer M: 5 mM sodium phosphate (pH 5.5), 150 mM NaCl, and 0.5 mM DTT.
14. Buffer N: 5 mM sodium phosphate (pH 5.5), 300 mM NaCl, and 0.5 mM DTT.
15. Buffer O: 5 mM sodium phosphate (pH 5.5), 800 mM NaCl, and 0.5 mM DTT.
16. Buffer P: 5 mM sodium phosphate (pH 5.5), 150 mM NaCl, 0.5 mM DTT, and 1 μM ZnCl$_2$.
17. 5 M NaCl.
18. 1 M DTT.

19. 2.5 *M* sodium sulfate.
20. 0.1 *M* ZnCl$_2$.

2.4.2. Buffers and Solutions for the Purification of Rat ShhN Expressed in Hi-5 Insect Cells from a Construct Encoding the N-terminal Fragment

1. Buffer Q: 5 m*M* sodium phosphate (pH 5.5), 150 m*M* NaCl, 0.5 m*M* PMSF, and 0.1% (w/v) Nonidet P-40.
2. Buffer R: 10 m*M* sodium phosphate (pH 6.5), 150 m*M* NaCl, 0.5 m*M* PMSF, 5 μ*M* pepstatin A, 10 μg/mL leupeptin, and 2 μg/mL E64. Pepstatin A, leupeptin, and E64 (Sigma).
3. Buffer S: 5 m*M* sodium phosphate (pH 5.5), 300 m*M* NaCl, and 0.1% (w/v) Nonidet P-40.
4. Buffer T: 5 m*M* sodium phosphate (pH 5.5), 800 m*M* NaCl, and 0.1% (w/v) Nonidet P-40.
5. Buffer U: 50 m*M* HEPES (pH 7.5).
6. Buffer V: 20 m*M* sodium phosphate (pH 7.2), 150 m*M* NaCl, and 1% (w/v) octyl-β-D-glucopyranoside. Octyl-β-D-glucopyranoside (US Biochemical Corp., Cleveland, OH).
7. Buffer W: 25 m*M* sodium phosphate (pH 3.0), 300 m*M* NaCl, and 1% (w/v) octyl-β-D-glucopyranoside.
8. Buffer X: 1 *M* HEPES (pH 7.5).
9. 10% (w/v) Triton X-100.

2.4.3. Buffers for the Purification of Human ShhN Expressed in Hi-5 Insect Cells from a Full-Length Construct

1. Buffer Y: 10 m*M* sodium phosphate (pH 6.5), 150 m*M* NaCl, and 0.5 m*M* PMSF.
2. Buffer Z: 10 m*M* sodium phosphate (pH 6.5), 150 m*M* NaCl, 0.5 m*M* PMSF, 5 μ*M* pepstatin A, 10 μg/mL leupeptin, 2 μg/mL E64, and 1% (w/v) Triton X-100.
3. Buffer AA: 5 m*M* sodium phosphate (pH 5.5), 150 m*M* NaCl, 0.5 m*M* PMSF, and 0.1% (w/v) Nonidet P-40.
4. Buffer BB: 5 m*M* sodium phosphate (pH 5.5), 300 m*M* NaCl, and 0.1% (w/v) Nonidet P-40.
5. Buffer CC: 5 m*M* sodium phosphate (pH 5.5), 800 m*M* NaCl, and 0.1% (w/v) Nonidet P-40.
6. Buffer DD: 20 m*M* sodium phosphate (pH 7.2), 150 m*M* NaCl, and 0.1% (w/v) hydrogenated Triton X-100. Protein-grade hydrogenated Triton X-100 (EMD Biosciences, Inc., San Diego, CA).
7. Buffer EE: 25 m*M* sodium phosphate (pH 3.0), 200 m*M* NaCl, and 0.1% (w/v) hydrogenated Triton X-100.
8. Buffer FF: 50 m*M* HEPES (pH 7.5) and 0.2% (w/v) hydrogenated Triton X-100.
9. Buffer GG: 20 m*M* sodium phosphate (pH 7.2), 150 m*M* NaCl, and 1% (w/v) octyl-β-D-glucopyranoside.
10. Buffer HH: 25 m*M* sodium phosphate (pH 3.0), 200 m*M* NaCl, and 1% (w/v) octyl-β-D-glucopyranoside.

2.4.4. Buffers and Solutions for the Preparation and Purification of N-terminally Fatty Acylated Human ShhN

1. Buffer II: 0.5 *M* sodium phosphate (pH 7.0).
2. Buffer JJ: 1 *M* sodium phosphate (pH 9.0).
3. Buffer KK: 5 m*M* sodium phosphate (pH 5.5), 150 m*M* NaCl, 1% (w/v) octyl-β-D-glucopyranoside, and 0.5 m*M* DTT.
4. Buffer LL: 5 m*M* sodium phosphate (pH 5.5), 1 *M* NaCl, 1% (w/v) octyl-β-D-glucopyranoside, and 0.5 m*M* DTT.
5. Buffer MM: 5 m*M* sodium phosphate (pH 5.5), 1 *M* NaCl, and 0.5 m*M* DTT.
6. 125 m*M* DTT.
7. 1.03 m*M* myristoyl coenzyme A.
8. 1 *M* hydroxylamine.
9. 5% (w/v) octyl-β-D-glucanopyranoside.
10. 150 m*M* NaCl and 0.5 m*M* DTT.
11. 0.1% (v/v) trifluroacetic acid and 5% (v/v) acetonitrile.
12. 0.1% (v/v) trifluoroacetic acid (TFA) and 85% (v/v) acetonitrile.

2.4.5. Buffers for the C3H10T1/2 Cell Assay (see **Note 2**)

1. Buffer NN: 20 m*M* sodium phosphate (pH 7.2) and 150 m*M* NaCl.
2. Buffer OO: 50 m*M* Tris–HCl (pH 7.5) and 150 m*M* NaCl.
3. Buffer PP: 10 m*M* diethanolamine (pH 9.5) and 0.5 m*M* MgCl$_2$.

2.4.6. Buffers and Solutions for the Preparation of mAb 5E1-Sepharose (see **Note 7**)

1. Buffer QQ: 1 *M* sodium borate (pH 8.4).
2. Buffer RR: 1 *M* ethanolamine (pH 8.0).
3. Buffer SS: 20 m*M* sodium phosphate (pH 7.2), 150 m*M* NaCl, and 0.02% (w/v) sodium azide.
4. 1 m*M* HCl.
5. 5 *M* NaCl.

3. Methods

The following sections describe the purification of human and rat ShhN purified from *E. coli* and Hi-5 cells. While each section describes the purification of a specific mammalian homolog expressed in a specific expression system, the sequence identity between human and rat ShhN (99%) is such that the methods described for one are likely to be equally applicable to the other. The biological activity of the various ShhN proteins can be assayed by measuring the induction of alkaline phosphatase (AP) in C3H10T1/2 cells (*see* **Note 2**). This assay provides a simple in vitro method to determine the relative activity of the purified proteins, and allows for batch-to-batch variability to be determined prior to use in complex ex vivo or in vivo systems.

3.1. Unmodified Human ShhN Expressed in *E. coli* from a Construct Encoding the N-terminal Fragment

The N-terminal fragment of human ShhN is purified from *E. coli* strain BL21/DE3/plysS (Stratagene, La Jolla, CA) containing plasmid p6H-SHH, a derivative of plasmid pET11d that carries the wild-type human *Shh* cDNA starting at Cys-24 and extending to Gly-197, followed by tandem termination codons, and cloned as an *NcoI*–*XhoI* fragment so that the *ShhN* cDNA is downstream of sequences encoding six consecutive histidine residues and an Asp-Asp-Asp-Asp-Lys enterokoinase cleavage site *(18)*.

1. Grow cells overnight at 37°C in TB-MGB medium containing 100 µg/mL ampicillin.
2. Inoculate (0.02 volumes) into fresh TB-MGB medium and 100 µg/mL ampicillin. Grow the cells at 37°C until the optical density at 550 nm reaches 0.6–0.8.
3. Add isopropylthiogalactoside (IPTG) to 0.5 m*M*.
4. 2–3 h after addition of the IPTG, harvest the cells by centrifugation. Cells can be used fresh or stored at −70°C for later use.
5. Record the volume of culture harvested; this information is required to determine the size of the sulfopropyl Sepharose column (*see* below). If the protein is required for in vivo animal studies, use pyrogen-free containers and buffers throughout. Pyrogen-free glassware can be prepared by soaking in 0.5 *M* NaOH at room temperature overnight and washing thoroughly with pyrogen-free water prior to use, or by baking overnight at 200°C. Wherever possible, new plastic containers and pipettes should be used.

3.1.1. Cell Breakage (2–8°C)

1. Resuspend cells (typically 500 g) in buffer A (1 g per 4 mL buffer).
2. Disrupt cells by passing through a high-pressure Gaulin homogenizer (e.g., Rannie, Copenhagen, Denmark at 700–900 psi) or an equivalent French pressure cell. Incubate on ice for 1 h, then disrupt for a second time as above.
3. Centrifuge disrupted cells at 19,000*g* for 30 min, decant, and keep the cell-free homogenate. Discard the pellet.

3.1.2. Sulfopropyl (SP) Sepharose Purification (2–8°C)

1. Pour a column of SP-sepharose resin; 1 mL packed resin per 100 mL original *E. coli* culture. Equilibrate the column with ≥10 column volumes (CV) of buffer B.
2. To the cell-free homogenate add 0.1 volumes of buffer C. Check the pH is ≤6.0.
3. Load the column under gravity feed, wash with 5 × 1 CV of buffer B, followed by 4 × 1 CV of buffer D, and then with 1 CV of buffer E.
4. Elute ShhN with 10 × 0.3 CV of buffer F. Determine which fractions to pool and the approximate percentage purity of the ShhN by running samples on reducing SDS-PAGE.
5. Pool appropriate fractions and determine the total protein content by measuring the absorbance at 280 nm, assuming a value of 1.0 = 1 mg/mL (*see* **Note 3**).

6. Filter-sterilize (0.2 μm) the pooled SP-sepharose fractions and proceed to the next step. If required, the SP-sepharose pool can be frozen at −70°C.

3.1.3. NTA-Ni²⁺ Agarose Purification (2–8°C)

1. Pour a column of NTA-Ni²⁺ Agarose resin; 1 mL packed resin per 20 mg ShhN (estimated from the percentage purity and total protein content of the SP-Sepharose pool above). Equilibrate the column with ≥10 CV of buffer G.
2. Mix the SP-Sepharose pool with sufficient 5 M NaCl, buffer H, buffer I, and 1 M DTT to bring the final concentration of NaCl, imidazole, sodium phosphate, and DTT to 1 M, 20, 50, and 1 mM, respectively, noting that the SP-sepharose pool already contains 800 mM NaCl, 25 mM sodium phosphate, and 0.5 mM DTT.
3. Load the column under gravity feed, wash with 5 × 1 CV of buffer G, then elute bound ShhN with 5 × 1 CV of buffer J, collecting into tubes that contain sufficient 1 M DTT to bring the fractions to 0.5 mM with respect to the added reductant.
4. Determine which fractions to pool by running samples on reducing SDS-PAGE, pool the appropriate fractions, and determine the total protein recovered by measuring the absorbance at 280 nm using the calculated Molar absorption coefficient of 26,030 Lmol⁻¹cm⁻¹ (1 mg/mL = 1.21 for the his-tagged ShNN protein) (*see* **Note 4**).

3.1.4. Phenyl Sepharose Purification (Room Temperature) (see **Note 5**)

1. Pour a column of phenyl sepharose (high sub) resin; 1 mL packed resin per 10 mg ShhN (estimated from the total protein content of the NTA-Ni²⁺ pool above). Equilibrate the column with ≥10 CV of buffer K.
2. Equilibrate the NTA-Ni²⁺ agarose pool to room temperature.
3. Slowly add an equal volume of 2.5 M sodium sulfate, 0.5 mM DTT, swirling gently while adding. The solution may become cloudy.
4. Load the column under gravity feed, wash with 2 × 1 CV of buffer K, then elute bound ShhN with 10 × 0.3 CV of buffer L.
5. Determine which fractions to pool, and the total protein recovered as in **Section 3.1.3**. The sample may be stored at −70°C if required. Typically, 2–3 g His-tagged ShhN is recovered per 500 g of cells, where the level of ShhN expression in the *E. coli* cells is ~5% of the total cellular protein.

3.1.5. Removal of the His-Tag with Enterokinase (28°C)

1. Equilibrate the phenyl Sepharose elution pool to room temperature, and add enterokinase at 1:1000 (w).
2. Mix gently and incubate at 28°C typically for 2 h. It is recommended that a pilot study be carried out, following the cleavage of the His-tagged ShhN protein by reducing SDS-PAGE. Samples should be monitored at $t = 0$, 30, 60, 120, and 240 min following the addition of the enterokinase to determine the minimal amount of time required for >95% cleavage. Overdigestion can lead, in part, to an undesired cleavage at the C-terminal side of Lys-32, leading to the formation of an inactive clipped form lacking the first nine amino terminal residues.
3. Following digestion, place on ice.

3.1.6. Purification of Detagged ShhN on NTA-Ni²⁺ Agarose (2–8°C)

1. Pour a column of NTA-Ni^{2+} agarose resin; 1 mL packed resin per 25 mg ShhN. Equilibrate the column with ≥10 CV of buffer G.
2. To the digested ShhN, add sufficient 5 M NaCl and buffer H to bring the final concentration of NaCl and imidazole to 1 M and 20 mM, respectively, noting that the digest already contains 400 mM NaCl.
3. Load the column under gravity feed and collect the flow through and 5 × 1 CV washes of buffer G into a tube containing sufficient 1 M DTT to bring the sample to 0.5 mM with respect to the added reductant. Determine the concentration of the detagged ShhN in the flow through and wash pool by measuring the absorbance at 280 nm using the calculated Molar absorption coefficient of 26,030 Lmol^{-1}cm^{-1} (1 mg/mL = 1.33 for the detagged protein) (*see* **Note 6**).
4. Apply 3 × 1 CV of buffer J to the column and collect into tubes containing sufficient 1 M DTT to bring the fractions to 0.5 mM with respect to the added reductant. Keep these fractions to recover any uncleaved His-tagged ShhN that bound to the column; in case the digestion failed to work satisfactorily and needs to be repeated.

3.1.7. Concentration of Detagged ShhN on SP-Sepharose (2–8°C)

1. Pour a column of SP-Sepharose fast flow resin; 1 mL packed resin per 20 mg ShhN. Equilibrate the column with ≥10 CV of buffer M.
2. Add 0.1 volumes of buffer C and 9 volumes of buffer M to the detagged ShhN NTA-Ni^{2+} agarose flow through and wash pool.
3. Load the sample onto the column under gravity feed, wash with 5 × 1 CV of buffer M, followed by 5 × 1 CV of buffer N. Elute the bound detagged ShhN with 10 × 0.3 CV of buffer O.
4. Determine which fractions to pool by running samples on reducing SDS-PAGE, pool the appropriate fractions, and determine the total protein recovered by measuring the absorbance at 280 nm using the calculated Molar absorption coefficient of 26,030 Lmol^{-1}cm^{-1} (1 mg/mL = 1.33).
5. Filter-sterilize (0.2 μm) the SP-sepharose pool. The sample may be stored at −70°C if required.

3.1.8. Dialysis (2–8°C)

1. As ShhN is a zinc-dependent protein (**19**), and to ensure full-occupancy of the bound metal ion, add 1 mol of ZnCl$_2$ (from a 0.1 M stock solution) per mole of protein and incubate for 1 h. It is recommended to first carry out a pilot addition to a small sample of the protein to ensure that the ZnCl$_2$ does not cause the protein to precipitate. This is especially important when working with high protein concentrations e.g., 10 mg/mL or above.
2. Dialyze the protein for 2–3 h against 4 L of buffer P. Then dialyze against 3 × 4 L changes of buffer M for at least 4 h per change.
3. Filter-sterilize (0.2 μm), and determine the final protein concentration as in **Section 3.1.7**. Aliquot the protein into appropriate size volumes, flash-freeze on liquid N$_2$,

and store at −70°C. If required, the protein can be concentrated up to 25 mg/mL prior to filter-sterilization and storage at −70°C. We recommend filter-sterilizing (0.2 μm) a portion of the last dialysis buffer against which the protein is dialyzed, flash-freezing on liquid N_2, and storing at −70°C. The buffer serves as a negative control in subsequent in vitro, ex vivo, or in vivo studies in which the protein is to be tested.

4. Analyze the purity of the detagged ShhN by reducing SDS-PAGE, and determine the intact mass by electrospray-ionization mass spectrometry. **Fig. 1A** (lane, ShhN) shows the purified protein analyzed by reducing SDS-PAGE. The intact mass of the protein by mass spectrometry was 19,560 Dalton, in good agreement with the calculated mass of 19,560.02 Dalton *(7,19)*.

3.2. Rat ShhN Expressed in Hi-5 Insect Cells from a Construct Encoding the N-terminal Fragment

The N-terminal fragment of rat Shh is purified from Hi-5 insect cells infected with a baculovirus expression vector pBluebac II (Invitrogen) that carries the wild-type rat ShhN N-terminal fragment (amino acid residues 25–198 in the rat gene sequence). Grow Hi-5 cells at 28°C to a density of 2×10^6 cells/mL in Sf-900 II serum-free medium in a 10 L bioreactor controlled for oxygen, then infect the cells with the rat ShhN-expressing baculovirus at a multiplicity of infection (MOI) of four virions per cell. Harvest the cells by centrifugation 48 h post-infection. The cells can be used fresh or stored at −70°C for later use. The N-terminus of rat ShhN starts at Cys-25 and not Cys-24 as in the human protein, since the signal sequence of the rat protein contains an additional amino acid residue. The rat sequence (amino acid residues 25–198) differs from the human sequence (amino acid residues 24–197) by only two residues; Ser-67 and Gly-196 in the human protein are replaced in the rat protein by threonine and aspartate, respectively. When expressed in Hi-5 cells, the majority of the protein is secreted into the culture medium since the construct lacks the autoprocessing domain responsible for attaching the cholesterol at the C-terminus. When purified, this soluble form had a similar specific activity (when measured in the C3H10T1/2 assay) as the soluble human ShhN purified from *E. coli*. However, a small fraction (~1%) of the protein found in the culture medium was significantly more potent (~100-fold). This form was found to be associated with membrane fragments (presumably liberated into the medium after baculovirus-induced lysis of the infected Hi-5 cells) and could be separated from the soluble form using size exclusion chromatography. The membrane fragment-associated forms have been shown to have a fatty acyl group attached to the N-terminal cysteine *(7)*. As the soluble form of rat ShhN secreted into the medium is susceptible to clipping and oxidation, we do not recommend purifying it from Hi-5 cell conditioned culture medium. Rather, we recommend purifying it from *E. coli* since the bacterial-expressed protein is of superior quality. Moreover, if the N-terminally fatty

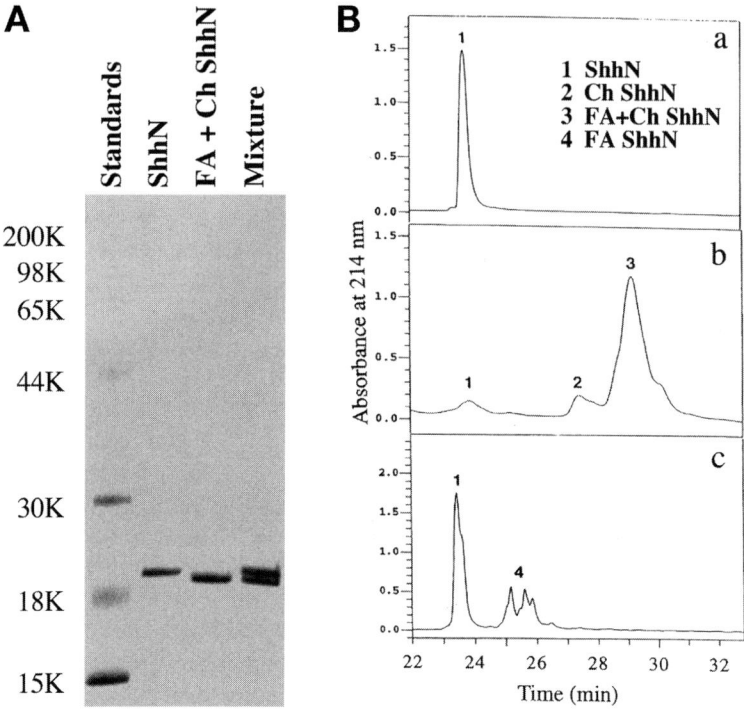

Fig. 1. (A) Reducing SDS-PAGE of unmodified, and palmitoyl- + cholesterol-modified human ShhN purified from *E. coli* and Hi-5™ cells, respectively. Samples containing 0.6 µg of unmodified ShhN purified from *E. coli* (lane: ShhN), and 0.6 µg of palmitoyl- + cholesterol-modified ShhN purified from Hi-5 cells (lane: FA + Ch ShhN) were electrophoresed either alone or in combination (lane: Mixture). The gel shows the difference in electrophoretic mobility of the two forms. **(B)** Reverse–phase (RP)–HPLC analysis of unmodified and hydrophobically modified forms of human and rat ShhN. Panel a, unmodified human ShhN purified from *E. coli*; panel b, palmitoyl- + cholesterol-modified human ShhN purified from Hi-5 cells; panel c, unmodified and fatty acylated rat ShhN purified from Hi-5 cells. The purified proteins were subjected to reverse-phase HPLC on a narrow bore Vydac C_4 column (2.1 mm internal diameter ×250 mm). The column was developed with a 35 min 0–80% acetonitrile gradient in 0.1% (v/v) TFA at 0.25 mL/min. The effluent was monitored using a photodiode array detector from 200 to 300 nm (data are shown at 214 nm). ShhN, unmodified protein; Ch ShhN, cholesterol-modified protein; FA + Ch ShhN, fatty acid- (i.e., palmitoyl) + cholesterol-modified protein; FA ShhN, fatty acid (i.e., myristoyl, palmitoyl, stearoyl, and arachidoyl)-modified protein.

acylated forms are required, we recommend purifying these from the Hi-5 cells themselves (described below) rather than from the conditioned culture medium as they represent a greater proportion of the purified product.

3.2.1. SP-Sepharose Purification (2–8°C)

1. Pour a column of SP-Sepharose fast flow resin; 1 mL packed resin per gram of Hi-5 cells. Equilibrate the column with ≥10 CV of buffer Q.
2. Thaw cells from −70°C (typically 40 g), and resuspend in 8 mL buffer R per gram of cells.
3. Add 0.1 volumes of 10% (w/v) Triton X-100, mix gently, and incubate on ice for 30 min to disrupt the cells. Centrifuge for 15 min at 1500g to pellet non-soluble material, decant, and add 0.1 volumes of buffer C to the supernatant.
4. Load the column under gravity feed, wash with 2 × 1 CV of buffer Q, followed by 2 × 1 CV of buffer S, prior to elution of the bound protein with 1 × 3 CV of buffer T.

3.2.2. Affinity Chromatography on mAb 5E1-Sepharose (2–8°C)

The SP-Sepharose-purified rat ShhN is then purified on a column of Sepharose to which the anti-human ShhN monoclonal antibody 5E1 is conjugated (*see* **Note 7**).

1. To the SP-Sepharose-purified rat ShhN, add 2 volumes of buffer U. Then add the 5E1-Sepharose resin; 1 mL settled resin per 12 g original Hi-5 cells. Rock the sample gently end-over-end for 2 h. Pack the slurry into a column, wash the resin with 10 × 1 CV of buffer V, then elute the ShhN protein with 10 × 0.3 CV of buffer W, collecting the fractions into tubes containing 0.1 volumes of buffer X.
2. Determine which fractions to pool by running samples on reducing SDS-PAGE. Pool the appropriate fractions, filter-sterilize (0.2 μm), and determine the total protein recovered by measuring the absorbance at 280 nm using the calculated Molar absorption coefficient of 26,030 Lmol^{-1}cm^{-1} (1 mg/mL = 1.33) (*see* **Note 8**). Typically, ~2 mg of ShhN is recovered per 40 g of cells, where ~30% of the final product is fatty acylated.

3.3. Human ShhN Expressed in Hi-5 Insect Cells from a Full-Length Construct

The N-terminal fragment of human Shh, with a palmitic acid group attached to Cys-24 and a cholesterol moiety attached to Gly-197, is purified from Hi-5 cells infected with recombinant baculovirus encoding the full-length human *Shh* gene. The cDNA for full-length *human Shh* was subcloned into the insect expression vector, pFastBac (Invitrogen), and recombinant baculovirus generated using the procedures supplied by the manufacturer and as described (*7*). Grow the Hi-5 cells at 28°C to a density of 2 × 10^6 cells/mL in Sf-900 II serum-free medium in a 10 L bioreactor controlled for oxygen, then infect the cells with the human Shh-expressing baculovirus at an MOI of three virions per cell. Harvest the cells by centrifugation 48 h post-infection, wash with buffer Y, and purify the ShhN protein immediately, since the protein is susceptible to proteolysis. At the time of harvest, over 95% of the ShhN is membrane-associated.

3.3.1. Cell Lysis (2–8°C)

1. Resuspend cells (typically 150 g) in buffer Z (1 g per 8 mL buffer), and incubate on ice for 30 min to disrupt the cells.
2. Centrifuge for 15 min at 14,000g, decant, keep the cell-free lysate, and discard the pellet.

3.3.2. Sulfopropyl (SP) Sepharose Purification (2–8°C)

1. Pour a column of SP-Sepharose fast flow resin; 1 mL packed resin per gram of Hi-5 cells. Equilibrate the column with ≥10 CV of buffer AA.
2. To the cell-free lysate add 0.1 volumes of buffer C.
3. Load the column under gravity feed, then wash with 6 × 0.3 CV of buffer AA, followed by 4 × 0.3 CV of buffer BB. Elute bound ShhN with 6 × 0.25 CV of buffer CC. Pool elution fractions 2–6 inclusive and proceed immediately to the next step.

3.3.3. Affinity Chromatography on mAb 5E1-Sepharose (2–8°C)

The SP-Sepharose-purified human ShhN is then purified on a column of Sepharose to which the anti-human Shh monoclonal antibody 5E1 is conjugated (*see* **Note 7**).

1. To the SP-Sepharose elution pool mix 2 volumes of buffer U. Then add the 5E1-Sepharose resin; 1 mL settled resin per 60 g original Hi-5 cells. Rock the sample gently end-over-end for 3 h. Collect the resin by centrifugation (760g for 30 min), resuspend in a small volume of buffer DD; and pack the slurry into a column. Wash the resin with 10 × 1 CV of buffer DD, then elute the ShhN protein with 10 × 0.25 CV of buffer EE, collecting the fractions into tubes containing 0.1 volumes of buffer X.
2. Determine which fractions to pool by running samples on reducing SDS-PAGE. Pool the appropriate fractions, and determine the total protein recovered by measuring the absorbance at 280 nm using the calculated Molar absorption coefficient of 26,030 Lmol^{-1}cm^{-1} (1 mg/mL = 1.29) (*see* **Note 9**). The sample may be stored at −70°C, if required.
3. Pool the peak fractions and mix with 1.3 volumes of buffer FF. Then add 5E1-Sepharose resin; 1 mL settled resin per 120 g original Hi-5 cells. Rock the sample gently end-over-end for 1 h. Pack the slurry into a column, wash with 3 CV of buffer GG, then elute the bound protein with 10 × 0.25 CV of buffer HH, collecting the fractions into tubes containing 0.1 volumes of buffer X.
4. Determine which fractions to pool by running samples on reducing SDS-PAGE. Pool the peak fractions, filter-sterilize (0.2 µm), determine the protein concentration as in **Section 3.3.3.**, aliquot, and store at −70°C (*see* **Note 10**). Analyze the

purity of the protein by reducing SDS-PAGE, and determine the intact mass by electrospray-ionization mass spectrometry. **Fig. 1A** (lane, FA + Ch ShhN) shows the purified protein analyzed by reducing SDS-PAGE, in which the lipid-modified form migrates with an apparent M_r of 19.5 kDa, 0.5 kDa smaller than that of the unmodified protein purified from *E. coli* (**Fig. 1A**; lane ShhN). While the apparent lower mass may be interpreted as proteolytic clipping, electrospray-ionization mass spectrometry shows that the protein is largely intact. The intact mass of the protein by mass spectrometry was 20,168 Dalton, in good agreement with the calculated mass of 20,167.14 Dalton for the protein carrying both palmitoyl and cholesterol groups *(7)*. Typically, ~200 µg of the purified palmitoyl- + cholesterol-modified ShhN is recovered per 10 L fermentation (*see* **Note 11**).

3.4. N-terminally Fatty Acylated Human ShhN

As relatively small amounts of ShhN carrying hydrophobic modifications at the N-terminus, or at both the N- and C-termini, can be purified from Hi-5 cells, and since these modified forms are insoluble in the absence of detergent, we have developed a method for producing mg quantities of soluble, N-terminally myristoylated Shh as a high potency surrogate molecule. The myristoylated protein is produced by modification of the detagged N-terminal fragment of human ShhN purified from *E. coli* (**Section 3.1.**) with myristoyl coenzyme A *(9)*. The following section describes the preparation and characterization of ShhN modified with myristoyl-CoA, as well as with palmitoyl-CoA, lauroyl-CoA, decanoyl-CoA, and octanoyl-CoA. The same methodology is used for modification with palmitoyl-CoA, lauroyl-CoA, decanoyl-CoA, and octanoyl-CoA as for myristoyl-CoA, except where stated otherwise. The fatty acyl coenzymes used to modify human ShhN were obtained from Sigma.

3.4.1. N-Terminal Modification

1. To a 3 mg/mL solution of detagged human ShhN formulated in buffer M, add sufficient buffer II, 125 m*M* DTT, water, and 1.03 m*M* myristoyl-CoA to bring the added components to the following concentrations: 0.8 mg/mL (41 µ*M*) ShhN, 40 m*M* sodium phosphate (pH 7.0), 25 m*M* DTT, and 410 µ*M* myristoyl-CoA (10-fold Molar excess over protein). For modification with palmitoyl-CoA and lauroyl-CoA, the concentration of the fatty acyl coenzyme A in the reaction mixture should also be 410 µ*M*; while for decanoyl-CoA and octanoyl-CoA, which have a significantly higher critical micelle concentration (CMC), the concentration should be 4.1 m*M* (100-fold Molar excess over protein).
2. Incubate the mixture at 28°C for 24 h. Monitor the extent of reaction by reverse-phase HPLC on a Vydac C$_4$ column (4.6 mm internal diameter × 250 mm) run at 1.4 mL/min with the following gradient where A = 0.1% (v/v) TFA, 5% (v/v) acetonitrile, and B = 0.1% (v/v) TFA, 85% (v/v) acetonitrile. 0–2 min = 100% A, 2–17 min = 0–100% B, 17–20 min = 100% B; and 20–25 min = 100% A. Protein is detected on-line at 280 nm. The reaction products at this stage contain fatty acid moieties

attached at either the α-amino group (singly myristoylated protein) or at both the α-amino group and the thiol group (doubly myristoylated protein) of the N-terminal cysteine (*see* **Note 12**).

3. Remove the thioester-linked acyl group from the side chain of Cys-24 by adding 0.1 volumes of buffer JJ, followed by 0.1 volumes (accounting for the increase in volume following the addition of buffer JJ) of 1 *M* hydroxylamine. Incubate for a further 18 h at 28°C, and check the extent of reaction by reverse-phase HPLC as described in **Section 3.4.1**. The reaction mixture should be devoid of any doubly myristoylated protein following incubation with hydroxylamine.

4. Add 0.25 volumes of 5% (w/v) octyl-β-D-glucopyranoside and incubate with gentle mixing at room temperature for 1 h (*see* **Note 13**).

3.4.2. Sulfopropyl (SP) Sepharose Purification (Room Temperature)

1. Pour a column of SP-Sepharose resin; 1 mL packed resin per 20 mg ShhN. Equilibrate with ≥10 CV of buffer KK.

2. Load the column under gravity feed, wash with 10×1 CV of buffer KK, then elute bound protein with 10×0.3 CV of buffer LL. Determine which fractions to pool by measuring the absorbance at 280 nm using the Molar absorption coefficient of 26,030 Lmol^{-1}cm^{-1} (1 mg/mL = 1.32 for myristoylated ShhN) (*see* **Note 14**). Pool the appropriate fractions and determine the total protein recovered as above. If required, the SP-Sepharose pool can be frozen at −70°C.

3.4.3. Bio-Scale S Purification (Room Temperature)

Unlike all other purification steps described in this chapter that require simple elution of bound protein, elution from the Bio-Scale S column requires an accurate linear NaCl gradient. Therefore, a chromatography system (e.g., Biorad's Biologic or GE Healthcare's AKTA), is required to deliver the gradient across the column.

1. Equilibrate a Bio-Scale S column with ≥10 CV of buffer KK. Dilute the SP-Sepharose pooled fractions with 4 volumes of buffer KK to reduce the NaCl concentration, and load onto the Bio-Scale S column (12 mg ShhN per mL of packed resin) at 160 cm/h (3 mL/min for a Bio-Scale S10 [10 mL] column with dimensions of 1.2 cm internal diameter × 8.8 cm) (*see* **Note 15**). Wash the column with 10 CV of buffer KK, then elute with a linear gradient of 150 m*M*–1 *M* NaCl in 20 CV, i.e., from 100% buffer KK to 100% buffer LL in 20 CV (200 mL for a Bio-Scale S10 column). Collect 0.1 CV fractions.

2. Determine which fractions to pool by analyzing samples by RP-HPLC as described in **Section 3.4.1**. Pool only those fractions containing the modified protein. Fractions can be stored at −70°C while analyzing representative samples by RP–HPLC.

3.4.4. Concentration and Dialysis (2–8°C)

1. Pour a column of SP-Sepharose resin; 1 mL packed resin per 20 mg fatty-acylated ShhN. Equilibrate with ≥10 CV of buffer KK.

2. Add 4 volumes of buffer KK to the pooled Bio-Scale S fractions to reduce the NaCl concentration, and load the column under gravity feed. Wash with 10×1 CV of buffer KK, then elute bound protein with 10×0.3 CV of buffer LL. Determine which fractions to pool by measuring the absorbance at 280 nm using the Molar absorption coefficient of 26,030 $Lmol^{-1}cm^{-1}$ (1 mg/mL = 1.32).

3. Add 1 volume of buffer MM to the concentrated myristoylated ShhN (to reduce the concentration of octyl-β-D-glucopyranoside to below its CMC), and dialyze against 8×4 L changes (changing approximately every 12 h) of 150 mM NaCl, 0.5 mM DTT to remove the octyl-β-D-glucopyranoside (*see* **Note 16**). Centrifuge the dialyzed sample for 15 min at 2000g, decant, and filter-sterilize (0.2 μm) the supernatant. Determine the protein concentration as described in **Section 3.4.4**. Aliquot, flash-freeze on liquid N$_2$, and store at $-70°C$ (*see* **Note 17**). We recommend filter-sterilizing (0.2 μm) a portion of the last dialysis buffer against which the myristolyated protein is dialyzed, flash-freezing it on liquid N$_2$, and storing at $-70°C$. The buffer serves as a negative control in subsequent in vitro, ex vivo, or in vivo studies in which the myristoylated protein is to be tested. The recovery of myristoylated ShhN is typically 40% of the initial material when starting with 30 mg unmodified protein. The activity in the C3H10T1/2 assay of human ShhN modified at the N-terminus with various fatty acyl groups is shown in **Fig. 2**. As can be seen, the potency increases with increasing chain length up to C14 (myristoylated ShhN) which is ~100-fold more potent than the unmodified protein. Increasing the chain length (C16) does not increase the activity further, possibly due to the decreased solubility of the palmitoylated protein. The myristoylated protein has also been shown to be more potent than the unmodified protein in ex vivo tissue explant assays *(10)*.

3.5. Purification of Human ShhN from Mammalian Cells and the Yeast Pichia Pastoris

In addition to the ShhN proteins purified from *E. coli* and Hi-5 cells described above, we have also purified human ShhN expressed in the yeast *Pichia pastoris* *(18)* and the EBNA-293 embryonic kidney cell line *(7)*. While the methods described above are applicable to the purification of ShhN from yeast and mammalian cells, the purified proteins showed extensive microheterogeneity, particularly with respect to the amount of post-translationally modified forms, as well as with respect to the amount of N-terminally clipped forms (*see* **Table 1**). Since the N-terminus of ShhN has been shown to be critical for activity *(18)*, we recommend using expression systems, such as *E. coli* (**Section 3.1.**) and Hi-5 cells (**Sections 3.2.** and **3.3.**), from which intact protein can be isolated. In addition, ShhN that is expressed in *P. pastoris* is secreted into the culture medium and is susceptible to oxidation, as has also been observed for ShhN secreted into Hi-5 cell conditioned culture medium. **Table 1** summarizes mass spectrometry data for ShhN purified from the various expression systems, as

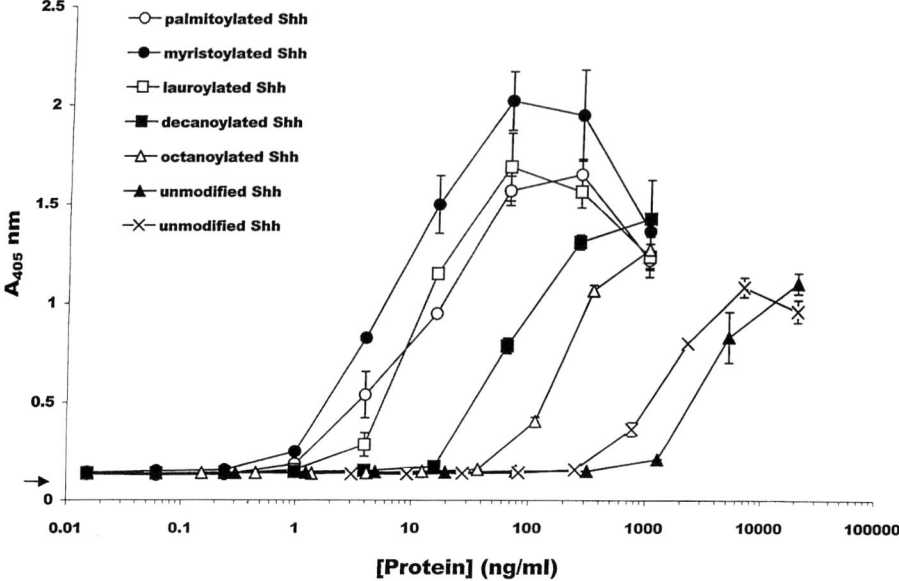

Fig. 2. C3H10T1/2 assay of palmitoylated, myristyolated, lauroylated, decanoylated, octanoylated, and unmodified human ShhN. Palmitoylated, lauroylated, decanoylated, and octanoylated human ShhN (formulated in 5 mM sodium phosphate (pH 5.5), 150 mM NaCl, 1% (w/v) octyl-β-D-glucopyranoside, and 0.5 mM DTT), and myristoylated human ShhN (formulated in 150 mM NaCl and 0.5 mM DTT), were assayed on C3H10T1/2 cells measuring the induction of AP. The numbers represent the mean of duplicate determinations. Serial threefold dilutions of the proteins were incubated with the cells for 5 d and the resulting levels of AP measured at 405 nm using the chromogenic substrate pNPP. The palmitoylated, myristoylated, lauroylated, and decanoylated proteins were assayed in one experiment with the unmodified protein shown as (p), while the octanoylated protein was assayed in an independent experiment with the unmodified protein shown as (×). The arrow on the *y*-axis denotes the background level of AP in the absence of added ShhN.

well as the *E. coli*-expressed protein modified at the N-terminus with various fatty acyl groups.

3.6. Engineering ShhN Variants with Improved Solubility and Pharmacokinetic Properties

While hydrophobic modifications increase the potency of ShhN when added to the N-terminus of the protein, the more hydrophobic modifications (e.g., palmitoyl and palmitoyl + cholesterol) compromise protein solubility. A practical

Table 1
Mass Data for ShhN Purified from *E. coli*, Hi-5™ Cells, EBNA-293 Cells, and *P. pastoris*. Samples were Fractionated by Reverse-Phase HPLC on a Narrow Bore C4 Column

Protein and expression system	Purified protein	Mass (Dalton)	
		Calculated	Measured
Human ShhN/*E. coli*	N-terminal construct		
	>95% intact, Unmodified[1]	19,560.02	19,560
Rat ShhN/Hi-5	N-terminal construct		
	~70% intact, unmodified[1]	19,632.08	19,632
	~30% intact, fatty acylated:		
	+myristoyl	19,842.05	19,842
	+palmitoyl	19,870.55	19,868
	+stearoyl	19,898.60	19,896
	+arachidoyl	19,926.66	19,925
Human ShhN/Hi-5	Full-length construct		
	>80% intact, +palmitoyl +cholesterol	20,167.74	20,168
	~15% as a mixture of:		
	Intact, +palmitoyl	19,798.49	19,796
	Intact, +cholesterol[3]		
	N-10, +cholesterol[3]		
	<5% unmodified[1]	19,560.02	19,560
Human ShhN/EBNA-293	Full-length construct		
	~30% intact, +palmitoyl +cholesterol	20,167.14	20,174[2]
	~60% as a mixture of:		
	intact, +cholesterol	19,928.64	19,934[2]
	N-10, +cholesterol	18,912.48	18,889[2]
	~10% as a mixture of:		
	intact, unmodified[1]	19,560.02	19,581[2]
	N-9, unmodified[1]	18,700.02	18,712[2]
Human ShhN/*P. pastoris*	N-terminal construct		
	~70% N-10, unmodified[1]	18,543.83	18,544[2]
	~25% as a mixture of intact and/or oxidized and/or with an N-terminal thiaproline adduct		
	~5% unidentified		
Human ShhN/*E. coli*	In vitro fatty acylated N-terminal construct		
	Intact, palmitoylated	19,798.43	19,798
	Intact, myristoylated	19,770.38	19,770
	Intact, lauroylated	19,742.33	19,742
	Intact, decanoylated	19,714.28	19,715
	Intact, octanoylated	19,686.23	19,686

Peaks were analyzed by electrospray–ionization mass spectrometry using a Micromass Quattro II triple quadrupole mass spectrometer or by MALDI-mass spectrometry using a Finnigan LaserMat mass spectrometer using α-cyano-4-hydroxycinnamic acid as the matrix. Average masses were used to calculate the expected masses.

[1]Unmodified refers to protein lacking fatty acyl or cholesterol groups.

[2]Masses measured by MALDI mass spectrometry *(7)* account for the lower accuracy of the measurement.

[3]Identified by retention time from RP–HPLC.

solution is to substitute the N-terminal cysteine with two isoleucine residues. The C24II ShhN mutant is 10-fold more potent than unmodified wild-type ShhN and is readily soluble in the absence of detergent *(9)*. Another problem encountered with ShhN is its relatively short serum half-life, making it difficult to evaluate in animal models. Two methods have been successfully developed to improve systemic exposure, which utilize PEGylation *(15)* and Fc fusion technologies *(16)*. In one method, polyethylene glycol is added to target sites that are introduced into the ShhN protein by surface cysteine mutations *(15)*. In the second method, the N-terminal fragment of *Shh* is genetically fused to an immunoglobulin Fc domain, introducing the properties of the Fc domain to alter the pharmacokinetic behavior *(16)*. These methods provide solutions for generating forms of ShhN that can be used in vivo.

4. Notes

1. For all buffers containing DTT, or PMSF, add freshly prepared.
2. ShhN is tested for function in a cell-based assay measuring AP induction in C3H10T1/2 cells *(20)*. AP is a marker for differentiation of the cell line into an osteoblast lineage and provides a simple bioassay for measuring changes in the potency of ShhN samples resulting from protein modifications. Maintain the C3H10T1/2 cell line at 37°C in a tissue culture incubator in Dulbecco's Modified Eagle Medium containing 10% fetal bovine serum. Prior to use in assays, passage the cells two to three times. The cells can then be split and used for up to 20 passages, at which point the cells lose responsiveness to ShhN. For assay, add cells in growth medium (5000 cells/100 µL/well) to 96-well tissue culture plates. Twenty-four hours later, add 100 µL purified ShhN protein in growth medium to each well and incubate for a further 5 d. Wash the plates twice with buffer NN (200 µL/well), and once with buffer OO, and then incubate at 37°C for 1 h in buffer PP (100 µL/well) to lyse the cells. Assay for AP activity using the chromogenic substrate pNPP and read at 405 nm. Dissolve a 20 mg pNPP tablet (Sigma) in 10 mL buffer PP, and add 100 µL per well. Read the plates (kinetic reading) for 1 h using a 96-well Molecular Devices Thermomax plate reader. Typical dose responses were in the range of 1–10 µg/mL for unmodified ShhN, and 0.01–0.1 µg/mL for the lipid–modified proteins. For lipid-modified ShhN, which we typically stored at 100 µg/mL, first dilute the samples 200-fold with normal growth medium and then subject to serial dilutions down the plates. If octyl-β-D-glucopyranoside is present in the test sample, the results are normalized for any potential effects of the detergent by including the same concentration of octyl-β-D-glucopyranoside (0.005% [w/v]) in the medium.
3. When measuring the protein concentration, it may be necessary to dilute the sample with buffer F to ensure the absorbance at 280 nm is ≤1.0, i.e., ≤90% of the light absorbed. Always blank the spectrophotometer with the buffer in which the protein is formulated.

4. As the imidazole (200 mM) in the NTA-Ni^{2+} agarose pool absorbs at 280 nm, blank the spectrophotometer with buffer J prior to measuring the absorbance.

5. The phenyl Sepharose purification step is omitted when purifying mutant forms of ShhN in which hydrophobic amino acids are introduced into the protein because of poor recovery from the column, e.g., C24II, where Cys-24 is replaced with two isoleucine residues (*see* **Section 3.6.**) and *(21)*.

6. As the imidazole (20 mM) in the NTA-Ni^{2+} agarose flow through and wash pool absorbs at 280 nm, blank the spectrophotometer with buffer G prior to measuring the absorbance.

7. 5E1-Sepharose resin is prepared by conjugating the anti-ShhN monoclonal anti-body 5E1 *(17)* to CNBr-activated Sepharose 4B. Quickly wash 2 g (dry weight) of CNBr-activated Sepharose with 200 mL of 1 mM HCl under vacuum on a glass fritted-filter unit. Then add 6 g (wet weight) of the washed resin to 12 mL of 2 mg/mL mAb 5E1 in buffer NN, followed by 850 μL of 5 M NaCl, and 1.2 mL of buffer QQ. Incubate the mixture for 8 h at 4°C with constant gentle mixing on a rocking platform. Then add 2.5 mL of buffer RR to quench the reaction, and incubate the slurry overnight at 4°C with constant gentle mixing. Wash the conjugated resin with 3 × 50 mL buffer SS, and store at 4°C in buffer SS.

8. Unlike human ShhN purified from *E. coli*, rat ShhN purified from Hi-5 cells is a mixture of unmodified and N-terminally fatty acylated proteins. **Fig. 1B**, panel c, shows that the unmodified protein (peak 1) elutes prior to a series of later-eluting peaks (peaks 4). Using a combination of peptide mapping and mass spectrometry, the proteins have been shown to have myristoyl (C14), palmitoyl (C16), stearoyl (C18), or arachidoyl (C20) groups attached to the α-amino group of Cys-25 *(7)*. The potency of fatty acylated rat ShhN in the C3H10T1/2 assay (as a mixture of the various forms) is significantly greater than for unmodified rat ShhN, or for human ShhN purified from *E. coli*. Similar increases in potency are seen for human ShhN purified from Hi-5 cells expressing the full-length construct (**Section 3.3.**), as well as for human ShhN modified specifically at the α-amino group of Cys-24 with various fatty acyl groups (**Section 3.4.**).

9. The use of hydrogenated Triton X-100 in the wash and elution allows for more accurate absorbance measurements than can be obtained with standard preparations of Triton X-100.

10. Human ShhN modified at the N-terminus with a palmitoyl group and with a cholesterol moeity at the C-terminus is very hydrophobic and consequently the protein should be handled with care. While readily soluble in detergent, or in 60% (v/v) acetonitrile, 0.1% (v/v) trifluoracetic acid, it precipitates and/or is lost on the surface of tubes when diluted out of these formulations. In fact, diluting the 1% (w/v) octyl-β-D-glucopyranoside-containing ShhN with 4 volumes of buffer NN (containing no detergent) results in quantitative precipitation of the protein, which we utilized as a buffer exchange step for some of the biochemical studies that are sensitive to detergent. However, the protein can readily be diluted into serum-containing growth medium, presumably because it is stabilized by the presence of a serum component.

11. Human ShhN purified from Hi-5 cells expressing the full-length construct is more hydrophobic than the unmodified protein purified from *E. coli*. Moreover, as the protein has both palmitoyl and cholesterol groups attached, it is also more hydrophobic than the N-terminally fatty-acylated forms of rat ShhN purified from Hi-5 cells (*see* **Note 8**). **Fig. 1B**, panel b, shows that the palmitoyl- plus cholesterol-modified protein (peak 3) is the major component, and that it elutes later in the acetonitrile gradient than either the small peak of unmodified protein (peak 1) or cholesterol-modified protein (peak 2), or the fatty acylated peaks of rat ShhN (peaks 4 in **Fig. 1B**, panel c).

12. The unmodified protein will elute first from the HPLC column, followed by the singly myristoylated protein, with the doubly myristoylated protein eluting last. The three species are separated to baseline resolution under these conditions.

13. Do not vortex, shake, or agitate vigorously and avoid the formation of bubbles. We recommend mixing by gentle inversion.

14. When measuring the protein concentration, it may be necessary to dilute the sample with buffer LL to ensure the absorbance at 280 nm is ≤1.0.

15. cm/h = flow rate (mL/h) divided by the cross-sectional area of the column (cm^2).

16. During the dialysis of myristoylated ShhN, a precipitate may be visible although the majority of the protein remains in solution.

17. For myristoylated, lauroylated, decanoylated, and octanoylated ShhN, the sample can be dialyzed against 150 mM NaCl, 0.5 mM DTT to remove the octyl-β-D-glucopyranoside since the proteins remain reasonably soluble in the absence of the detergent. However, for palmitoylated ShhN, the protein should be dialyzed against 150 mM NaCl, 0.5 mM DTT containing 1% (w/v) octyl-β-D-glucopyranoside since it is insoluble in the absence of the detergent. Due to the insolubility of the palmitoylated form, we recommend using the myristoylated protein as a surrogate for the more physiologically relevant form of the protein carrying both palmitoyl and cholesterol groups.

Acknowledgments

We thank the hedgehog project teams at Biogen Idec, Inc. and Curis, Inc. for contributing to the work described in this chapter.

References

1. Marigo, V., Roberts, D. J., Scott, M. K. L., et al. (1995) Cloning, expression, and chromosomal location of SHH and IHH: two human homologues of the *Drosophila* segment polarity gene Hedgehog. *Genomics* **28**, 44–51.

2. Echelard, Y., Epstein, D. J., St-Jacques, B., et al. (1993) Sonic hedgehog, a member of a family of putative signaling molecules, is implicated in the regulation of CNS polarity. *Cell* **75**, 1417–1430.

3. Lee, J. J., Ekker, S. C., von Kessler, D. P., Porter, J. A., Sun, B. I., and Beachy, P. A. (1994) Autoproteolysis in Hedgehog protein biogenesis. *Science* **266**, 1528–1537.

4. Bumcrot, D. A., Takada, R., and McMahon, A. P. (1995) Proteolytic processing yields two secreted forms of Sonic hedgehog. *Mol. Cell. Biol.* **15,** 2294–2303.
5. Porter, J. A., Ekker, S. C., Park, W.-J., et al. (1996) Hedgehog patterning activity: role of a lipophilic modification mediated by the carboxy-terminal autoprocessing domain. *Cell* **86,** 21–34.
6. Porter, J. A., Young, K. E., and Beachy, P. A. (1996) Cholesterol modification of Hedgehog signaling proteins in animal development. *Science* **274,** 255–259.
7. Pepinsky, R. B., Zeng, C., Wen, D., et al. (1998) Identification of a palmitic acid-modified form of human Sonic hedgehog. *J. Biol. Chem.* **273,** 14, 037–14,045.
8. Valentini, R. P., Brookhiser, W. T., Park, J., et al. (1997) Post-translational processing and renal expression of mouse Indian hedgehog. *J. Biol. Chem.* **272,** 8466–8473.
9. Taylor, F. R., Wen, D., Garber, E. A., et al. (2001) Enhanced potency of human Sonic hedgehog by hydrophobic modification. *Biochemistry* **40,** 4359–4371.
10. Kohtz, J. D., Lee, H. Y., Gaiano, N., et al. (2001) N-terminal fatty-acylation of Sonic hedgehog enhances the induction of rodent ventral forebrain neurons. *Development* **128,** 2351–2363.
11. Kohtz, J. D., Baker, D. P., Corte, G., and Fishell, G. (1998) Regionalization within the mammalian telencephalon is mediated by changes in responsiveness to Sonic hedgehog. *Development* **125,** 5079–5089.
12. Bhardwaj, G., Murdoch, B., Wu, D., et al. (2001) Sonic hedgehog induces the proliferation of primitive human hematopoietic cells via BMP regulation. *Nat. Immunol.* **2,** 172–180.
13. Pola, R., Ling, L. E., Silver, M., et al. (2001) The morphogen Sonic hedgehog is an indirect angiogenic agent upregulating two families of angiogenic growth factors. *Nat. Med.* **7,** 706–711.
14. Pascual, O., Traiffort, E., Baker, D. P., Galdes, A., Ruat, M., and Champagnat, J. (2005) Sonic hedgehog signaling in neurons of adult ventrolateral nucleus tractus solitarus. *Eur. J. Neurosci.* **22,** 389–396.
15. Pepinsky, R. B., Shapiro, R. I., Wang, S., et al. (2002) Long-acting forms of Sonic hedgehog with improved pharmacokinetic and pharmacodynamic properties are efficacious in a nerve injury model. *J. Pharm. Sci.* **91,** 371–387.
16. Shapiro, R. I., Wen, D., Levesque, M., et al. (2003) Expression of Sonic hedgehog-Fc fusion protein in *Pichia pastoris*. Identification and control of post-translational, chemical, and proteolytic modifications. *Protein Exp. Purif.* **29,** 272–283.
17. Ericson, J., Morton, S., Kawakami, A., Roelink, H., and Jessell, T. L. (1996) Two critical periods of Sonic hedgehog signaling required for the specification of motor neuron identity. *Cell* **87,** 661–673.
18. Williams, K. P., Rayhorn, P., Chi-Rosso, G., et al. (1999) Functional antagonists of Sonic hedgehog reveal the importance of the N-terminus for activity. *J. Cell. Sci.* **112,** 4405–4414.

19. Day, E. S., Wen, D., Garber, E. A., et al. (1999) Zinc-dependent structural stability of human Sonic hedgehog. *Biochemistry* **38,** 14,868–14,880.

20. Nakamura, T., Aikawa, T., Iwamoto-Enomoto, M., et al. (1997) Induction of osteogenic differentiation by Hedgehog proteins. *Biochem. Biophys. Res. Commun.* **237,** 465–469.

21. Pepinsky, R. B., Rayhorn, P., Day, E. S., et al. (2000) Mapping Sonic hedgehog-receptor interactions by steric interference. *J. Biol. Chem.* **275,** 10,995–11,001.

2

Application of Sonic Hedgehog to the Developing Chick Limb

Eva Tiecke and Cheryll Tickle

Abstract

Here, we describe methods for applying Sonic hedgehog (Shh) to developing chick limbs. The *Sonic hedgehog* gene is expressed in the polarizing region, a signaling region at the posterior margin of the limb bud and application of Shh-expressing cells or Shh protein to early limb buds mimics polarizing region signaling. The polarizing region (or zone of polarizing activity) is involved in one of the best known cell–cell interactions in vertebrate embryos and is pivotal in controlling digit number and pattern. At later stages of limb development, the application of Shh protein to the regions between digit primordia can induce changes in digit morphogenesis.

Key Words: Sonic hedgehog; chick embryo; polarizing region; limb development; digit morphogenesis; bead.

1. Introduction

Chicken embryos are readily manipulated through a window in the shell and have proved a very powerful model for studying vertebrate limb development *(1)*. The polarizing region (also known as the zone of polarizing activity) was first discovered through a grafting experiment carried out in chick wing buds by John Saunders *(2)*. When tissue from the posterior margin of an early chick wing bud was grafted to the anterior margin of a second wing bud, 6 digits developed instead of 3, with additional digits 432 arising in mirror-image symmetry with the normal set of digits 234 giving the complete pattern of 432234 (reading from anterior to posterior). Extensive grafting experiments have shown that signaling from the polarizing region is dose-dependent and long range; digit character, for example, depends on distance from the polarizing region *(3)*. Thus, when an additional polarizing region was grafted not right at the anterior

From: *Methods in Molecular Biology: Hedgehog Signaling Protocols*
Edited by: J. Horabin © Humana Press Inc., Totowa, NJ

margin but further posterior, toward the mid-point of the wing bud apex, the pattern of digits that developed was 4334 and no digit 2 was formed. In contrast, attenuating signaling by the polarizing region by, for example, irradiating the graft, resulted in failure to induce formation of the most posterior digit, giving partial duplications such as 32234 and 2234 *(4)*, as did grafting small numbers of polarizing region cells *(5)*. These data from experiments on chick wing buds support a model in which the polarizing region produces a long-range diffusible molecule that spreads across the limb bud, from posterior to anterior setting up a concentration gradient. This would result in cells at different distances from the polarizing region being exposed to different concentrations of morphogen thus providing them with information about their position across the antero-posterior axis of the limb bud. The polarizing region also controls digit number and there is a positive feedback loop between signaling by the polarizing region and the apical ectodermal ridge, the thickened epithelium that rims the tip of the limb bud.

The induction of additional digits from the anterior of the chick wing bud is an excellent assay for polarizing activity. This assay was used to map polarizing activity in chick wings throughout development *(6,7)* and also to identify the polarizing region of mammalian limb buds, including those of human embryos *(8)*. In addition, tissue from several different regions of embryos has been shown to possess polarizing activity, including the node in both chick and mouse, and the genital tubercle in mouse *(9,10)*. The chick wing bud assay was also adapted to test defined chemical substances for polarizing activity. The first success was achieved using small pieces of filter paper soaked in retinoic acid *(11)*. This technique was then refined by examining the effectiveness of a number of different carriers for retinoic acid *(12)*. Formate derivatised AG1-X2 beads (Bio-Rad, USA) were found to be most effective and release retinoic acid over at least 24 h. AG1-X2 beads were then used to characterize the effects of retinoic acid on chick wing development *(13,14)* and have been used extensively to apply retinoic acid to other regions of vertebrate embryos.

The *Sonic hedgehog* gene was one of the first genes found to be expressed in the polarizing region of the chick limb (*[15]*; **Fig. 1A,B**). Riddle et al. showed that the distribution of *Shh* transcripts along the posterior margin of the chick wing bud correlated very closely with the maps of polarizing activity *(7)*. Furthermore, they found that pellets of cells expressing Shh grafted to the anterior margin of the chick wing bud produced duplicated digit patterns. Soon after, it was shown that beads soaked in the amino-terminal cleavage product of Shh (Shh-N) could also produce digit duplications in the chick wing *(16)*. Beads soaked in Shh-N were used to characterize signaling in the chick wing, and this confirmed similarities with polarizing region signaling *(17)*. Thus, beads soaked in different concentrations of Shh implanted at the anterior margin of chick wing buds led to dose-dependent changes in digit pattern (**Fig. 1C**). There

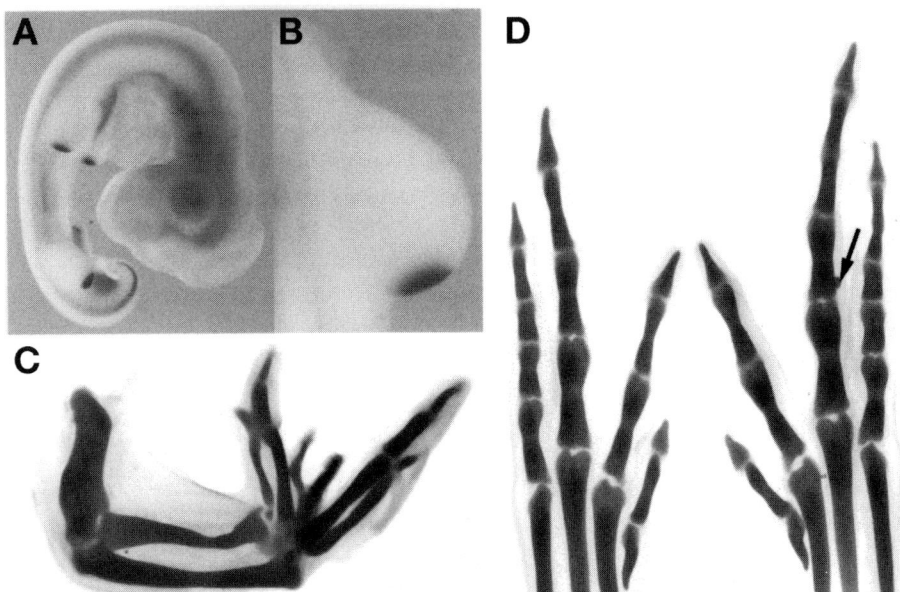

Fig. 1. (**A**) Stage 21 chick embryo showing *Shh* expression in the nervous system, limb bud, and branchial arches. (**B**) Close up of stage 21 wing bud showing Shh expression in the posterior part of the limb bud where the polarizing region is located. (**C**) Alcian Green staining of 10 d chick wing with a digit pattern of 432234 obtained after implanting a Shh-soaked bead. (**D**) Alcian Green staining of 10 d chick leg showing an extra phalanx in digit 3 obtained after implanting a Shh-soaked bead (arrowed) at stage 28.

are indications that Shh diffuses across the chick limb, as indicated by a cell differentiation assay *(18)*, while Shh protein has been detected by immuno-histochemistry *(19)*. Nevertheless, cells expressing membrane tethered Shh still induce complete digit duplications *(17)*. Here, we will describe how to apply Shh protein on beads and graft pellets of Shh-expressing cells to chick limb buds. Each method has its own merits; cells process the Shh protein, while Shh-soaked beads can be readily removed and thus permit experiments that investigate the importance of the timing of Shh signaling *(17)*.

Pellets of Shh-expressing cells and beads soaked in Shh have also been implanted into early mouse limb buds in organ culture *(20)*. These manipulations can be used to explore short-term effects on gene expression. In addition, both normal and mutant mouse limb buds can be treated.

Shh-soaked beads have also been used to probe downstream events that are regulated by Shh in the early limb bud. High-level expression of the vertebrate *patched* (*ptc*) gene, which acts as a receptor for Shh, was seen in response to application of Shh-soaked beads in chick wing buds *(21)*. However, in a poly-dactylous chicken mutant, *talpid³*, Shh beads did not induce high level *ptc (22)*. Also, dissection of downstream consequences of Shh signaling can be accomplished by simultaneous or sequential challenge with beads soaked in other factors *(23)*. The polarizing region overlaps with the posterior necrotic zone of the early chick wing bud and implantation of Shh-soaked beads was found to increase cell death in the posterior necrotic zone *(24)*. There is also extensive cell death in the mesenchyme between digit primordia at later stages in chick limb development, but beads soaked in Shh implanted in the mesenchyme rescued cell death. More unexpectedly, these Shh-soaked beads placed in interdigital mesenchyme also led to the production of an additional phalanx in neighboring digit(s) showing that digit morphogenesis is plastic even at relatively late stages (*[24–26]*; **Fig. 1D**). For further protocols in chick limb bud development, *see* ref. *(27)*.

2. Materials

1. Fertilized chicken eggs. Newly laid eggs can be stored in an incubator at 15°C for up to 1 wk. To start development of the embryos, place in a humidified 38°C incubator and incubate until they reach the desired stage of development; staging according to Hamburger and Hamilton *(28)*.
2. Shh protein. Recombinant mouse N-terminus Shh protein (Shh-N; R & D systems, USA order number 461-SH.025). Alternatively, cells expressing Shh can be used—both transient and stable transfection of cells; chick embryo fibroblasts *(15)*; QT6 cells *(29)*; and COS7 cells *(17)*.
3. Beads. CM Affigel-blue beads (Bio-Rad order number 153-7304) stored at 4°C. Wash 500 µL aliquots of beads to be used for Shh application several times in phosphate-buffered saline (PBS) and store at 4°C in PBS.
4. HEPES buffered DMEM containing 1% antibiotics/antimycotic (GIBCO; Invitrogen UK, USA).
5. PBS.
6. Nile blue sulfate (34059; BDH).
7. Trichloroacetic acid (TCA).
8. Alcian green (34160; BDH).
9. Methyl salicylate.
10. Tools: Several tools are needed to perform surgery on chick limb buds. Pair of blunt forceps to pierce the egg; scissors to enlarge the window in the egg shell; a pair of fine forceps to remove the membranes over the embryo; and a sharp tungsten needle to make the incision in the limb bud. Tungsten needles can be made by fixing a length of 250 or 500 µm diameter tungsten wire into glass tubing with Araldite. The tip is "sharpened" electrolytically by applying a DC potential of 10 V across the wire in 1 *M* NaOH. Sellotape or clear tape is used to cover the egg while enlarging the window and to seal over the window after manipulation.

3. Methods

3.1. Dissolving Shh Protein

1. Recombinant Shh protein comes in powdered form and should be stored at −20°C. Dissolve the powder in PBS containing 0.1% bovine serum albumin (BSA). The protein is made up as a stock solution at a concentration of 10 mg/mL (*see* **Notes 1** and **2**). To do this, the required amount of PBS containing 0.1% BSA is pipetted around the sides of the tube containing the Shh protein.
2. Put the tube in a 50 mL falcon tube and centrifuge for 5 min at 228*g* in a Beckman GS-6 bench top centrifuge (288*g*).
3. Re-pipette the liquid around the sides of the tube to dissolve any remaining powder and centrifuge the tube again. This step is repeated until all the protein is dissolved. Siliconized tips and tubes should always be used.
4. Once the protein is dissolved, divide the solution into 1 μL aliquots in siliconized eppendorf tubes. Store at −80°C.

3.2. Preparation of Beads

1. Place an aliquot of CM Affigel-blue beads in a bacteriological petri dish containing PBS (*see* **Note 3**). Using a calibrated eyepiece graticule to measure the size, select beads with a diameter of 200–250 μm. These will be used as carriers to apply Shh.
2. Suck up 20–30 selected beads in a small volume of PBS using a Glison pipette and place in a drop in the centre of a bacteriological 35 mm tissue culture plate. All the liquid is then carefully removed from the beads using a Gilson pipette and some tissue paper.
3. Thaw an aliquot of Shh solution (1 μL) and add it to the beads (**Fig. 2A**). In addition, make a circle of small drops (25 μL) of PBS around the edge of the petri dish. This keeps the petri dish humid and prevents evaporation of the Shh solution.
4. Wrap the dish in parafilm and leave the beads to soak for at least 30 min at 4°C. The beads and protein solution can be stored for up to 1 wk in the sealed petri dish.

3.3. Windowing the Egg

1. Eggs are incubated for 2–3 d (*see* **Note 4**). First, eggs are gently rotated around the long axis to ensure that the embryo is floating freely on top of the yolk.
2. Disinfect all tools by spraying them with 70% ethanol or wiping them with ethanol soaked tissue paper. Make sure the tools are dry before using them and take care not to transfer any ethanol into the egg.
3. Lay the egg on its side and, using blunt forceps, make a small hole in the shell at the rounded end of the egg (**Fig. 3A**), where the air sac is.
4. Make a second hole in the shell on the uppermost side of the egg, with the blunt forceps and remove a small piece of eggshell without breaking the thick white shell membrane which lies underneath (**Fig. 3B**).

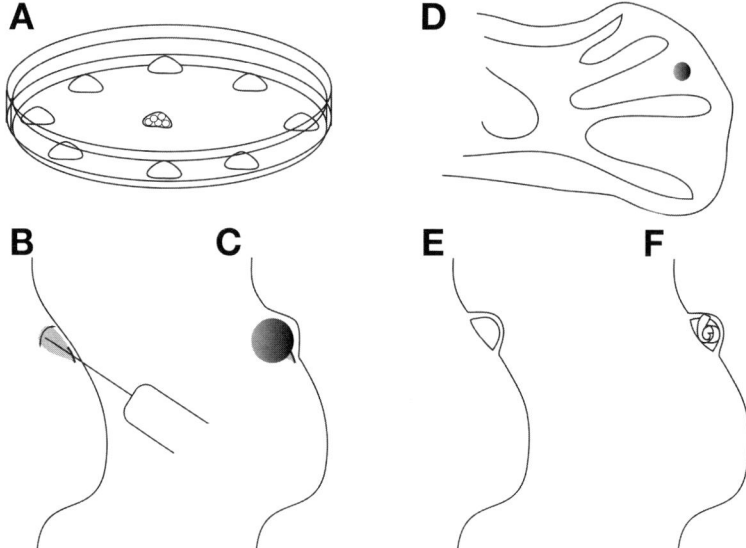

Fig. 2. (**A**) Selected beads are soaked in a small volume of Shh solution in a bacterio-logical petri dish. A circle of drops of PBS is placed around the edge of the dish to keep the chamber moist. (**B**) To implant a Shh-soaked bead, a small incision is made (black line). The space behind this is the hollowed out (gray area) using a sharp needle. (**C**) The bead will be contained in the hollow and not be able to escape through the incision. (**D**) Diagram shows the position in which to implant a Shh-soaked bead to alter digit morphology. It is important to place the bead close to the ectoderm rimming the inter-digital mesenchyme. (**E**) To graft Shh-expressing cells, a cut is made just under the ectoderm, this is then stretched into a loop. (**F**) The graft of Shh-expressing cells is held in place by the loop.

5. Put a drop of medium containing antibiotics/antimycotic onto the shell membrane and pierce the membrane gently; this will allow the embryo to drop (**Fig. 3C**). You will notice the membrane change color.
6. Put a piece of Sellotape over the hole and enlarge it carefully using scissors to make a window. The hole can then be resealed using Sellotape and eggs can be returned to the incubator until the desired stage is reached.

3.4. Implanting Beads

3.4.1. Early Limb Buds

1. Reopen the window by cutting the Sellotape with a pair of scissors.
2. Tear the two transparent membranes (vitelline membrane and amnion) over the embryo in the region of the limb bud using a pair of sharp forceps.

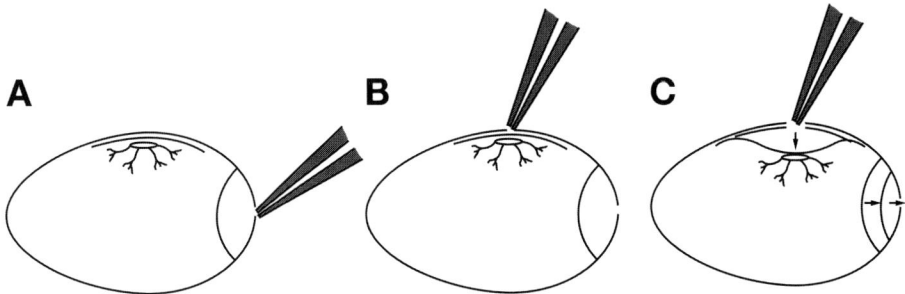

Fig. 3. Windowing an egg (**A**) Make a hole at the blunt end of the egg, where the airsac is. (**B**) Carefully make a hole in the uppermost side of the egg, leaving the underlying membrane intact. (**C**) Carefully enlarge the hole and pierce the membrane. This will cause the airsac to empty (*see* arrows) and the embryo to drop.

3. With a sharp tungsten needle, make a small incision in the ectoderm and underlying mesenchyme at the desired position and make a hollow in the mesenchyme underneath the intact ectoderm near the incision (**Fig. 2B**; *see* **Note 5**).
4. Impale a Shh-soaked bead on the tip of a sharpened tungsten needle and remove it from the drop of protein solution, let it dry a little so that it shrinks slightly, then poke it into the hollow so that the bead is well embedded in the mesenchyme. Keep the needle and bead in this position while the bead gradually rehydrates. The bead will swell and therefore be unable to escape through the small incision.
5. Remove the needle by sliding it back through a fine pair of forceps to leave the bead in place (**Fig. 2C**).
6. Cover the window in the egg with Sellotape, place the egg back at 38°C and incubate until the embryo has developed to the desired stage.

3.4.2. Late Limb Buds

1. Incubate eggs until the embryos are stages 26–28. Reopen the window as described in **Section 3.4.1.** to gain access to the limbs.
2. Tear the membranes over the back of the embryo and gently maneuver the limb on to the top of the membranes (*see* **Note 6**). The membranes should support the limb bud, if not, elevate the limb with forceps.
3. Make an incision and implant the bead as in **Section 3.4.1.** (*see* **Note 7**). It is crucial to keep the embryos moist and using antibiotic/antimyotic solution in PBS is best at these stages.

3.5. Preparing Grafts of Shh-Expressing Cells

There are several different ways of preparing cells for grafting. The simplest method is to graft sheets of Shh-transfected cells (**Sections 3.5.1.** and **3.5.2.**).

Alternatively, cell pellets (**Sections 3.5.3.** and **3.5.4.**) or aggregates (**Sections 3.5.5.** and **3.5.7.**) can be made.

1. Grow Shh-transfected cells to confluence and then scrape off the cells as a sheet using a rubber policeman.
2. Cut the sheets into pieces and lightly stain in medium containing 0.01% Nile blue sulphate to aid visualization.
3. Alternatively, trypsinize transfected cells and then pellet by centrifugation.
4. Incubate the cell pellet for 30 min at 37°C before removing from the tube. Cut into pieces and stain with 0.01% Nile blue sulphate *(15)*.
5. Pellets can also be made by placing a 3 µL drop of transfected cells (5×10^7 cells/mL) on the lid of a petri dish. Then flip the lid upside down to form a hanging drop.
6. After about 3 h, cut the aggregates of cells that form into small pieces for grafting *(17)*.
7. Pellets have also been made by seeding transfected cells at high density on bacteriological petri dishes and allowing the cells to form aggregates overnight *(30,31)*.

3.6. Grafting Shh-Expressing Cells into Early Limb Buds

1. Window eggs, reopen them as described above, and tear away the membranes covering the limb bud (as in **Section 3.4.1.**).
2. Place Shh-cell pellets directly underneath the apical ectodermal ridge. Make a cut along the base of the apical ectodermal ridge at the desired position at the margin of the limb bud with a sharpened tungsten needle.
3. Use a needle with a thick tip to pierce through the bud and ease the apical ectodermal ridge away from the mesenchyme to make a loop.
4. Transfer the piece of cell pellet into the egg using a Gilson pipette and maneuver it under the loop using the thick needle. The apical ectodermal ridge then contracts back to hold the piece of cell pellet in place (**Fig. 2E,F**).

3.7. Skeletal Analysis After Implanting a Shh-Soaked Bead or Grafting Shh-Expressing Cells

1. After 10-d incubation, reopen the window, cut the membranes surrounding the embryo, and transfer the embryo into a dish containing PBS.
2. Remove any excess membrane and, if desired, the internal organs from the embryo. Place the embryo or limbs in a glass vial (*see* **Note 8**) containing 5% TCA and fix overnight at room temperature.
3. The next day, transfer the embryo or limbs to 70% ethanol, 1% HCl for 2 h.
4. Stain 3 h to overnight in 1% Alcian green, 70% ethanol, and 1% HCl.
5. If the color is too dark, staining can be differentiated (reversed) by washing in 70% ethanol and 1% HCl.
6. Dehydrate in successive steps of ethanol for 2 h each (70, 90, and two washes in 100%).
7. Clear the tissue in methyl salicylate and photograph in a glass petri dish. The specimen can be gently flattened using a square of glass, cut from a microscope slide using a diamond pencil.

4. Notes

1. It is important not to try and dissolve the Shh protein powder by pipetting the added liquid up and down because the solution will froth.
2. Beads soaked in 8 µg/µL Shh from R&D systems have given full digit duplications. Each time a new solution of Shh is made, it should be tested for duplicating activity. Beads soaked in lower concentrations of Shh can also produce duplications.
3. Use bacteriological petri dishes rather than wettable tissue culture dishes so that the PBS forms a nice drop instead of spreading out.
4. It is best to window eggs at 2–3 d of incubation even if manipulations are not carried out until later in development. This also provides an opportunity for adjusting incubation times to allow "younger" embryos to catch up.
5. Chicken embryos usually lie on their left side in the egg, with the right side uppermost. Therefore, one sees the dorsal side of the right limb bud and anterior is toward the head and posterior toward the tail.
6. Be careful not to damage the vascularized allantois.
7. In order to produce changes in digit morphogenesis, Shh-soaked beads should be placed near the ectoderm rimming the interdigital mesenchyme.
8. It is crucial to use glass vials as methyl salicylate will melt plastic.

Acknowledgments

Cheryll Tickle is supported by The Royal Society and both Eva Tiecke and Cheryll Tickle are supported by the MRC. We would like to thank Angie Blake for help with preparation of the manuscript. Dr Allyson Clelland for supplying **Fig. 1C** and Dr Juan-Jose Sanz-Ezquerro for help with **Section 3.4.2.** and supplying **Fig. 1D**.

References

1. Tickle, C. (2004) The contribution of chicken embryology to the understanding of vertebrate limb development. *Mech. Dev.* **121,** 1019–1029.
2. Saunders, J. W. and Gasseling, M. T. (1968) Ectodermal–mesodermal in the origin of limb symmetry. In *Epithelial-Mesenchymal Interactions* (Fleischmeyer, R. and Billingham, R. E., eds), Williams and Wilkins, Baltimore, pp. 78–97.
3. Tickle, C., Summerbell, D., and Wolpert, L. (1975) Positional signaling and specification of digits in chick limb morphogenesis. *Nature* **254,** 199–202.
4. Smith, J. C., Tickle, C., and Wolpert, L. (1978) Attenuation of positional signaling in the chick limb by high doses of gramma-radiation. *Nature* **272,** 612–613.
5. Tickle, C. (1981) The number of polarizing region cells required to specify additional digits in the developing chick wing. *Nature* **289,** 295–298.
6. MacCabe, A. B., Gasseling, M. T. Jr., and Saunders. J. W. (1973) Spatiotemporal distribution of mechanisms that control outgrowth and anteroposterior polarization of the limb bud in the chick embryo. *Mech. Ageing. Dev.* **2,** 1–12.
7. Honig, L. S. and Summerbell, D. (1985) Maps of strength of positional signaling activity in the developing chick wing bud. *J. Embryol. Exp. Morphol.* **87,** 163–174.

8. Fallon, J. F. and Crosby, G. M. (1977) Polarizing zone activity in limb buds in amniotes. In *Vertebrate Limb and Somite Morphogenesis* (Hinchliffe, J. R., Balls, M., and Ede, D. A., eds), Cambridge University Press, Cambridge, pp. 55–70.

9. Hornbruch, A. and Wolpert, L. (1986) Positional signaling by Hensen's node when grafted to the chick limb bud. *J. Embryol. Exp. Mophol.* **94,** 257–265.

10. Dollé, P., Izpisúa-Belmonte, J. C., Brown, J. M., Tickle, C., and Duboule, D. (1991) HOX-4 genes and the morphogenesis of mammalian genitalia. *Genes Dev.* **5,** 1767–1777.

11. Tickle, C., Alberts, B., Wolpert, L., and Lee, J. (1982) Local application of retinoic acid to the limb bud mimics the action of the polarizing region. *Nature* **296,** 564–566.

12. Eichele, G., Tickle, C., and Alberts, B. (1984) Microcontrolled release of biologically active compounds in chick embryos: beads of 200-microns diameter for the local release of retinoids. *Anal. Biochem.* **142,** 542–555.

13. Tickle, C., Lee, J., and Eichele, G. (1985) A quantitative analysis of the effect of all-trans-retinoic acid on the pattern of chick wing development. *Dev. Biol.* **109,** 82–95.

14. Eichele, G., Tickle, C., and Alberts, B. M. (1985) Studies on the mechanism of retinoid-induced pattern duplications in the early chick limb bud: temporal and spatial aspects. *J. Cell Biol.* **101,** 1913–1920.

15. Riddle, R. D., Johnson, R. L., Laufer, E., and Tabin, C. (1993) Sonic hedgehog mediates the polarizing activity of the ZPA. *Cell* **75,** 1401–1416.

16. Lopez-Martinez, A., Chang, D. T., Chiang, C., et al. (1995) Limb-pattern activity and restricted posterior localization of the amino-terminal product of Sonic hedgehog cleavage. *Curr. Biol.* **5,** 791–796.

17. Yang, Y., Drossopoulou, G., Chuang, P. T., et al. (1997) Relationship between dose, distance and time in Sonic Hedgehog-mediated regulation of anteroposterior polarity in the chick limb. *Development* **124,** 4393–4404.

18. Zeng, X., Goetz, J. A., Suber, L. M., Scott, W. J. Jr., Schreiner, C. M., and Robbins, D. J. (2001) A freely diffusible form of Sonic hedgehog mediates long-range signalling. *Nature* **411,** 716–720.

19. Gritli-Linde, A., Lewis, P., McMahon, A. P., and Linde, A. (2001) The whereabouts of a morphogen: direct evidence for short- and graded long-range activity of Hedgehog signaling peptides. *Dev. Biol.* **236,** 364–386.

20. Zuniga, A., Haramis, A. P., McMahon, A. P., and Zeller, R. (1999) Signal relay by BMP antagonism controls the SHH/FGF4 feedback loop in vertebrate limb buds. *Nature* **401,** 598–602.

21. Marigo, V., Scott, M. P., Johnson, R. L., Goodrich, L. V., and Tabin, C. J. (1996) Conservation in Hedgehog signaling: an induction of a chicken patched homolog by Sonic hedgehog in the developing limb. *Development* **122,** 1225–1233.

22. Lewis, K. E., Drossopoulou, G., Paton, I. R., et al. (1999) Expression of ptc and gli genes in talpid3 suggests bifurcation in Shh pathway. *Development* **126,** 2397–2407.

23. Drossopoulou, G., Lewis, K. E., Sanz-Ezquerro, J. J., et al. (2000) A model for anteroposterior patterning of the vertebrate limb based on sequential long- and short-range Shh signaling and Bmp signaling. *Development* **127,** 1337–1348.

24. Sanz-Ezquerro, J. J. and Tickle, C. (2000) Autoregulation of Shh expression and Shh induction of cell death suggest a mechanism for modulating polarising activity during chick limb development. *Development* **127,** 4811–4823.

25. Dahn, R. D. and Fallon, J. F. (2000) Interdigital regulation of digit identity and homeotic transformation by modulated BMP signaling. *Science* **289,** 438–441.

26. Sanz-Ezquerro, J. J. and Tickle, C. (2003) Fgf signaling controls the number of phalanges and tip formation in developing digits. *Curr. Biol.* **13,** 1830–1836.

27. Ros, M. A., Simandl, B. K., Clark, A. W., and Fallon, J. (2000) Methods for manipulating the chick limb bud to study gene expression, tissue interactions and patterning. In *Developmental Biology Protocols III* (Tuan, R. S. and Lo, C. W. eds), Humana, Totowa, NJ.

28. Hamburger, V. and Hamilton, H. L. (1951) A series of normal stages in the development of the chick embryo. *J. Morph.* **88,** 49–92.

29. Duprez, D., Fournier-Thibault, C., and Le Douarin, N. (1998) Sonic Hedgehog induces proliferation of committed skeletal muscle cells in the chick limb. *Development* **125,** 495–505.

30. Duprez, D. M., Kostakopoulou, K., Francis-West, P. H., Tickle, C., and Brickell, P. M. (1996) Activation of Fgf-4 and HoxD gene expression by BMP-2 expressing cells in the developing chick limb. *Development* **122,** 1821–1828.

31. de la Pompa, J. L. and Zeller, R. (1993) Ectopic expression of genes during chicken limb pattern formation using replication defective retroviral vectors. *Mech. Dev.* **43,** 187–198.

3

Manipulation of Hedgehog Signaling in *Xenopus* by Means of Embryo Microinjection and Application of Chemical Inhibitors

Thomas Hollemann, Emmanuel Tadjuidje, Katja Koebernick, and Tomas Pieler

Abstract

Xenopus embryos provide a powerful model system to investigate the complex molecular mechanisms, which are controlled by or control the activity of the Hedgehog (Hh) signaling pathway. The use of synthetic mRNA or antisense oligonucleotide (morpholino) microinjection into blastomeres of early embryos or by simply treating the embryos with small organic inhibitors, has already led to an idea of the network in which the Hh pathway is embedded. More needs to be done in order to achieve a detailed understanding of how the different players of the Hh signaling pathway are integrated to control different genetic programs, such as axis formation in early embryos or cell differentiation during retinogenesis.

Key Words: Patched; smoothened; Hedgehog-interacting protein (HIP); 7-dehydrocholesterol reductase (DHCR7); AY9944; mevinolin; statin; hydroxymethyl-glutaryl coenzyme A reductase (HMGR).

1. Introduction

During early development, patterning events within the embryo are mediated by morphogens, which provide positional information for the definitive fate of uncommitted precursor cells. Important examples for morphogens are provided by members of the secreted Hedgehog (Hh) family proteins. These signaling molecules have been highly conserved from *Drosophila* to humans, but are not found in lower bilaterians like *Caenorhabditis elegans*. Hh proteins are involved in pattern formation and cell specification, e.g., in the neural tube and in appendices like the wing in *Drosophila* and the limbs in vertebrates. In addition, it has

From: *Methods in Molecular Biology: Hedgehog Signaling Protocols*
Edited by: J. Horabin © Humana Press Inc., Totowa, NJ

been shown that Hh proteins are involved in stem cell maintenance and axon guidance. As a consequence, mutations in elements of the Hh signaling pathway have been found to result in a variety of severe congenital developmental defects; in addition, they are also often associated with the manifestation of cancer. Despite the huge body of knowledge on the various biological functions of Hh signaling, our understanding of the exact molecular mechanisms that mediate Hh signal transduction is still fragmentary (reviewed in Refs. *[1–6]*). In *Xenopus laevis*, Hh proteins, namely Sonic, Banded (*Xbhh*, Indian in mammals), and Cephalic (*Xchh*, Desert in mammals) Hh, have been identified by Ekker and colleagues *(7)*. More recently, we have reinvestigated the embryonic expression of *Xbhh* and *Xchh* and could show that *Xbhh* is expressed only very weakly in the notochord of tadpole stage embryos (NF st. 34), whereas *Xchh* transcripts were mainly detected in the endodermal germ layer *(8)*.

Manipulation of the Hh-signaling pathway has been achieved by a panel of molecular tools, which either activate the pathway (such as overexpression of Hh proteins and of the co-receptor, Smoothened) or repress the pathway (such as application of dominant-negative Patched or of the Hedgehog-interacting protein (HIP), or by treatment of embryos with small organic molecules like cyclopamine and jervine) *(7,9,10)*. In addition, Hh mutant phenotypes have been observed in embryos that were treated with proximal or distal inhibitors of the cholesterol biosynthesis pathway (hydroxymethyl-glutaryl coenzyme A reductase (HMGR) inhibitors, such as the statins or 7-dehydrocholesterol reductase (DHCR7) inhibitors such as AY9944, respectively), which may interfere with the chemical modification of the Hh protein in the sending cell, or with the reception of the signal in the receiving cell *(11*, **Fig. 1**). In this chapter, we describe how such manipulations of Hh signaling can be performed using Xenopus embryos as an experimental system (*see* **Note 1**).

2. Materials

1. Adult *Xenopus laevis* frogs are purchased from NASCO (Wisconsin, USA).
2. Chorionic gonadotropin (HCG, cat. no. CG-10, Sigma, USA).
3. L-15 (Leibovitz; GibcoBRL, Germany).
4. L-cysteine hydrochloride monohydrate (cat. no. 30129, Fluka, Germany).
5. Penicillin/streptomycin solution (with 10,000 U penicillin and 10 mg streptomycin per ml, cat. no. P-0781, Sigma).
6. Mevinolin (cat. no. M2147, Sigma).
7. AY9944 (cat. no. S693162, Sigma).
8. Cyclopamine (cat. no. C988400, TRC, Germany).
9. Jervine (cat. no. J211000, TRC).
10. Modified Barth's Solution (MBS): 88 mM NaCl, 2.4 mM NaHCO$_3$, 1.0 mM KCl, 10 mM HEPES, 0.82 mM MgSO$_4$, 0.41 mM CaCl$_2$, and 0.33 mM Ca(NO$_3$)$_2$ (pH 7.4).

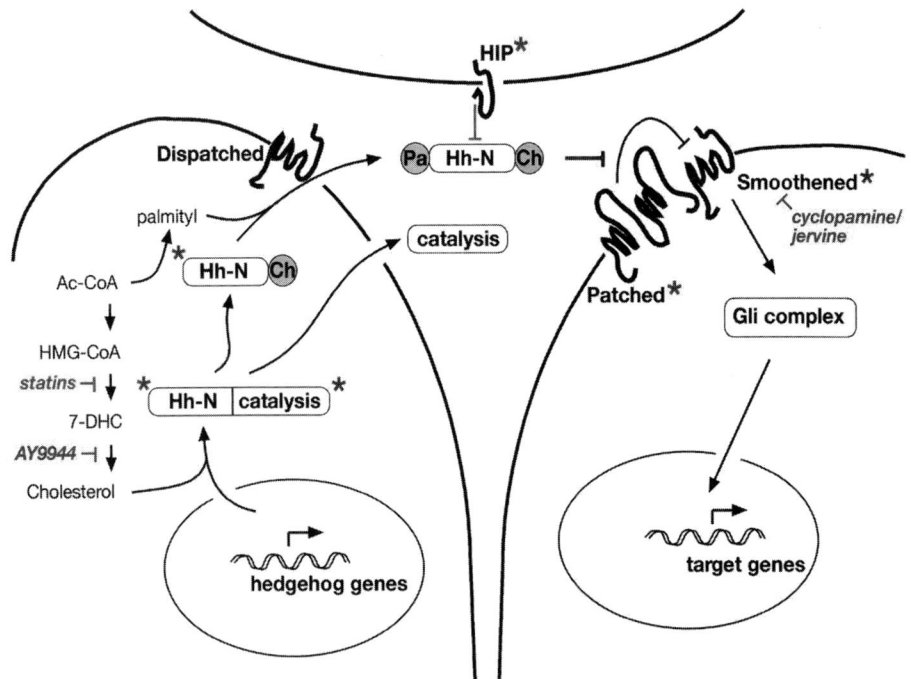

★ Overexpression of activating constructs

★ Overexpression of dominant-negative constructs

Fig. 1. Molecular tools to dissect the Hedgehog (Hh) signaling pathway. Micro-injection of synthetic Hh mRNA leads to an activation of the pathway, as does the injection of Smoothened or a constitutive active form of Smoothened (Smo-M2), whereas the overexpression of a dominant-negative form of patched (PtcDLoop2) or of the Hedgehog-interacting protein represses the pathway, as do small organic molecules, which bind to and inhibit Smo activity. Within the sending cell, Hh modification and secretion seem to depend on the availability of cholesterol. Lipid raft formation in both cells may equally be necessary for proper secretion and reception of the signal.

11. MEMFA buffer: 0.1 M MOPS (pH 7.4), 2 mM EGTA, 1 mM MgSO$_4$, and 3.7% formaldehyde.
12. Nile blue stain (saturated solution of Nile blue sulfate in 50 mM phosphate buffer (pH 7.8) filtered through 3 mm filter).
13. Ca^{2+}/Mg^{2+}-free MBS: 88 mM NaCl, 2.4 mM NaHCO$_3$, 1.0 mM KCl, 10 mM HEPES (pH 7.4).
14. Prehybridization mix: 50% formamide (Merck, Germany), 5× SSC (20× SSC: 3 M NaCl, 0.3 M sodium citrate (pH 7.0) with NaOH), 1 mg/mL Torula RNA (Sigma), 100 µg/mL Heparin (Sigma), 1× Denhardts (100× Denhardts: 2% BSA (Fluka),

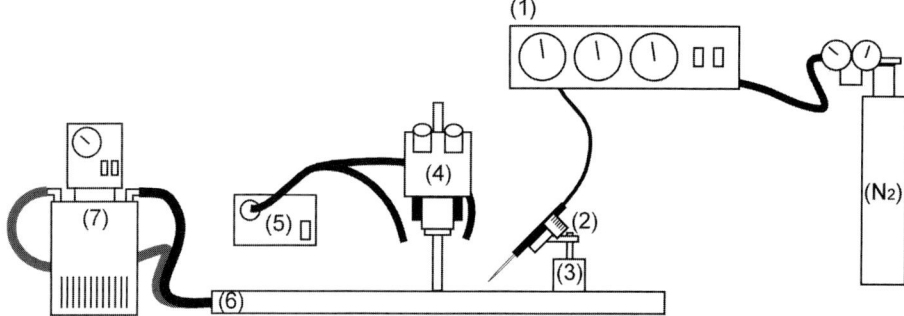

Fig. 2. Schematic of the microinjection set-up (*see* text for details).

2% PVP (Sigma), 2% Ficoll 400 (Pharmacia, Sweden), 0.1% Tween 20 (Sigma), 0.1% CHAPS (Sigma), and 10 mM EDTA in DEPC H$_2$O (Merck).
15. L-15/BSA solution: 65% L-15 Leibovitz medium (Sigma-Aldrich), 0.1% BSA (pH 7.4).
16. Gastromaster and the replacement microsurgery tips (XENOTEK Engineering, Belleville, IL, USA).
17. Microinjection set-up (as illustrated in **Fig. 2**): pneumatic picopump (PV 820) for injection is from World Precision Instruments, micromanipulator (M3301R) and magnetic stand (M1; WPI, Inc., USA), stereo-microscope (Stemi 2000; Zeiss, Germany), cold light equipped with fiber optics (KL1500; Schütt, Germany), metal plate (self-made) is cooled by a *(7)* refrigerating circulator (Biometra, Germany), and incubator (WB22K, 10–99°C; Mytron, Germany).
18. Micropipette Puller (PN-30; Narishige, Japan).

3. Methods

In general, we use pigmented embryos for injection experiments, although for the whole mount *in situ* hybridization which follows, best results are obtained using albino embryos. Pigmented embryos are easier to inject, since the orientation of the dorsal–ventral axis is easy to see.

3.1. Preparation and Culturing of Xenopus Embryos

1. Obtain eggs from adult *Xenopus* females 6 to 8 h after injection with human chorionic gonadotropin (500–1000 U/frog). Spawning can be stimulated by a warm water bath (30°C) and/or a gentle massage of the primed frogs.
2. Prepare the sperm by mincing in 100 µL of 1× MBS on ice 1/5 of a testis using a fine pair of scissors. Dilute to 0.1× MBS with 900 µL H$_2$O prior to fertilization.
3. Fertilize the eggs in a petri dish with a solution of freshly prepared sperm for 2 min and cover with 0.1× MBS.
4. Remove the jelly coat of the fertilized eggs 40 min after fertilization by swirling in 2% cysteine (pH 8.0) for 2 to 3 min (longer treatment will damage the embryos).

Wash extensively in 0.1× MBS. If any jelly coat was not removed, repeat the cysteine step.

5. For staging and orientation prior to injection, albino embryos need to be stained with Nile blue after dejellying, or prior to use.

3.2. Preparation of Synthetic Messenger RNA

In order to achieve efficient translation, mRNAs for injection should contain a Cap-structure, which can be introduced using either the Stratagene in vitro RNA Transcription Kit supplemented with RNasin (Promega) and Cap-nucleotides (Biolabs) or Ambion's mMessage Machine Kit.

1. In the first case, react the following mixture for 2 to 4 h at 37°C: 2 µL 10× RNA polymerase buffer, 10 µL of 2× nucleotide triphosphate mix (6 mM ATP, 6 mM CTP, 3 mM GTP, 6 mM UTP, and 9 mM m7(5′)Gppp(5′)G cap analog (NEB)), 2 µL of 200 mM DTT, 0.5 µL of 20 U/µL RNasin (Promega), 2 µL linearized DNA template (1 µg), 1.5 µL of 20 U/µL of the respective RNA polymerase (*see* **Notes 2** and **3**).
2. After RNA synthesis, digest the DNA template by adding 1 µL RNase free DNase I (Boehringer, Mannheim) and incubate the reaction for 15 min at 37°C.
3. Purify the RNA with the help of the RNeasy Kit (Qiagen).
4. Elute the RNA in a small volume (25 µL H$_2$O) and check 1/20 of the newly synthesized RNA on a freshly prepared 1% agarose-gel in 1× TBE. If desired, dilute the remaining Cap-RNA with RNase-free H$_2$O.
5. Store synthesized RNA at −80°C.

3.3. Microinjection of Synthetic mRNA into Xenopus Embryos

Depending on the construct (gene of interest), 1 or up to 2000 pg in vitro synthesized capped mRNA can be microinjected into blastomeres of 1- to 32-cell stage embryos in 5 to 10 nL H$_2$O. Fate maps can be found in Refs. *(12–14)*. The correct injection volume needs to be adjusted under a microscope with the help of an eyepiece micrometer. To do so, measure the diameter of the drop and calculate the respective volume ($v = 4/3\pi r^3$). After injection, the embryos are transferred into 1× MBS and kept at 15°C. Two hours after injection, embryos are transferred to 0.1× MBS (*see* **Note 4**).

3.4. Gain-of-Function Assays

3.4.1. Overexpression of Hh Proteins by mRNA Microinjection

3.4.1.1. WHOLE EMBRYOS

1. Overexpression of *Xshh* in whole embryos is achieved by injection of 500 pg of the corresponding synthetic capped mRNA into one blastomere of two-cell stage embryos.

2. Injection can also be done at the four-cell stage or later if one needs to restrict the overexpression to a specific group of cells.
3. To trace the injected side, 50–100 pg β-galactosidase-encoding synthetic mRNA is coinjected.

3.4.1.2. ANIMAL CAP EXPLANTS

1. For overexpression of *Xshh* in animal cap explants, inject 500 pg capped RNA into both blastomeres (250 pg per blastomere) of two-cell stage embryos, to ensure a homogenous distribution of the injected RNA. *Xshh* induces its own expression and that of targets like *patched* only in animal caps, which are neuralized. Therefore, *Xshh* mRNA is usually injected in combination with *chordin* mRNA (25 pg per blastomere) (*see* **Note 5**).
2. When different combinations of RNA are injected, adjust the RNA load to the same total with β-galactosidase mRNA. This is especially relevant when injected caps are to be analyzed by RT-PCR, to prevent RNA concentration effects.
3. Excise animal cap explants at stage 8 with a gastromaster (Xenotek, USA) and culture in 0.5× MBS containing penicillin/streptomycin on top of 0.8% agarose in small plastic dishes.
4. Harvest explants when control embryos (of the same batch) have reached the desired stage.

3.4.2. Overexpression of Constitutively Active Smoothened Receptor (Smo-M2) by RNA Microinjection

The 7-pass transmembrane protein Smoothened (Smo) is essential for the transmission of the Hh signal into the cell. In the absence of Hh ligands, the Hh receptor Patched (Ptc) inhibits the signaling activity of Smo *(15)*. A single amino acid mutant of Smo identified in human basal cell carcinoma, Smo-M2 *(16,17)*, is understood to constitutively activate Smo independent of the Hh ligand.

1. For overexpression, inject 1.5 ng synthetic mRNA encoding XSmo-M2, the corresponding *Xenopus* mutant (containing the amino acid substitution W508L), into one cell of two-cell stage embryos in combination with 100 pg β-galactosidase mRNA as lineage tracer.
2. Allow embryos to develop to the desired stage.

3.5. Loss-of-Function Assays (see Note 6)

3.5.1. Overexpression of HIP by RNA Microinjection

The HIP binds to Hh proteins in the extracellular space. In *Xenopus*, the expression domain of HIP (*Xhip*) has been reported to be located in close proximity to *Shh*, *Wnt-8*, and *Fgf-8*-positive cells, and *Xhip* has been described as a multifunctional antagonist of Hh, Fgf, and Wnt-8 pathways *(8)*.

3.5.1.1. WHOLE EMBRYOS

1. To overexpress *Xhip* in whole embryos, inject 750 pg synthetic, capped mRNA into one blastomere of two-cell stage embryos, in combination with 100 pg β-galactosidase mRNA as lineage tracer.

3.5.1.2. ANIMAL CAP EXPLANTS

1. For animal cap explants assays, inject a total of 1 ng *Xhip* mRNA into both blastomeres (500 pg for each) of two-cell stage embryos.
2. When different combinations of RNA are injected, adjust the RNA load to the same total with β-galactosidase mRNA. When animal caps injected with increasing amounts of mRNA are to be analyzed by RT-PCR, this is especially important since high amounts of RNA are more stable than low amounts.
3. Excise animal cap explants at stage 8.
4. Culture the animal cap explants in 0.5× MBS containing penicillin/streptomycin on top of 0.8% agarose-coated dishes.
5. Harvest explants when control embryos (of the same batch) have reached the desired stage.

3.5.2. Overexpression of PtcΔLoop2 by RNA Microinjection

PtcΔLoop2 is a truncated version of the Hh receptor *Patched1* that lacks the second extracellular loop to which the Hh protein normally binds. It was first described as a Hh-insensitive form of Patched whose expression blocks Shh-induced signal transduction in the neural tube *(18)*.

Overexpression of PtcΔLoop2 in whole embryos is performed by injection of 2 ng synthetic capped mRNA into one blastomere of two-cell stage embryos, in combination with 100 pg β-galactosidase mRNA as linage tracer.

3.6. Treatment of Xenopus Embryos with Hh Signaling Inhibitors

3.6.1. Cyclopamine/Jervine Treatment

Cyclopamine and jervine are naturally occurring steroidal alkaloids that cause cyclopia by blocking the Sonic Hh signaling pathway. Jervine is a close structural analog of cyclopamine, which inhibits Hh signaling by binding to the signal transducer Smoothened.

1. Prepared stock solutions of cyclopamine (20 m*M*) and jervine (10 m*M*) in 100% ethanol and stored at −20°C.
2. For efficient drug absorption, remove the vitelline membrane of the embryos manually with a pair of good forceps (Dsumont no. 5) prior to treatment.
3. Treat embryos in small dishes of 1-cm diameter in which approx 500 μL of solution can be used for 10 embryos. This allows application of high concentrations while saving the chemical.

4. Treat control embryos with an equivalent dilution of ethanol (1:100 for cyclopamine and 1:50 for jervine).
5. Completely cover the dishes with aluminum foil. To prevent photodegradation of the drugs, treatment should be done in darkness.
6. From stages 8.5 to 9 onwards, culture embryos in dishes that have been coated with 0.8% agarose, in 0.1× MBS in the presence of cyclopamine and/or jervine at a final concentration of 200 μ*M* each.
7. Change solutions daily until control embryos have reached the desired stage.
8. At tadpole stage, cyclopamine- and/or jervine-treated embryos show cyclopic eyes (the two eyes tend to fuse to the midline). Examples are shown in **Fig. 3**.

3.6.2. Mevinolin/AY9944 Treatment

Cholesterol modification of Hh proteins is a prerequisite for the production of an active signal, and for the proper spatial distribution of the signal. Mevinolin is a derivative of a chemical substance called statin. Statins inhibit the activity of HMGR, which catalyzes the rate-limiting step in the biosynthesis of sterols. AY9944 (*trans*-1.4-*bis*-2 chlorobenzylaminomethyl cyclohexane) inhibits the activity of DHCR7, which catalyzes the production of cholesterol from its direct precursor 7-dehydrocholesterol (*see* **Note 7**).

1. Prepare stock solutions of mevinolin and AY9944 (25 m*M*) in 100% ethanol and 100% DMSO, respectively. Store at −20°C.
2. Working concentrations are 125 and 250 μ*M* in 0.1× MBS for mevinolin and AY9944, respectively. Treatment from earlier stages (two-cell stage) or with higher concentrations of mevinolin (250 μ*M*) does not alter the observed phenotypes.
3. Treat embryos from stages 8.5 to 9 onwards.
4. Treat control embryos with equivalent dilutions of 100% ethanol (1:200) and 100% DMSO (1:100), respectively.
5. Change solutions daily until the desired stage of development is reached.
6. Mevinolin treatment is highly reproducible, and the manipulated embryos easily grow to late tadpole stage.
7. Examples of neural phenotypes caused by mevinolin and AY9944 treatment are shown in **Fig. 3**.

4. Notes

1. The experimental protocols described above clearly illustrate the availability of a multitude of experimental protocols to either positively or negatively modulate Hh signaling activities in *Xenopus* embryos and in embryonic explants derived from such manipulated embryos. The major advantage is the relative ease in generating large numbers of experimental embryos, allowing for the routine use of more than one approach for the generation of either loss- or gain-of-function effects in the context of a given biological problem. A considerable limitation can be seen in the rather global modulation of Hh signaling in all these protocols.

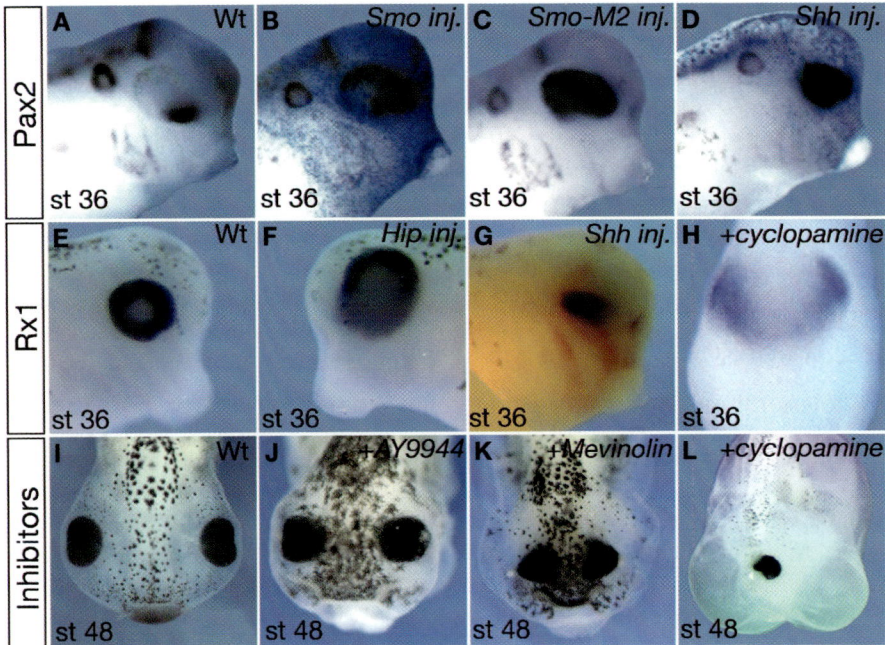

Fig. 3. Phenotypic effects of manipulating Hh signaling in *Xenopus* embryos. (**A–D**) *Pax2* expression within the optic stalk is positively regulated by Hedgehog (Hh) signaling. Ectopic activation of the Hh-pathway by either microinjection of 1.5 ng synthetic mRNA for *Smo* (**B**), or for *Smo*-M2 (**C**), or of 500 pg mRNA for *Shh* (**D**) into one cell of a two-cell stage embryo expands the *Pax2* domain into the remaining eye cavity. (**E–H**) *Rx1* expression is negatively regulated by Hh signaling. *Rx1* transcripts are found in retinal cell of the eye cup (**E**). The *Rx1* expression domain can be either expanded by Hedgehog-interacting protein overexpression (**F**) and cyclopamine treatment (**H**), or it can be reduced by injection *Shh* mRNA (**G**). (**I–L**) Shh-dependent separation of the eye field can be blocked by treating embryos with small organic molecules that interfere with different degrees of efficiency with cholesterol synthesis and hedgehog signal transduction: (**J**) AY9944: inhibitor of 7-dehydrocholesterol-reductase, (**K**) Mevinolin: inhibitor of HMG-CoA-reductase, and (**H,L**) Cyclopamine: inhibitor of Smoothened.

2. The pCS2+ expression vector has been utilized to effectively produce synthetic messenger RNA. In addition to the basic vector, modified versions are available containing either a nuclear localization signal (pCS2+-NLS), antibody-binding domain (pCS2+-Myc or Flag) or the ligand-binding domain of the glucocorticoid receptor (GR-LBD) for the respective fusion constructs.

3. Do not use H_2O treated with pyrocarbonate to prepare synthetic RNA; we obtained best results using the RNAse-free water of the RNAeasy Kit from Qiagen.
4. The development of Xenopus embryos can be decelerated when the embryos are cultured at lower temperatures. However, temperatures below 12°C may significantly increase the rate of gastrulation errors.
5. In experiments where relative activities of individual proteins are to be compared by means of RNA injection, a second non-interfering RNA should be added to make sure that all injections contain the same total amount of RNA, since injected RNAs are more stable at higher concentrations.
6. Attempts to make use of antisense morpholino oligonucleotides in order to knock down the activity of individual Hh signaling components have so far been mostly unsuccessful in our hands, but the availability of such tools would define a further significant improvement with respect to the experimental approaches applicable to the functional analysis of Hh signaling in vertebrate embryos.
7. AY9944 treatment is often compromised by massive embryonic death after stage 34/35, rendering it difficult to analyze late phenotypes. However, tadpole stage embryos can be raised if the AY9944 treatment is terminated at around stage 32/33. AY9944 treatment earlier than stages 8.5 to 9 results in massive death during neuralization, and a higher concentration (500 μM from stages 8.5 to 9) kills the embryos very early (around stages 28–30).

References

1. Briscoe, J. and Ericson, J. (2001) Specification of neuronal fates in the ventral neural tube. *Curr. Opin. Neurobiol.* **11**, 43–49.
2. Ingham, P. W. and McMahon, A. P. (2001) Hedgehog signaling in animal development: paradigms and principles. *Genes Develop.* **15**, 3059–3087.
3. Jacob, J. and Briscoe, J. (2003) Gli proteins and the control of spinal-cord patterning. *EMBO Reports* **4**, 761–765.
4. Lum, L. and Beachy, P. A. (2004) The Hedgehog response network: sensors, switches, and routers. *Science* **304**, 1755–1759.
5. Kalderon, D. (2005) Hedgehog signaling: an Arrestin connection? *Curr. Biol.* **15**, 175–178.
6. Ashe, H. L. and Briscoe, J. (2006) The interpretation of morphogen gradients. *Development* **133**, 385–394.
7. Ekker, S. C., McGrew, L. L., Lai, C. J., et al. (1995) Distinct expression and shared activities of members of the Hedgehog gene family of Xenopus laevis. *Development* **121**, 2337–2347.
8. Cornesse, Y., Pieler, T., and Hollemann, T. (2005) Olfactory and lens placode formation is controlled by the Hedgehog-interacting protein (Xhip) in Xenopus. *Dev Biol.* **277**, 296–315.
9. Koebernick, K., Hollemann, T., and Pieler, T. (2003) A restrictive role for Hedgehog signaling during otic specification in Xenopus. *Dev. Biol.* **260**, 325–338.
10. Perron, M., Boy, S., Amato, M. A., et al. (2003) A novel function for Hedgehog signaling in retinal pigment epithelium differentiation. *Development* **130**, 1565–1577.

11. Tadjuidje, E. and Hollemann, T. (2006) Cholesterol homeostasis in development: the role of Xenopus 7-dehydrocholesterol reductase (Xdhcr7) in neural development. *Dev. Dyn.* **235**(8), 20.

12. Hirose, G. and Jacobson, M. (1979) Clonal organization of the central nervous system of the frog. I. Clones stemming from individual blastomeres of the 16-cell and earlier stages. *Dev. Biol.* **71**, 191–202.

13. Jacobson, M. and Hirose, G. (1981) Clonal organization of the central nervous system of the frog. II. Clones stemming from individual blastomeres of the 32- and 64-cell stages. *J. Neurosci.* **1**, 271–284.

14. Dale, L. and Slack, J. M. (1987) Fate map for the 32-cell stage of Xenopus laevis. *Development* **99**, 527–551.

15. Chen, Y. and Struhl, G. (1996) Dual roles for patched in sequestering and transducing Hedgehog. *Cell* **87**, 553–563.

16. Hynes, M., Ye, W., Wang, K., et al. (2000) The seven-transmembrane receptor Smoothened cell-autonomously induces multiple ventral cell types. *Nat. Neurosci.* **3**, 41–46.

17. Xie, J., Murone, M., Luoh, S. M., et al. (1998) Activating Smoothened mutations in sporadic basal-cell carcinoma. *Nature* **391**, 90–92.

18. Briscoe, J., Chen, Y., Jessell, T. M., and Struhl, G. (2001) A Hedgehog-insensitive form of patched provides evidence for direct long-range morphogen activity of Sonic hedgehog in the neural tube. *Mol. Cell* **7**, 1279–1291.

4

Isolation of Rat Telencephalic Neural Explants to Assay Shh GABAergic Interneuron Differentiation-Inducing Activity

Rina Mady and Jhumku D. Kohtz

Abstract

Multiple assays for Shh activity using cell lines, primary cultures, and explanted tissue have been described. We first described the use of E11.5 rat dorsal telencephalic explants to assay Shh ventralizing and differentiation-inducing activity in Kohtz et al. *(1)*. Using this assay, we subsequently showed that N-lipid modification is critical for Shh activity in the telencephalon *(2)*. In vivo assays for lipid-modified Shh support the results of our E11.5 telencephalic neural explant assay *(2)*. More recently, the method of isolating telencephalic explants was improved by an intraocular grid, increasing both its accuracy and reproducibility *(3)*. Shh induces the expression of the following ventral telencephalic markers: MASH-1, the Dlx's, and Islet 1/2. Therefore, this assay for Shh induction of GABAergic interneurons defines a competent, but naïve region within the E11.5 dorsal telencephalon, allowing the study of GABAergic interneuron induction and differentiation from an unspecified progenitor population.

Key Words: Sonic hedgehog; forebrain; telencephalon; GABAergic; neural explants; Dlx; differentiation.

1. Introduction

We describe a detailed procedure for the isolation and culture of a precisely defined region within the rat E11.5 embryonic telencephalon that responds to lipid-modified Shh signaling. On isolation, this region does not express ventral genes spontaneously without the addition of Shh and, therefore, has not been exposed to Shh signaling in the embryo. In this assay, Shh induces the differentiation of neurons expressing members of the Dlx gene family (Dlx-1, -2, -5, and -6) and an embryonic form of glutamate decarboxylase (GAD-ES). Step-by-step photographs from a dissection microscope describe the dissection of rat E11.5

From: *Methods in Molecular Biology: Hedgehog Signaling Protocols*
Edited by: J. Horabin © Humana Press Inc., Totowa, NJ

embryonic telencephalic neural explants. The subsequent culture and immunohistochemistry of Shh-treated explants is also described in detail. Although technically challenging at first, the use of the intraocular grid increases the reproducibility of this assay.

2. Materials

1. Leibovitz's L-15 with L-glutamine (Cellgro, Herndon, VA; cat. no. 10-045-CV).
2. 1 Lumsden BioScissor (Dr Andrew Lumsden, King's College, London).
3. 2 Microdissection Tweezers .05 × .01 mm (Roboz, Gaithersburg, MD; cat. no. RS-4978).
4. Tungsten Wire (A-M Systems, Inc., Carlsborg, WA; cat. no. 7170).
5. Explant culture medium, syringe-filtered: 25 mL DMEM/F12 50/50 mix with L-glutamine (Cellgro; cat. no. 10-090-CV), 250 µL penicillin–streptomycin, 100× (Gibco, Rockville, MD; cat. no. 15140-148), 250 µL L-glutamine, 200 nM-100× (Gibco; cat. no. 25030-149), 25 µL Mito+ Serum Extender (BD San Jose, CA; cat. no. 355006), 250 µL N-2 (100×) supplement (Gibco; cat. no. 17502-048), 500 µL B-27 (50×) serum free supplement (Gibco; cat. no. 17504-044).
6. 10-mm tissue culture inserts, 0.02 µm membrane (NUNC, Rochester, NY; cat. no. 162243).
7. 4-well plate (NUNC; cat. no. 176740).
8. 24-well plate (NUNC; cat. no. 143982).
9. 96-well plate (NUNC; Cat. No. 249946).
10. 100 × 15 mm Petri dish (BD; cat. no. 351029).
11. 3 mL sterile transfer pipette (BD; cat. no. 357575).
12. 20 mL sterile syringe (BD; cat. no. 309661).
13. 25-mm syringe filter, 0.2 µm membrane (Pall Corporation, Ann Arbor, MI; PN 4192).
14. Blocking solution: 20 mL PBS + 0.5% Triton, 5 mL goat serum, 50 µL 5% sodium azide. Stored at 4°C.
15. FluorSave Reagent (Calbiochem, San Diego, CA; cat. no. 345789).
16. Leica MZ12.5 dissecting microscope.
17. Custom intraocular grid (Leica, Bannockburn, IL).

3. Methods

The methods given below outline (1) the basic dissection of rat E12 embryos, (2) isolation of the dorsal telencephalon from the brain, (3) extracting neural explants from the dorsal telencephalon, (4) culturing neural explants, and (5) staining neural explants on filters.

3.1. Rat E11.5 Embryo Collection

1. Place embryos into the lid of a petri dish filled with L-15 medium.
2. Using two microdissection tweezers, tear through fat, and uterine wall to expose amniotic sac.
3. Carefully tear open the amniotic sac and cut the umbilical chord.

4. Collect whole embryo using a sterile plastic transfer pipette with the tip cut off.
5. Place embryo in a dish with ice-cold L-15 medium, on ice.
6. Repeat steps 2–5 for each embryonic sac.

3.2. Rat E11.5 Embryo Dissection

1. Using Lumsden BioScissors (*see* **Note 1**), remove the hind limbs and branchial arches from each embryo (**Fig. 1A,B**).
2. Next, separate the brain from the trunk by cutting through the otic vesicle (**Fig. 1A,B**).
3. Place limbs, branchial arches, trunks, and heads into a 4-well plate containing ice-cold L-15 medium.
4. Place trunks, limbs, and branchial arches into separate eppendorf tubes and centrifuge samples for 4 min at 250 *g* at 4°C.
5. Remove excess L-15 media and freeze samples on dry ice. Stored at −80°C. These samples can be used for RNA or protein isolation at a later date.

3.3. Rat E11.5 Brain Dissection

1. Perform dissection at magnification ×2.5 on the Leica MZ12.5.
2. Prop up the brain so that it is standing upright (**Fig. 1C**).
3. Immobilize the brain by grasping around the hindbrain with microdissecting tweezers.
4. Using BioScissors, cut along telencephalic/diencephalic border (blue dashed line, **Fig. 1D,E**).
5. Next, cut slightly downward on each side through the eyes (red dashed line, **Fig. 1D**).
6. Flip and flatten dorsal tissue as shown (green arrow, **Fig. 1D,F**)
7. Separate the dorsal telencephalon from the ventral telencephalon. There is a slight indentation where the dorsal and the ventral halves meet. Using BioScissors, cut through this border, following it around to completely separate the dorsal portion of the telencephalon (green dashed line, **Fig. 1F**).
8. The ventral telencephalon is still attached to the mid/hindbrain. Separate the ventral forebrain from the mid/hindbrain.
9. Store the ventral forebrain and mid/hindbrain tissue on ice in separate wells of a 4-well plate containing L-15 medium. Later the tissue can be centrifuged and stored at −80°C. These samples can be used for RNA or protein isolation at a later date.

3.4. Tungsten Needles

1. Wear safety goggles.
2. Fill a 50 mL beaker with 1 *M* KOH.
3. Hook up a pair of electrode clamps to a power supply.
4. Insert a long piece of wire (or straightened jumbo paperclip) into the anode clamp. This wire is placed in the 1 *M* KOH.
5. Into the cathode clamp, insert a piece of tungsten wire (0.5 in.).
6. Hold the cathode clamp so that only the tungsten wire is in the KOH and run the power supply at 50 V.

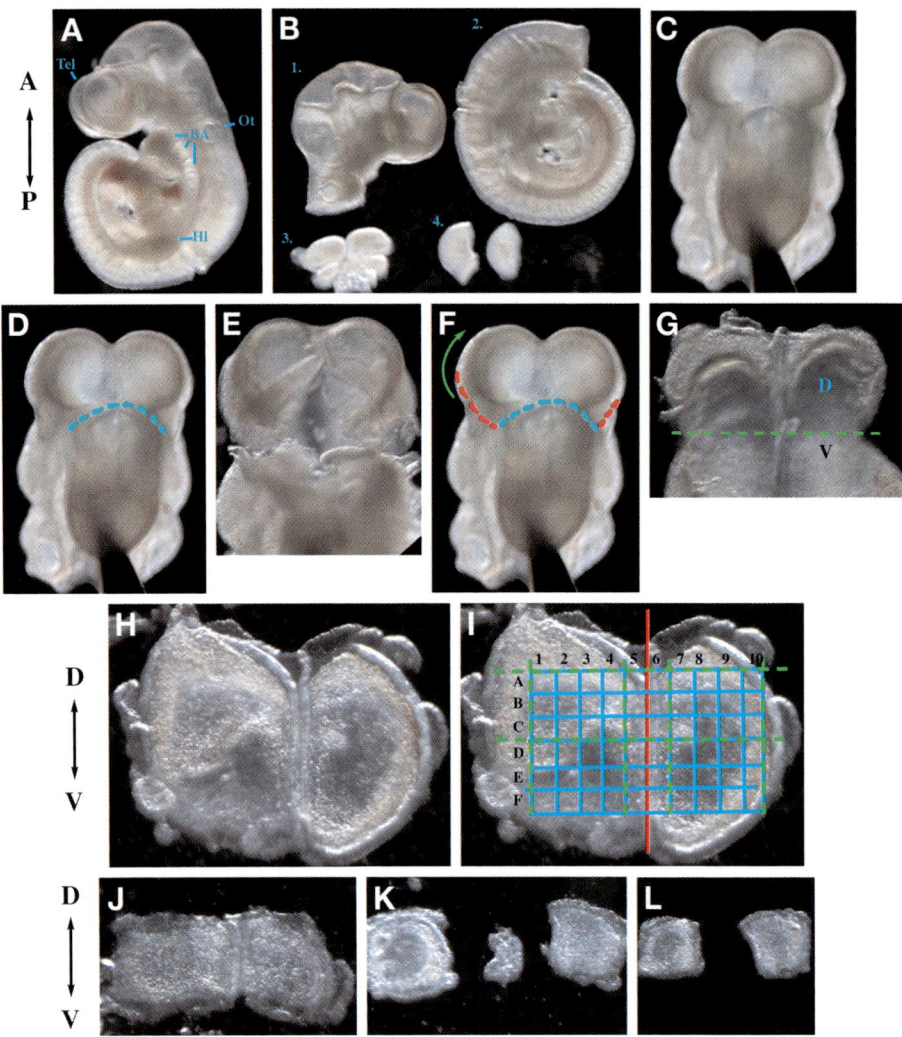

Fig. 1. Detailed schematic of rat E11.5 telencephalic neural explant assay. (**A**) Whole embryo, saggital view. (**B**) Pieces obtained after initial dissection: 1. Head, 2. Trunk, 3. Branchial arches, 4. Hind limbs. (**C**) Head, dorsal view, held steady at the level of the mid/hindbrain using tweezers. (**D**) The first cut in the head at the telencephalon/diencephalon border (dashed blue line). (**E**) Head shown after dashed blue cut. (**F**) The cuts along the sides of the cortex are shown in dashed red lines, and flipped (green arrow). (**G**) Dorsal (D) and ventral (V) telencephalon still attached after blue and red cuts and flipped with ventricular side facing. (**H**) Dorsal telencephalic piece. (**I**) View of dorsal telencephalic piece with intraocular grid superimposed. The grid is blue, except where cuts are made, where it is green and blue dashed lines. The midline is marked by the red line. Rows are labeled with letters, and columns with numbers.

7. Without breaking the circuit, move the tungsten wire up and down very quickly in the KOH to sharpen the wire.
8. Continue sharpening for about 2 min or until a visible thinning of the wire can be seen.
9. Using a wire clipper, clip the tungsten wire to the desired length and lock it into a bacterial loop holder.
10. The optimal thickness, length, and strength of the tungsten needle should be determined for each person. Successful dissections depend on generating the ideal needle (*see* **Note 1**).

3.5. 4 × 3 Dorsal Tissue Dissection

1. Perform dissection of isolated dorsal tissue at magnification ×8.0 on Leica MZ12.5 (**Fig. 1G,H**).
2. An intraocular grid (Leica) representing 125 μm × 125 μm squares (*see* **Note 2**) at this magnification is used to guide the dissection of the neural explants from the dorsal telencephalon (**Fig. 1H**; *see* **Note 1**).
3. The midline (*see* **Note 3**) is lined up between columns 5 and 6 (red line, **Fig. 1H**).
4. Using the tungsten needle, make the first cut above row A, and second cut between rows C and D (green dashed lines, **Fig. 1H**).
5. The next two cuts are made parallel to the midline, 625 μm away from the midline on each side (green dashed lines, line of column 1, right line of column 10, **Fig. 1H**).
6. The midline is cut away by making parallel cuts 125 μm on either side of the midline between columns 4 and 5 and 6 and 7 (green dashed lines, **Fig. 1H,J**).
7. The final neural explants should measure approx 500 μm × 375 μm. (**Fig. 1K**). Explants should be stored in a 4-well plate containing L-15 on ice until dissections are finished, and culturing begins. The dissection takes approx 3–4 h per 12 embryos.

3.6. Culturing Neural Explants

1. Using ethanol wiped forceps, place tissue culture inserts into wells of 24-well plate.
2. Add 200-μL syringe-filtered culture medium (<2 wk old) into each insert.
3. Place neural explants in tissue culture inserts.
4. Prepare medium containing N-lipid-modified ShhN (Curis, Inc., myristoylated ShhN: .5–1 μg/mL; *see* **Note 4**) in 600 μL.
5. Remove as much media from each insert as possible without drying out explants.
6. Add 300 μL culture medium into each tissue culture insert and 300 μL culture medium into each outer well.
7. Orient the explants using tungsten needles so that the neuroectodermal layer is at the surface.
8. Culture explants in 37°C, humidified CO_2 incubator for 3 d (*see* **Note 5**).

(**J**) Dorsal telencephalic explant after top, bottom and side cuts are made, before midline cut (green dashed lines). (**K**) 4 × 3 explant pieces adjacent to dorsal midline piece. (**L**) Final trimmed 4 × 3 dorsal telencephalic explants. Tel, telencephalon, Ot, otic vesicle, BA, branchial arches, Hl, hind limb, D, dorsal, V, ventral.

3.7. Staining Explants on Filters

1. Remove all media from outer wells.
2. Rinse explants in tissue culture inserts twice with 1× PBS.
3. Fix explants in fresh 4% paraformaldehyde for 15 min at 4°C, rocking.
4. Rinse three times with 1× PBS.
5. Wash four times with PBS + 0.5% Triton for 10 min rocking at room temperature.
6. Transfer explants into a 96-well plate using a disposable pipette.
7. Block explants in blocking solution for 1 h at room temperature or overnight at 4°C.
8. Add primary antibody (1 µg/mL) in blocking solution to explants. Cover plate with lid and wrap parafilm around the edges.
9. Leave rocking overnight at 4°C.
10. Rinse two times in PBS + 0.5% Triton.
11. Rinse two times in PBS + 0.5% Triton for 30 min rocking at room temperature.
12. Add secondary antibody (Jackson Immuno Cy2 and Cy3 antibodies, 1:250) in blocking solution to explants for 1 h, rocking at room temperature, and wrapped in foil.
13. Rinse two times in PBS + 0.5% Triton.
14. Rinse two times in PBS + 0.5% Triton for 30 min rocking at room temperature. (Before second wash can add DAPI in the ratio of 1:1000 for 15 min).
15. Using a disposable pipette, transfer explants onto a microscope slide.
16. Use a tungsten needle to help line up the explants on each slide.
17. Dry the edges of each slide with a Kimwipe.
18. Add approx three drops of FluorSave Reagent onto each slide and seal with a coverslip.
19. Air dry slides for at least 1 h, preferably overnight before viewing for confocal microscopy. Seal edges with nail polish for long-term storage of slides at 4°C.

4. Notes

1. Part of the difficulty of cutting tissues from the telencephalon is that they are rounded and difficult to flatten. The Bioscissor is excellent for cutting the round edges. However, once the piece in **Fig. 1H** is achieved, this tissue must be flattened for further manipulation and aligning with the grid. Flattening is done with the fatter edge of the tungsten needle. Too much flattening can destroy the tissue, resulting in poor survival after culturing. Too little flattening will result in inaccurate dissection, and ventral contamination.
2. The addition of the grid has made it possible to perform the telencephalic explant assay more reproducibly. However, one problem often encountered with this assay is the inadvertent and unwanted inclusion of a region that has already been exposed to Shh signaling. This region is what we call "ventral contamination" and will express Dlx genes without the addition of exogenous Shh protein. If the first cut made above row A is accidentally shifted downward, at the level of row B, the entire dissection will be shifted, and ventral contamination will result. If the lateral cuts at column 1 and 10 are too wide, ventral contamination can also result. Ventral contamination will generate false positives (expression of Dlx/Islet) in ShhN (–) control

explants, confounding the analysis. Only by practicing dissections several times, can ventral contamination can be avoided. After practice, one out of eight 4 × 3 pieces may be ventrally contaminated, and this is taken into account in the analysis.
3. The midline serves as an excellent guide for orientation under the microscope, and for making cuts parallel to it; sometimes, it is useful to mark the midline with a shallow dent of the tungsten needle without cutting all the way through.
4. N-lipid modified ShhN must be used, ShhN without modification has a very low potency in these assays.
5. Explants can also be isolated for RT-PCR as described *(1)*. A minimum of two explants should be used in each sample.

Acknowledgments

The authors wish to thank HaeYoung Lee, who initially helped to define the 4 × 3 explant, Dr M. Placzek for showing us how to make good tungsten needles, and Dr A. Lumsden for designing and providing us with Lumsden Bioscissors. This work is supported by RO1 HD044745 and R21 HD049875 (NICHHD) to J. D. K.

References

1. Kohtz, J. D., Baker, D. P., Corte, G., and Fishell, G. (1998) Regionalization within the mammalian telencephalon is mediated by changes in responsiveness to Sonic Hedgehog. *Development* **125,** 5079–5089.
2. Kohtz, J. D., Lee, H. Y., Gaiano, N., et al. (2001) N-terminal fatty-acylation of Sonic Hedgehog enhances the induction of rodent ventral forebrain neurons. *Development* **128,** 2351–2363.
3. Feng, J., White, B., Tyurina, O. V., et al. (2004) Synergistic and antagonistic roles of the Sonic Hedgehog N- and C-terminal lipids. *Development* **131,** 4357–4370.

5

Genetic Analysis of the Vertebrate Hedgehog-Signaling Pathway Using Muscle Cell Fate Specification in the Zebrafish Embryo

Sudipto Roy

Abstract

Over the recent years, a large number of embryological studies with the zebrafish have provided substantial evidence of its usefulness for the investigation of the genetic and cellular basis of vertebrate development. With regard to the Hedgehog (Hh) pathway, forward as well as reverse genetic approaches in this organism have not only validated the roles of evolutionarily conserved players of the signaling cascade, but have also contributed to the isolation of several novel components that had remained unidentified through screens in other animal models. Here, the author describes a whole mount antibody labeling method that allows the detection of three unique muscle cell fates in the zebrafish embryo, which are induced by distinct levels and timing of Hh-signaling activity. This technique provides a rapid and convenient assay that can be utilized for the evaluation of effects of loss- or gain-of-function of any gene on the levels of Hh pathway activation during embryogenesis.

Key Words: Hedgehog; zebrafish; slow muscle; muscle pioneer; fast muscle; antibody labeling.

1. Introduction

The transparent zebrafish embryo is well suited for high-resolution cell biological analysis of developmental processes. Moreover, the feasibility of large-scale genetic and small molecule screens, gene "knock-down" studies with antisense morpholino oligonucleotides as well as gene misexpression using stable transgenics or transient overexpression with mRNA and DNA constructs, are the further advantages that this organism provides for the analysis of the genetic regulation of embryonic development. Until recently, much of our understanding of the Hedgehog (Hh) pathway in vertebrates had heavily relied on the characterization of the functions of the mammalian homologs of signaling

From: *Methods in Molecular Biology: Hedgehog Signaling Protocols*
Edited by: J. Horabin © Humana Press Inc., Totowa, NJ

components that were described originally in *Drosophila*. Chemical and retroviral insertional mutagenesis in the zebrafish have now resulted in the identification of loss-of-function alleles of almost all of the evolutionarily conserved components, including Sonic Hh (Shh) *(1)*, Patched2 *(2)*, Dispatched1 *(3)*, Smoothened (Smo) *(4–6)*, Gli1 *(7)*, Gli2 *(8)*, Suppressor of fused (Su[fu]) *(2)*, as well as in the Hh-interacting protein (Hip) *(2)*—a vertebrate-specific Hh receptor, first identified through biochemical studies in mammals. Such forward genetic screens have also been instrumental in the discovery of two new players in the Hh pathway—the zinc finger and coiled-coil protein Iguana (Igu) *(9,10)* and the Epidermal Growth Factor and Complement Subcomponents C1r/C1s, (EGF)-related sea urchin protein, Bone Marphogenetic Protein-1 (CUB) domain containing molecule Scube2 *(11,12)*. In parallel, complementary reverse genetic studies have not only provided independent evidence of the requirement of the Su(fu) protein in vertebrate Hh signal transduction *(13)*, but have also helped to clarify the roles played by the homologs of the serine– threonine kinase fused (Fu) and the kinesin-like protein, Costal2 (Cos2) *(13, 14)*. In addition, this approach has identified β-arrestin 2 as a novel regulator of Smo activity in the vertebrate embryo *(15)*.

It is anticipated that future work with the zebrafish will help to refine our understanding of the interactions among the already known constituents of the Hh pathway, and also continue to identify and link new molecules to the current framework of the vertebrate Hh signal transduction network. The availability of a convenient assay in the developing embryo to assess the levels of Hh-signaling activity is an essential element for such kinds of analyses. The majority of work on vertebrate Hh signaling has utilized induction of cell fates in the ventral neural tube of the chick and mouse embryo as a biological read out of the measure of graded Hh pathway activity during development. Besides the conserved effects of Hh on ventral neural tube patterning in the zebrafish embryo, we and others have shown that a unique set of muscle cell types are induced in the myotome in response to distinct levels and timing of Hh activity that emanates from the axial midline (notochord and the floor plate of the neural tube).

The zebrafish myotome consists of two muscle cell lineages—slow twitch and fast twitch *(16)*. Cells closest to the source of Hh secretion, the midline, are induced to form slow-twitch muscle fibers by low levels of Hh; subsequently, these cells differentiate into a superficial layer on the surface of the myotome and are referred to as superficial slow fibers (SSFs). Slow muscle precursors that are exposed to the highest levels of Hh, continue to reside next to the midline and mature into muscle pioneer (MP) cells *(13)*. Subsequent to the specification of these two cell identities within the slow lineage, submaximal levels of Hh activity are required for the formation of the medial fast fibers (MFFs) *(13)*. Loss-of-function mutations in components that stimulate the pathway, such as *shh (1,17)* and *smo (4)*, result in the loss of some or all of these cell types, depending on the extent of reduction of the levels of signaling activity. Conversely, abrogating the function of negative regulators of the pathway, like protein kinase A *(18)*, Su(fu) *(13)*, Igu *(9)*, or Cos2 *(14)*, induces ectopic signaling, and the concomitant specification of supernumerary Hh-dependent muscle fates. Each kind of muscle cell fate, again, is affected to a different extent, since the loss of a specific negative regulator results in the de–repression of the pathway to a specific degree. Thus, the segregation of these distinct kinds of muscle fiber types is a reliable indicator of the levels of Hh-signaling activity in the developing zebrafish embryo. Furthermore, the added advantage of being readily amenable for forward and reverse genetic analyses make the muscle system an ideal alternative paradigm for the study of the vertebrate Hh pathway.

In this chapter, the author describes the method for performing whole-mount antibody labeling against two homeodomain-containing proteins—Prox1 and Engrailed (Eng), which facilitates unequivocal identification of the three kinds of Hh-dependent muscle cells in the zebrafish myotome. All cells of the slow fiber lineage are mononucleate and express the Prox1 protein in their nuclei *(19)*. The slow-twitch MP cells, in addition, express high levels of the Eng proteins *(13,19,20)*. By contrast, the MFFs comprise multinucleate muscles that express low levels of Eng proteins exclusively, but not Prox1 *(13,19,20)*.

2. Materials

1. Zebrafish embryos (*see* **Note 1**).
2. Zebrafish embryo culture medium: 0.3 g sea salt/L deionized water. Add few drops of methylene blue solution (0.1% stock in water) to prevent fungal growth.
3. Pronase (Sigma, USA): 20 mg/mL stock (in deionized water), stored at −20°C.
4. Fixative: 4% paraformaldehyde, 4% sucrose, 0.15 mM CaCl$_2$ in 0.1 M sodium phosphate buffer (pH 7.3). Heat at 65°C for 30 min in a water bath. Make sure that paraformaldehyde has completely dissolved by swirling. Cool to room temperature and store at 4°C. Use within 2-d preparation.
5. Methanol: absolute methanol and grades of 75, 50, and 25% in 1× phosphate buffered saline (PBS).
6. Ice-cold acetone stored at −20°C.
7. PBDT: 1× PBS, 1% bovine serum albumin (BSA), 1% dimethylsulfoxide, 0.5% Triton X-100. Agitate on a shaker to dissolve BSA and Triton X-100 and stored at 4°C.
8. Primary antibodies: anti-Eng monoclonal antibodies mAb 4D9 (Developmental Studies Hybridoma Bank (DSHB), University of Iowa, USA) and rabbit anti-Prox1 polyclonal antibodies (Covance Research Products, USA) (*see* **Notes 2** and **3**).
9. Secondary antibodies: anti-mouse IgG coupled to Cy3 and anti-rabbit IgG coupled to FITC (Jackson Immunoresearch Laboratories, USA) (*see* **Note 4**).
10. Normal goat/sheep serum (Vector Laboratories, USA).
11. Blocking solution: PBDT containing 2% normal goat/sheep serum.
12. 50 and 70% glycerol (in deionized water).

13. Glass cavity dishes (Heinz Herenz, Germany).
14. Dumont forceps and mounted dissection needles (Fine Science Tools, USA).
15. Microscope slides, coverslips (22 mm × 22 mm) and nail varnish.

3. Methods

3.1. Culturing Zebrafish Embryos

Sort freshly fertilized zebrafish eggs into Petri dishes containing embryo medium (approx 60–70 eggs per dish) and grow in an incubator at 28.8°C for 24 h post-fertilization (hpf), by which time all Hh-dependent muscle cell fates get specified (*see* **Note 1**).

3.2. Embryo Dechorionation and Fixation

1. Under a dissection microscope, manually remove the chorions using two pairs of Dumont forceps. Transfer dechorionated embryos into 1.5 mL eppendorf tubes. Aspirate excess embryo medium and rinse with 1× PBS. Chemical dechorionation with pronase is recommended for large batches of embryos. For this, transfer embryos into a 50 mL Falcon tube in a volume of approx 5–6 mL embryo medium. Add pronase to a final concentration of 1 mg/mL and agitate the tube gently on a shaker. The chorions will begin to fall off from the embryos within a period of 10–20 min. Stop pronase activity quickly by replacing the pronase containing embryo medium with fresh embryo medium, and rinse —two to three times with embryo medium before transferring into 1.5 mL eppendorf tubes. Prolonged exposure to pronase solution should be avoided as this will result in the digestion of the embryos themselves. Rinse the embryos with 1× PBS.
2. Remove PBS and add 1–mL cold fixative to the embryos in the eppendorf tubes. Invert tubes —two to three times for proper mixing and leave at room temperature for 2 h (*see* **Note 5**).
3. Remove fixative and rinse embryos with 1× PBS. Wash embryos with PBS for 1 h with —two to three changes. Agitate the tubes on a nutator during this period.
4. Remove PBS from the final washing step and add 1 mL absolute methanol. Let the tube stand for 5 min. Remove methanol and add another 1 mL fresh methanol. Store the tubes at −20°C. The embryos can be left in methanol at −20°C for a few months (*see* **Note 6**).

3.3. Primary Antibody Labeling

1. Transfer eppendorf tubes containing the fixed embryos in methanol from −20°C to room temperature.
2. Remove methanol and add approx 1 mL of 75% methanol:PBS. Keep tubes lying on their sides for 5 min. Repeat this procedure for 50% methanol:PBS and 25% methanol:PBS.
3. Pour out embryos in 25% methanol:PBS from eppendorfs into glass cavity dishes. If some embryos remain stuck to the tube, flush them with 25% methanol:PBS to dislodge them into suspension and pour them out into the dishes. It is advisable

to use the glass dishes for all subsequent steps of the antibody labeling reaction as the embryos are less adherent to glass than the plastic of the eppendorf tubes.

4. Remove as much of 25% methanol:PBS as possible. Flood the dishes with 1× PBS. The embryos tend to float around at this stage. Squirt PBS on the floating embryos to make them sink.

5. Let embryos soak in PBS for 5 min. Remove PBS and flood with fresh PBS, repeating two more times with 5 min of soak time each. These PBS washes are necessary to remove all traces of methanol.

6. Remove as much of the PBS as possible from the final wash step. Add 1 mL ice-cold acetone (stored at −20°C) to the embryos (*see* **Note 7**). If some embryos float around, squirt more acetone to make them sink. Immediately transfer the dishes to −20°C for 7 min.

7. Remove the dishes from −20°C and discard all of the acetone. Allow residual acetone to evaporate; at the same time, do not allow embryos to become completely dry. Flood dishes with 1× PBS. If embryos float around, make them sink by squirting more PBS over them. Let the embryos soak in PBS for 5 min. Remove PBS and wash embryos with fresh PBS, repeating the procedure two more times with 5 min of soak time each.

8. Remove as much of PBS as possible from the final wash. Add 250 μL blocking solution to each dish. Make sure all embryos have sunk to the bottom of the dish. The presence of Triton X-100 in the PBDT in this and all subsequent steps reduces surface tension and usually prevents embryos from floating around. The volume of blocking solution used varies, depending on the number of embryos being stained and is typically 250 μL and can be as much as 500 μL. Seal the dishes with parafilm to prevent evaporation, transfer to an orbital shaker, and agitate gently for 1 h at room temperature.

9. Remove as much of the blocking solution as possible. Add primary antibody diluted in PBDT. Typical and minimal volume of diluted primary antibody is in 200 μL PBDT per dish. It can vary up to 400 μL depending on the number of embryos being used for a particular experiment. Use mAb 4D9 at the ratio of 1:50 and anti-Prox1 at the ratio of 1:5000. Ensure that all embryos are submerged in the antibody solution.

10. Seal the dishes with parafilm and place them on an orbital shaker at low speed in the cold room (i.e., 4°C) overnight.

3.4. Secondary Antibody Labeling

1. Remove dishes from 4°C and discard primary antibody solution. Flood the dishes with room temperature PBDT and rinse. Add more PBDT (approx 1–1.5 mL) and wash for 30 min at room temperature on a shaker with moderate agitation. Repeat three times. This 2-h washing step ensures minimal nonspecific background reaction from the primary antibodies.

2. After the final wash, remove PBDT and add secondary antibodies—anti-mouse-Cy3 and anti-rabbit-FITC—diluted in PBDT, to each dish. Typical and minimal volume of diluted secondary antibodies is 200 μL, but can vary up to 400 μL depending on number of embryos being processed for each reaction. The secondary antibodies are used at a dilution in the ratio of 1:100–1:200.

3. Seal the dishes with parafilm and agitate gently on a shaker at room temperature for 5 h. Since FITC and Cy3 are light-sensitive fluorescent dyes, all incubation steps using these reagents must be done in the dark either by covering the dishes with aluminum foil or placing them in a light tight box. The secondary antibody reaction can also be performed overnight at 4°C.

4. Remove as much of the secondary antibody solution as possible. Flood with room temperature PBDT and rinse. Add more PBDT and wash for 30 min on a shaker at room temperature with gentle agitation. Repeat three times for efficient removal of nonspecific labeling by the secondary antibodies. Keep the dishes covered from light at all times.

5. Remove PBDT after last wash. Add about 1 mL of 50% glycerol:deionized water and swirl the dish to get all the embryos into the glycerol solution. Cover with aluminum foil and leave at 4°C for 2 h.

6. Remove 50% glycerol and replace with approx 1 mL of 70% glycerol:deionized water and incubate at 4°C for 2 h. After the 2-h incubation period in 70% glycerol, the embryos can be dissected from their yolk and mounted on slides for confocal microscopy. For long-term storage, the embryos can be transferred into eppendorf tubes and kept at −20°C (*see* **Note 8**).

3.5. Mounting Stained Embryos for Confocal Microscopy

For visualizing the muscle cells using a confocal microscope, the trunk and tail region of the embryos need to be dissected from the head and yolk portion and mounted laterally (**Fig. 1**). Proper mounting is essential for obtaining good quality images.

1. Using two pairs of dissection needles gently break the yolk ball, and tease out as much of the yolk particles as possible. There is no need to remove the yolk extension. Break the embryo at the junction between the head and the trunk.

2. Place two strips of scotch tape on a microscope slide, approx 0.5-cm apart. Place a drop of glycerol in the space between the strips of tape and transfer the dissected trunk and tail fragment into the drop of glycerol. Gently lower a coverslip such that its edges lie over the tape strips. The two strips of tape act like spacers and prevent excessive flattening of the embryo by the pressure of the coverslip. Make sure that the preparation is lying laterally. If the preparation has disoriented during the mounting procedure, gently tap the edges of the coverslip to restore it back to the lateral position. Seal the edges of the coverslip with nail varnish.

3. Use a laser scanning confocal microscope to view the samples. For FITC (green emission) and Cy3 fluorophores (red emission), excitation wave lengths of 488 and 543 nm, respectively, should be used (*see* **Note 9**).

4. The superficial location of the SSFs in the myotome allows them to be visualized with the minimal amount of Z-sectioning; by contrast, the MP and MFF cell types require slightly deeper scans in the region immediately adjacent to the notochord. Confocal micrographs illustrating the disposition of the different Hh-dependent

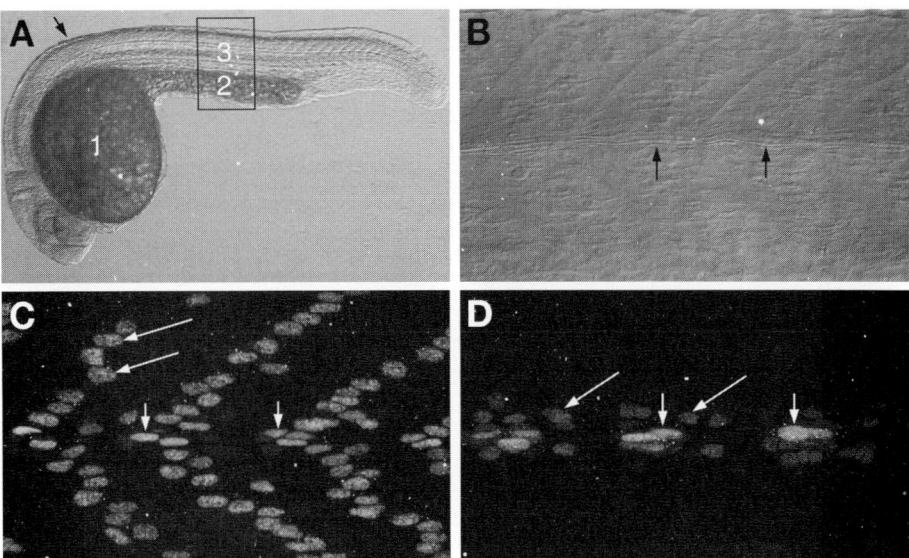

Fig. 1. Muscle fiber types in the myotome of a wild-type embryo. (**A**) DIC image of a 24 h post-fertilization embryo. The arrow points to the junction between the neck and the trunk. 1, yolk ball; 2, yolk extension; 3, myotomal segments. (**B**) DIC image of the myotomal region circumscribed by the box in A. The position of the MP cells along the horizontal myoseptum is indicated (arrows). (**C**) Confocal image of Prox1-expressing SSFs (green; long arrows) and the Eng- and Prox1-expressing MP cells (yellow; short arrows). (**D**) Confocal image of the Eng-expressing MFFs (red; long arrows) and the Eng- and Prox1-expressing MP cells (yellow; short arrows). (C) and (D) represent projection images of multiple Z-scans performed in the superficial (C) and medial (D) sections of the myotome, respectively. All panels in this and the following figure are oriented anterior to the left and dorsal to the top.

muscle fiber types in the wild-type embryo and mutants with differing degrees of loss or gain of Hh-signaling activity are depicted in **Figs. 1** and **2**, respectively.

4. Notes

1. For general zebrafish husbandry, staging and anatomy of the embryo, strategy for raising fry, mating adult fish, obtaining fertilized eggs, forward and reverse genetic screens, and other related protocols, refer to more exhaustive zebrafish methods books (*21–24*).

2. The partially purified IgG concentrate form of mAb 4D9 produces the best results with minimal background. By contrast, the anti-Prox1 antibodies produce some amount of background labeling in the form of speckles. This can be reduced to a large extent by first preabsorbing the antibody on fixed zebrafish embryos. For

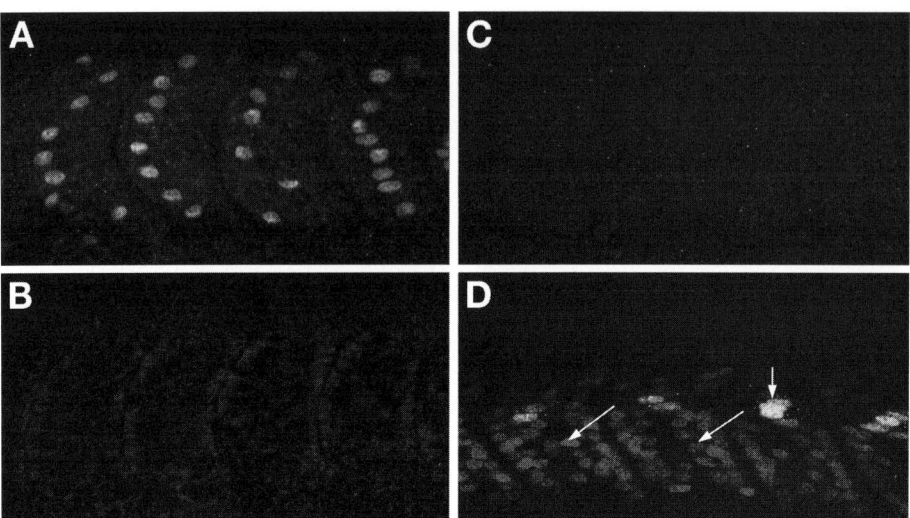

Fig. 2. Effects of loss and gain of Hh-signaling activity on the specification of muscle cell fates. **(A)** Superficial view of the myotome of an embryo homozygous for a mutation in the *shh* gene, *sonic you* (*syu*), showing complete absence of MP cells and a dramatic reduction in the number of SSFs (cf. **Fig. 1C**). Note also the alteration in shape of the myotome from the chevron shape in the wild-type embryo to the U-shape in the mutant on account of the loss of MP cells. **(B)** Medial view of the myotome of the embryo depicted in **(A)**, showing complete absence of the MFFs (cf. **Fig. 1D**). Shh mutant embryos only have a partial reduction in Hh signaling due to the redundancy among paralogous *hh* genes (*tiggy winkle hh* (*twhh*) and *echidna hh* (*ehh*)) that are expressed in the midline of the zebrafish embryo. Consequently, cells requiring high levels of Hh, like the MPs and the MFFs are not specified, but the SSFs are induced, albeit in fewer numbers. **(C)** Projection image of Z-scans through the entire width of the myotome (superficial and medial) of an embryo homozygous for a mutation in the *smo* gene, *slow-muscle-omitted* (*smu*), showing complete absence of all Hh-dependent muscle cells (cf. **Fig. 1C,D**). Unlike the multiple *hh* paralogs, vertebrate genomes contain a single *smo* gene that is absolutely essential for all Hh-signaling activity. Consequently, loss-of-function of *smo* results in the complete inhibition of the signaling pathway. **(D)** Medial view of the myotome of an *igu* mutant embryo, showing supernumerary numbers of Eng-expressing MFFs (red; long arrows) (cf. **Fig. 1D**). The MP cells are indicated (short arrow). Loss of the *igu* gene product triggers constitutive Hh-signaling downstream of Smo through ectopic induction of Gli1 activity.

this, use approx 250 μL packed volume of fixed 5 hpf (approx 50% epiboly) embryos in an eppendorf tube and add 500 μL of 1:50 diluted anti-Prox1 antibodies (in 1× PBS). Rotate the tube on a nutator overnight at 4°C. Remove and store the supernatant at 4°C. Use this at a final dilution in the ratio of 1:5000.

3. Two other useful primary antibodies for detecting the slow-twitch muscle fibers are the monoclonals, mAb F59 and mAb S58, both available from the DSHB. These antibodies recognize the slow myosin heavy-chain protein, although mAb F59 exhibits some cross-reactivity with fast muscle fibers, in post-24 hpf embryos. The S58 antibody does not work on paraformaldehyde fixed embryos. Instead, fix embryos in Carnoy's fixative (60% ethanol, 30% chloroform, and 10% glacial acetic acid) for 2 h at room temperature and then wash with grades of ethanol—95, 85, 70, 50, and 30% (in deionized water) for 10 min each, followed by washes in PBDT. Proceed from step 8 of *"Primary Antibody Labeling"* as previously described. S58 is a mouse IgA antibody. Therefore, an anti-mouse IgA secondary antibody (FITC-coupled anti-mouse IgA; Sigma) should be used.
4. The FITC and Cy3 labels can be used interchangeably on the secondary antibodies. In addition, combinations of secondary antibodies conjugated with other kinds of fluorophores (for example, Alexa 488, Cy5, etc.) can also be utilized.
5. We routinely fix embryos for 2 h at room temperature for all kinds of antibody-labeling reaction with fixative that has been prepared for not more than 2 d. Use of stale fixative dramatically affects the quality of antibody labeling and should be avoided. Fixation can also be performed overnight at 4°C. However, this may affect the quality of certain antibody labelings. We have noted that mAb 4D9 performs best on embryos fixed for 2 h at room temperature.
6. The cold methanol treatment is essential for proper fixation and for good quality immunolabeling. For this reason, the embryos should be left at −20°C at least for 3–4 h before proceeding with the subsequent steps of the protocol.
7. The acetone cracking step helps to permeabilize the embryos and facilitates proper antibody penetration.
8. The glycerol grades clear the embryos by replacing water and make them optically more transparent. Moreover, 70% glycerol is an ideal mounting medium for microscopy.
9. We routinely use a Zeiss Meta confocal microscope for all of our fluorescence imaging purposes. Use of a 40× oil immersion lens system is recommended for acquisition of the confocal images depicted in **Figs. 1** and **2**.

Acknowledgments

The work is supported by the Institute of Molecular and Cell Biology and the Agency for Science, Technology and Research (A*STAR), Singapore, in the author's laboratory on Hh signaling in the zebrafish embryo.

References

1. Schauerte, H. E., van Eeden, F. J., Fricke, C., Odenthal, J., Strahle, U., and Haffter, P. (1998) Sonic hedgehog is not required for the induction of the medial floor plate cells in the zebrafish. *Development* **125,** 2983–2993.
2. Koudijs, M. J., den Broeder, M. J., Keijser, A., et al. (2005) The zebrafish mutants *dre, uki,* and *lep* encode negative regulators of the Hedgehog signaling pathway. *PLos Genet.* **1(2),** e19.

3. Nakano, Y., Kim, H. R., Kawakami, A., Roy, S., Schier, A. F., and Ingham, P. W. (2004) Inactivation of *dispatched 1* by the *chameleon* mutation disrupts Hedgehog signaling in the zebrafish embryo. *Dev. Biol.* **269,** 381–392.

4. Barresi, M. J., Stickney, H. L., and Devoto, S. H. (2000) The zebrafish *slow-muscle-omitted* gene product is required for Hedgehog signal transduction and the development of slow muscle identity. *Development* **127,** 2189–2199.

5. Varga, Z. M., Amores, A., Lewis, K. E., et al. (2001) Zebrafish *Smoothened* functions in ventral neural tube specification and axon tract formation. *Development* **128,** 3497–3509.

6. Chen, W., Burgess, S., and Hopkins, N. (2001) Analysis of the zebrafish *Smoothened* mutant reveals conserved and divergent functions of Hedgehog activity. *Development* **128,** 2385–2396.

7. Karlstrom, R. O., Tyurina, O. V., Kawakami, A., et al. (2003) Genetic analysis of zebrafish *gli1* and *gli2* reveals divergent requirements for *gli* genes in vertebrate development. *Development* **130,** 1549–1564.

8. Karlstrom, R. O., Talbot, W. S., and Schier, A. F. (1999) Comparative synteny cloning of zebrafish *you-too*: mutations in the Hedgehog target *gli2* affect ventral forebrain patterning. *Genes Dev.* **13,** 388–393.

9. Wolff, C., Roy, S., Lewis, K. E., et al. (2004) *Iguana* encodes a novel zinc-finger protein with coiled-coil domains essential for Hedghog signal transduction in the zebrafish embryo. *Genes Dev.* **18,** 1565–1576.

10. Sekimizu, K., Nishioka, N., Sasaki, H., Takeda, H., Karlstrom, R. O., and Kawakami, A. (2004) The zebrafish *iguana* locus encodes Dzip1, a novel zinc-finger protein required for proper regulation of Hedgehog signaling. *Development* **131,** 2521–2532.

11. Kawakami, A., Nojima, Y., Toyoda, A., et al. (2005) The zebrafish secreted matrix protein You/Scube2 is implicated in long range regulation of Hedgehog signaling. *Curr. Biol.* **15,** 480–488.

12. Woods, I. G. and Talbot, W. S. (2005) The *you* gene encodes an EGF-CUB protein essential for Hedgehog signaling in zebrafish. *PLoS Biol.* **3,** e66.

13. Wolff, C., Roy, S., and Ingham, P. W. (2003) Multiple muscle cell identities induced by distinct levels and timing of Hedgehog activity in the zebrafish embryo. *Curr. Biol.* **13,** 1169–1181.

14. Tay, S. Y., Ingham, P. W., and Roy, S. (2005) A homologue of the *Drosophila* kinesin-like protein Costal2 regulates Hedgehog signal transduction in the vertebrate embryo. *Development* **132,** 625–634.

15. Wilbanks, A. M., Fralish, G. B., Kirby, M. L., Barak, L. S., Li, Y. X., and Caron, M. G. (2004) β-arrestin 2 regulates zebrafish development through the Hedgehog signaling pathway. *Science* **306,** 2264–2267.

16. Devoto, S. H., Melancon, E., Eisen, J. S., and Westerfield, M. (1996) Identification of separate slow and fast muscle precursor cells *in vivo*, prior to somite formation. *Development* **122,** 3371–3380.

17. Lewis, K. E., Currie, P. D., Roy, S., Schauerte, H., Haffter, P., and Ingham, P. W. (1999) Control of muscle cell-type specification in the zebrafish embryo by Hedgehog signaling. *Dev. Biol.* **216,** 469–480.

18. Hammerschmidt, M., Bitgood, M. J., and McMahon, A. P. (1996) Protein kinase A is a common negative regulator of Hedgehog signaling in the vertebrate embryo. *Genes Dev.* **10,** 647–658.
19. Roy, S., Wolff, C., and Ingham, P. W. (2001) The *u-boot* mutation identifies a Hedgehog-regulated myogenic switch for fiber-type diversification in the zebrafish embryo. *Genes Dev.* **15,** 1563–1576.
20. Hatta, K., Bremiller, R., Westerfield, M., and Kimmel, C. B. (1991) Diversity of expression of Engrailed-like antigens in zebrafish. *Development* **112,** 821–832.
21. Westerfield, M. (ed.) (2000) *The Zebrafish Book,* University of Oregon Press, Oregon.
22. Nusslein–Volhard, C. and Dahm, R. (ed.) (2002) *Zebrafish: A Practical Approach,* Oxford University Press, Oxford.
23. Detrich, H. W. 3rd, Zon, L. I., and Westerfield, M. (ed.) (2004) *Zebrafish: Cellular and Developmental Biology,* Elsevier Academic Press, San Diego.
24. Detrich, H. W. 3rd, Zon, L. I., and Westerfield, M. (ed.) (2004) *Zebrafish: Genetics, Genomics and Informatics,* Elsevier Academic Press, San Diego.

6

Efficient Manipulation of Hedgehog/GLI Signaling Using Retroviral Expression Systems

Maria Kasper, Gerhard Regl, Thomas Eichberger, Anna-Maria Frischauf and Fritz Aberger

Abstract

Efficient manipulation of Hedgehog (HH)/GLI signaling activity is crucial to the analysis of molecular events underlying HH/GLI-regulated cell fate determination and tumor growth. In this article, we describe the use of retroviral expression systems as a valuable tool to activate or repress Hh-pathway activity in a broad spectrum of mammalian cells—including human cells—either by forced expression of the major Hedgehog-effectors GLI1 and GLI2 or by expression of the short-hairpin RNAs-targeting GLI mRNAs. We focus on two distinct retroviral systems that allow efficient and sustainable expression of GLI proteins in primary cells and cell lines of human origin: (i) a Moloney Murine Leukemia Virus-based and (ii) an HIV-derived lentivirus expression system, which allows transduction of both dividing and quiescent cells.

Key Words: Retroviral gene expression; lentivirus; Hedgehog signal transduction; GLI proteins; RNA interference.

1. Introduction

The Hedgehog/GLI (HH/GLI) signal transduction pathway plays a fundamental role in the development of vertebrates by controlling critical biological processes, such as proliferation, survival, differentiation, and pattern formation in many different cell types and tissues (reviewed in Refs. [1–3]). Recent studies have implicated aberrant HH/GLI pathway activation in the growth and maintenance of a variety of malignancies in man and have also shown that targeted pathway blockade using specific small molecule inhibitors may hold promise in future cancer therapies (4–10).

Activation of HH/GLI signaling is triggered by binding of biologically active HH protein to its receptor patched (PTCH), which eventually leads to (transcriptional)

From: *Methods in Molecular Biology: Hedgehog Signaling Protocols*
Edited by: J. Horabin © Humana Press Inc., Totowa, NJ

activation of members of the GLI family of zinc finger transcription factors, with GLI1 and GLI2 being the major Hh-effectors in the control of Hh-target gene expression. Methods that allow manipulation of GLI activity in a multitude of cell lines and primary cells thus represent a valuable tool for the detailed analysis of the molecular events controlled by Hedgehog signaling in normal development and disease.

In this article, we describe the use of two retroviral expression systems to efficiently manipulate Hh-pathway activity by forced expression of the GLI1 and GLI2 oncogenes in a multitude of dividing and quiescent cells and cell lines. We also present a lentiviral vector system that can be used for conditional RNAi-mediated inactivation of GLI function.

Retroviral transduction is a versatile method for efficient and stable gene delivery into a broad spectrum of primary cells or cell lines of various species. Here, we elaborate on two distinct expression systems: (i) a Moloney Murine Leukemia Virus (MoMuLV)-based and (ii) a HIV-derived lentiviral system. The latter offers the advantage that it can transduce both proliferating and non-dividing cells, while the MoMuLV-derived viruses can deliver transgenes to dividing cells only, which poses a problem for studies of differentiated or quiescent cells.

2. Materials
2.1. Cell Culture and Cell Lines

1. Phoenix-Ampho or Phoenix-Eco stable packaging cell lines for MoMuLV-derived retrovirus production *(11,12)*: use Phoenix-Ampho cell line (ATCC product# SD 3443) for transduction of most mammalian and human cells, and Phoenix-Eco line (ATCC product# SD 3444) for transduction of murine and rat cells (for ordering details, see http://www.stanford.edu/group/nolan/).
2. 293FT cell line for lentivirus production (Invitrogen, Carlsbad, CA).
3. Dulbecco's modified Eagle's medium (DMEM), high glucose (PAA, Pasching, Austria; cat. no. E15-843) supplemented with 10% fetal bovine serum (FBS) (Invitrogen), 1× MEM, 1× L-glutamine, 1× penicillin–streptavidin, and 0.5 mg/mL Geneticin (G418, for 293FT cells only).
4. MEM non-essential amino acids solution (100× stock) (Invitrogen). Stored at 4°C.
5. L-glutamine (100× stock (200 mM), Invitrogen). Store aliquots at −20°C.
6. Penicillin–streptavidin solution (100× stock, Invitrogen). Store aliquots at −20°C.
7. Geneticin (Invitrogen). Store powder at 4°C.
8. Trypsin solution (0.25 %) and 1 mM ethylenediamine tetra-acetic acid (Invitrogen). Store aliquots at −20°C.
9. Kohrsolin FF (Roth, Karlsruhe, Germany); use at a final concentration of 1% to inactivate and dispose of infectious viral particles.

2.2. Virus Production

1. Retroviral, lentiviral, and packaging plasmid constructs (*see* **Note 1**).
   Successful virus production requires appropriate packaging systems. Lentiviral pLVTHM needs second generation systems. This vector is also suitable for conditional shRNA-mediated knockdown in combination with tTR-KRAB repressor expressing lentivirus *(13)*. pLL3.7 *(14)* works with third generation systems. For production of MoMuLV-based SIN-IP retrovirus *(15)*, we recommend the use of Phoenix cell lines, a second generation packaging system that stably expresses gag, pol, and envelope protein for production of ecotropic or amphotropic retroviruses *(11,12*; **Table 1**).

   Plasmid maps and detailed sequences of the constructs listed above can be obtained from the following web-sites: http://tronolab.epfl.ch/ (for information on pLVTHM, psPAX2, pMD.G, pMDL g/p RRE, and pRSV-Rev). http://web.mit.edu/ccr/labs/jacks/protocols/rnairesources.htm (for information on pLL3.7).
2. 2× HBS (HEPES-buffered saline): 50 mM HEPES, 280 mM NaCl, and 1.5 mM Na$_2$HPO$_4$, adjust pH exactly to 6.95 with HCl (note that pH is a critical parameter for efficient complex formation, also try 7.00 and 7.05). Sterile filter the solution through 0.2 µm filter and store at 4°C.
3. 2 M CaCl$_2$. Filter sterilize through 0.2 µm filter and store at 4°C.
4. 12 mL polystyrol tubes.
5. 10 mL syringes, Luer Lock.
6. Rotilabo 0.45 µm low protein-binding filter for 10 mL syringes (Roth, Karlsruhe, Germany, cat. no. P665.1).

2.3. Retroviral Transduction (for Safety Risks see Note 2)

1. Polybrene (hexadimethrine bromide) (Sigma, St Louis, MO) dissolved in double-distilled water at 5 mg/mL. Store aliquots in screw-cap glass tubes for up to 6 months at 4°C. Note that polybrene is toxic and care should be taken to avoid any direct contact or inhalation.

3. Methods

3.1. Cloning of GLI Genes Into Retroviral Vectors

Efficient packaging of virus DNA and thus, transduction efficiency and expression of the transgene of interest strongly depend on the total size of recombinant retroviral DNA comprising the transgene (i.e., distance between 5′ LTR and 3′ LTR). For the vectors described here, we noticed a significant decrease in virus titers if inserts exceeded a length of 5 kb. As the coding sequence for GLI transcription factors is in the range of 3.3 (GLI1) to 4.8 kb (full-length human GLI2), we recommend the use of short tags (e.g., MYC- or HA-tag) rather than constructs containing an additional internal ribosomal entry site driving expression of GFP or a drug-resistance marker. For human GLI1 *(18)*

Table 1
Virus Production Packaging Systems

	Second generation	Third generation	Phoenix
Packaging plasmid	psPAX2	pMDL g/p	–
	(http://tronolab.epfl.ch/)	RRE *(16)*	
Rev-expressing plasmid	–	pRSV-Rev *(16)*	–
Envelope plasmid	pMD.2G *(17)*	pMD.2G *(17)*	–
Retroviral transfer vectors	pLVTHM *(13)*	pLL3.7 *(14)*	SIN-IP-GFP *(15)*

and constitutively active N-terminally truncated GLI2 *(19,20)* (about 3.8 kb), we got high viral titers and transgene expression by fusing EGFP (Clontech, Mountain View, CA) to the N-terminus of either transcription factor (*see* **Fig. 2A** and **B**). The biological activity of the EGFP-GLI fusion proteins is comparable with wild-type GLI1 and GLI2, respectively.

3.2. Growth of Viral Producer Cell Lines

Phoenix-Ampho and Phoenix-Eco are second generation producer cell lines for the production of helper-free ecotropic and amphotropic MoMuLV-based retroviruses (to be used with SIN-IP retroviral vectors). The lines were generated by stable integration of viral gag-pol and env genes into highly transfectable 293T cells (*11,12*; *see* **Note 3**). Phoenix-Ampho cells produce viral particles that transduce and deliver genes into *dividing* cells of most mammalian species (including human), while Phoenix-Eco cells are to be used with *dividing* murine or rat cells.

For production of GLI1, GLI2 or shRNA expressing lentiviruses (e.g., LL3.7-GLI1, LL3.7-GLI2, pLVTHM-GLI2-shRNA1), we recommend the use of the 293FT producer cell line (Invitrogen), which is a clonal isolate derived from transformed embryonic kidney cells 293. The 293FT strain is a fast-growing, highly transfectable variant of the 293 cell line, which stably expresses the large T antigen, contributing to high viral titers.

1. Grow Phoenix and 293FT cells in DMEM, supplemented with 10% FBS. For normal growth and expansion, Phoenix and 293FT cells should be split 1:4 and 1:5, respectively, at 80% confluence every 2–3 d. Except for transfection, 293FT cells should be grown in the presence of 0.5 mg/mL Geneticin to maintain the capacity of high-titer virus production. Note that Phoenix-Ampho/-Eco and 293FT cells should never reach confluence, since this dramatically reduces transfection efficiency. Under optimal conditions, we routinely achieve transfection efficiencies between 70 and 95%.

Table 2
Reagents for Viral Packaging in Phoenix and 293FT cells

Phoenix packaging cells	
Solution A	30 µg SIN-IP-vector
	100 µL 2 M CaCl$_2$
	add H$_2$O to 800 µL
Solution B	800 µL 2× HBS (pH 7.0)

293FT cells for lentivirus production		
	Second generation packaging	**Third generation packaging**
Solution A	5 µg pMD2G	5 µg pMD2G
	15 µg psPAX2	10 µg pMDLg/pRRE
		5 µg pRSV-Rev
	20 µg pLVTHM	20 µg pLL3.7
	100 µL 2 M CaCl$_2$	100 µL M CaCl$_2$
	add H$_2$O to 800 µL	add H$_2$O to 800 µL
Solution B	800 µL 2× HBS (pH 7.0)	800 µL 2× HBS (pH 7.0)

3.3. Transfection and Virus Production (Protocol Applies to Phoenix and 293FT Producer Lines Unless Stated Otherwise)

1. Day 1: 18–24 h prior to transfection split a 75 cm^2 flask with 90% confluent 293FT or Phoenix cells into three 100-mm petri dishes (resulting in about 30–35% confluence). Culture cells at 37°C in 10 mL DMEM with 10% FBS. Omit Geneticin from 293FT culture medium.
2. Day 2: replace culture medium with fresh DMEM supplemented with 10% FBS (omit Geneticin from 293FT culture medium) approximately 2–3 h prior to transfection. At this point, cells should be at 95% confluence.
3. Immediately before transfection, equilibrate plasmid solutions, buffers and sterile water to room temperature before mixing the solutions. This increases the reproducibility of calcium–phosphate–DNA complex formation.
4. Prepare solutions A and B in polystyrol tubes as described below, according to the viral packaging and producer lines to be used (i.e., Phoenix or 293FT cells). The protocol applies to one 100-mm culture dish (**Table 2**).
5. Vortex Solution B at 2000 rpm (IKA MS2 shaker) and dropwise add Solution A **close** to the bottom of the tube. At this point, the solution should become turbid due to complex formation.
6. Incubate the mixture for 30–40 min at room temperature.
7. Before adding the solution to 293FT/Phoenix cells, resuspend or briefly vortex the precipitate.
8. Add the solution carefully and dropwise onto the cell monolayer and gently agitate the culture dish to evenly distribute the complexes. We recommend processing of no more than two plates at a time, since 293-derived cells are very sensitive to temperature shifts and readily shrink and detach.

9. Day 3: discard the medium and add 7-mL fresh culture medium (*see* **Note 4**). Note that the supernatant already contains infectious virus particles.

10. Day 4: harvest the first virus supernatant and add another 7 mL fresh medium to the cells. Filter the virus supernatant through a 45 µm sterile filter (low protein binding) using a 10 mL syringe. Purified virus supernatant can be directly used for trans-duction or stored for 2–4 d at 4°C. For long-term storage of virus, we recommend snap-freezing of small aliquots on dry ice and storage at −80°C (*see* **Note 5**). To increase the multiplicity of infection (MOI), e.g., for shRNA-mediated knock-down, concentrate the virus by spinning the supernatant at 90,000 g (e.g., SORVALL Ultra Pro80, SW-28 rotor) for 2 h at 4°C and resuspend the virus pellet in an appropriate volume of serum-free medium or phosphate-buffered saline containing 1% BSA. We found that GLI transgene expression from SIN-IP or LL3.7 retro-viruses works well without concentration in a variety of cell types (**Fig. 1**), though a higher MOI may be necessary for efficient RNAi-mediated knockdown.

11. Day 5: harvest the second virus supernatant and proceed as described for day 4. Due to lower stability of virus particles at 37°C, the virus titer may be improved by incubating cells for 24 h at 32°C prior to harvesting. Discard the virus producing cells and autoclave all virus contaminated waste (for propagation and selection of stable virus-producing Phoenix cells, *see* **Note 6**)

3.4. Viral Transduction

1. For transduction with SIN-IP-GLI1/2 retrovirus, target cells must proliferate. We got best results when cells were at 60–70% confluence at the time of transduction. By contrast, lentiviral particles produced by 293FT cells can be applied to prolif-erating as well as quiescent cells.

2. We recommend seeding cells into appropriate multi-well plates as this facilitates the subsequent centrifugation step.

3. Prior to virus transduction pre-treat cells for 10–20 min with polybrene (*see* **Note 7**) by adding polybrene to the culture medium. For most cells and cell lines tested, we found a polybrene concentration of 8 µg/mL to give best results though, due to possible cytotoxicity, lower concentrations may be required for certain cell types.

4. In the meantime transfer the virus to a 12 mL polystyrol or glass tube and add polybrene to the same concentration as used for the pre-treatment (*see* above). Briefly vortex the virus-polybrene solution.

5. If using 6-well plates (35-mm dishes), replace the culture medium with 1.6 mL virus-polybrene solution per well and carefully seal the multi-well plate with parafilm.

6. To improve transduction efficiencies, spin the cultures in a heated centrifuge at 600 × g for 1 h at 37°C using a swing-out microtiter plate rotor.

7. Remove the virus-containing supernatant and add fresh medium. Dispose of the virus supernatant by adding kohrsolin FF to a final concentration of 1% and autoclave all virus contaminated waste.

3.5. GLI Transgene and GLI Target Gene Expression Analysis

If GFP or GFP-fusion proteins are expressed from the retroviral constructs, measure the amount of GFP-positive cells (**Fig. 1**) (e.g., by flow-cytometry), to

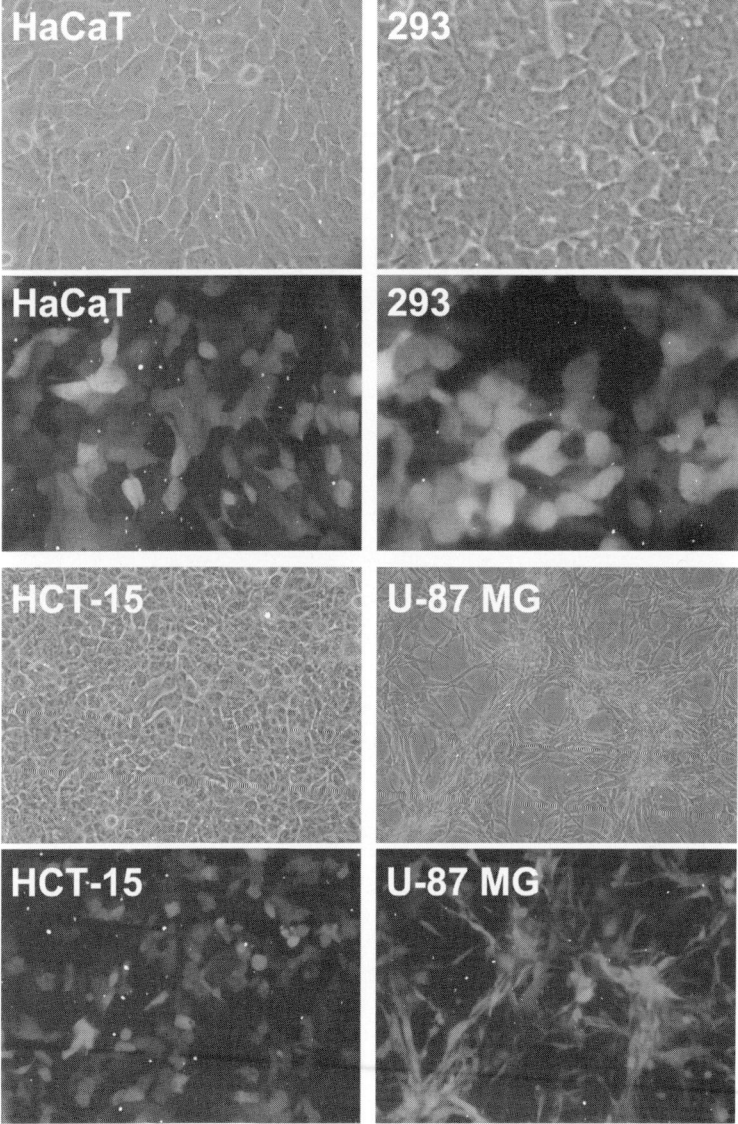

Fig. 1. Efficient and stable transduction of human cell lines using retroviral expression vectors. HaCaT keratinocytes *(22)* human embryonic kidney cells (HEK 293, ATCC: CRL-1573), the colon cancer cell line HCT-15 (ATCC: CCL-225) and the glioblastoma/ astrocytoma cell line U-87 MG (ATCC: HTB-14) were transduced with LL3.7 lentivirus expressing EGFP. To demonstrate stable gene transfer, transduced cells were cultured for several passages (*n* > 3), frozen in liquid nitrogen and regrown for another two passages. U-87 MG cells were analyzed 72-h post-infection. Fluorescence of EGFP was monitored to visualize transduced cells. Corresponding bright-field images are shown above the fluorescence images.

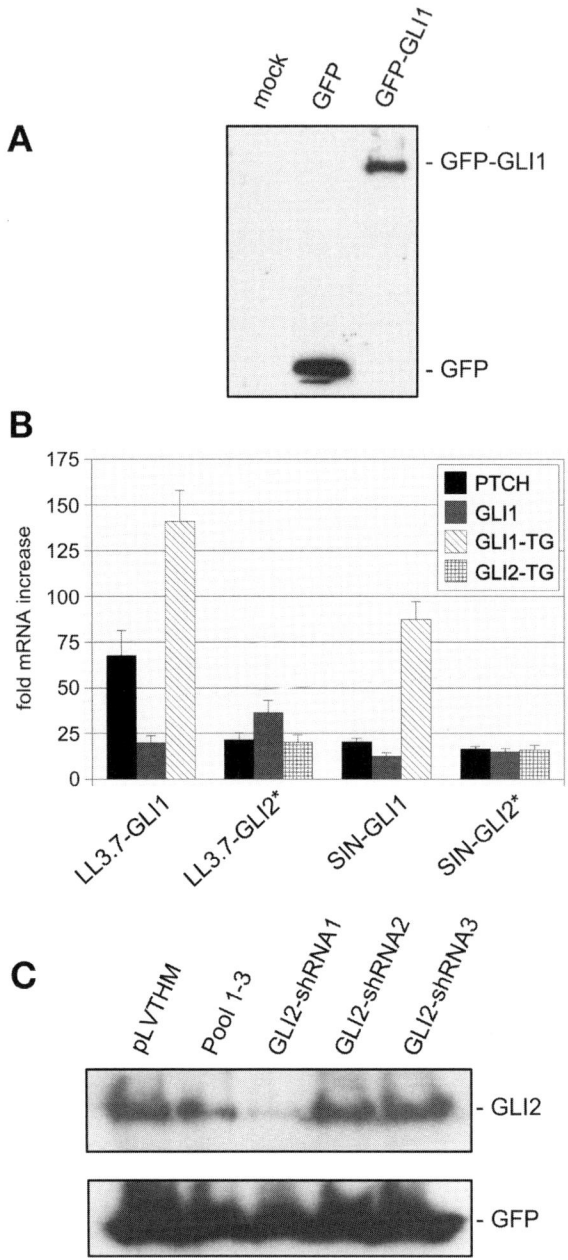

Fig. 2. Retroviral expression of GLI proteins and target gene activation. (**A**) Western blot analysis of the DAOY medulloblastoma cells (ATCC: HTB-186) transduced with LL3.7 lentivirus expressing N-terminally GFP-tagged GLI1 or GFP, only. GFP-GLI1 and GFP protein were detected with mouse anti-GFP antibody (B-2) (Santa Cruz

determine the transduction efficiency. Note that integration of the viral DNA into the host genome takes 8–12 h and transgene expression is detectable only 18–24 h post-transduction. Lentiviral titers can also be determined by quantifying the level of p24 antigen in ELISA assays. The amount of p24 can then be converted to transducing units per milliliter (for details, *see* http://tronolab.epfl.ch/).

To monitor the expression of GLI transgenes and their biological activity, we routinely analyze transgene mRNA and protein levels as well as the transcriptional response of known GLI target genes, such as PTCH and endogenous GLI1 (*see* **Fig. 2A** and **B**; for primer sequences *see* *[20]*). To identify functional pLVTHM-shRNA constructs, we monitor RNAi-mediated knockdown of GLI protein expression by Western blot analysis (*see* **Fig. 2C**).

For detection of human GLI1 and GLI2 proteins on Western blots, we obtained best results with the following antibodies from Santa Cruz Biotechnology (Santa Cruz, CA):

polyclonal goat anti-GLI1 (C-18) (cat. no. sc-6152);
polyclonal goat anti-GLI2 (N-20) (cat. no. sc-20290);
polyclonal rabbit anti-GLI2 (H-300) (cat. no. sc-28674);

4. Notes

1. Efficient transfection of packaging cells is a critical step in virus production and to a large extent depends on the quality of the plasmid preparation. We got best results with Qiagen Plasmid Maxi Kits (Qiagen, Hilden, Germany). To avoid recombi-

---

Biotechnology). (**B**) qRT-PCR analysis of DAOY cells transduced with lentivirus LL3.7 (LL3.7-GLI1, LL3.7-GLI2* or control LL3.7-EGFP) or SIN-IP retrovirus expressing human GLI1 (SIN-GLI1) or constitutively active GLI2* (SIN-GLI2*). As read-out for the biological activity of GLI1 and GLI2 transgenes, transcriptional activation of the Hh/GLI targets PTCH and GLI1 are shown. Note that endogenous GLI1 transcript can be distinguished from GLI1 transgene mRNA (GLI1-TG) by primers specific to the 3′ UTR of endogenous GLI1 mRNA. Fold mRNA induction represents the fold mRNA increase in GLI1 or GLI2 expressing cells when compared with the cells transduced with GFP-expressing control virus. The lower fold increase in GLI2-TG mRNA levels when compared with the increase in GLI1-TG levels is due to higher background levels of endogenous GLI2 mRNA. (**C**) RNAi-mediated knockdown of GLI2 by shRNA expressed from the lentiviral plasmid pLVTHM. 293 cells were transiently co-transfected with GLI2 expression plasmid and a different lentiviral shRNA expression construct as indicated. The knockdown efficiency of GLI2-shRNA1 was >80%. GLI2-shRNA2 and 3 had no effect on GLI2 expression. Pools 1–3: cells were transfected with an equimolar amount of GLI2-shRNAs 1–3. GLI2-shRNA1 targets human GLI2 mRNA (GenBank accession number DQ086814) at positions 4736–4754 (for detailed instructions on how to clone shRNAs into pLVTHM, *see* http://tronolab.epfl.ch/).

nation of viral vector sequences in bacterial host cells, we strongly recommend to transform the retroviral transfer vectors into *rec*A1 E. coli strains, such as STBL2 (Invitrogen).

2. Work with retroviruses, in particular amphotropic viruses, represents a potential biosafety risk, as amphotropic viral particles efficiently transduce cells of human origin. Work with replication-incompetent retroviruses requires at least Biosafety level 2 measures. Depending on your local guidelines, retroviral expression of certain disease-related genes, such as human oncogenes, may have to be carried out in level 3 facilities.

3. To prevent loss of viral packaging genes and to maintain the high virus-producing efficiency of Phoenix lines, cells should be selected every few months for a period of 1 wk in DMEM containing 10% serum, diphtheria toxin (1 μg/mL) (Sigma), and hygromycin (300–500 μg/mL) (Sigma).

4. The medium has to be added slowly onto the monolayer, as 293-derived cells readily detach from the culture dish. The cells are also temperature sensitive and tend to shrink and detach if left at room temperature for more than a couple of minutes. We found that adding as little as 7 mL medium to 100-mm dishes leaves producer cells intact and yields high viral titers.

5. We got best results with snap freezing the virus supernatant on dry ice instead of liquid nitrogen, which in our hands resulted in lower transduction efficiencies.

6. Stable SIN-IP-GLI virus expressing Phoenix cells can be selected for puromycin resistance. To do so, seed Phoenix cells to 30% confluence on day 1 and proceed to day 3 as described in **Section 3.2**. After transfection with viral transfer constructs, Phoenix cells can be directly frozen in complete DMEM containing 10% DMSO or further passaged. To maintain high-viral titers, transfected Phoenix producer cells should be selected in 1.25 μg/mL puromycin and split before reaching confluence.

7. Polybrene is a cationic polymer that significantly increases retroviral gene transfer by enhancing adsorption of the virus to the cell surface. Recent evidence suggests that cationic polymers modulate transduction via membrane charge neutralization and virus aggregation *(21)*.

Acknowledgment

We are particularly grateful to Dr Graham W. Neill for help with optimization of virus production protocols. pLVTHM, pLL3.7, and SIN-IP-GFP plasmids were kindly provided by Profs Didier Trono, Luc van Parijs, and Paul Khavari. We also thank Prof Garry P. Nolan for the use of Phoenix-Ampho and Phoenix-Eco cell lines. This work was supported by grant P16518-B14 of the Austrian Science Fund (FWF), by the Austrian genome project Gen-AU and by the research focus "Life Sciences and Health" of the University of Salzburg.

References

1. Ruiz, I. A. A., Palma, V., and Dahmane, N. (2002) Hedgehog-Gli signalling and the growth of the brain. *Nat. Rev. Neurosci.* **3,** 24–33.

2. Ingham, P. W. and McMahon, A. P. (2001) Hedgehog signaling in animal development: paradigms and principles. *Genes Dev.* **15,** 3059–3087.

3. McMahon, A. P., Ingham, P. W., and Tabin, C. J. (2003) Developmental roles and clinical significance of hedgehog signaling. *Curr. Top Dev. Biol.* **53,** 1–114.

4. Berman, D. M., Karhadkar, S. S., Hallahan, A. R., et al. (2002) Medulloblastoma growth inhibition by hedgehog pathway blockade. *Science* **297,** 1559–1561.

5. Berman, D. M., Karhadkar, S. S., Maitra, A., et al. (2003) Widespread requirement for Hedgehog ligand stimulation in growth of digestive tract tumours. *Nature* **425,** 846–851.

6. Karhadkar, S. S., Bova, G. S., Abdallah, N., et al. (2004) Hedgehog signalling in prostate regeneration, neoplasia and metastasis. *Nature* **431,** 707–712.

7. Watkins, D. N., Berman, D. M., Burkholder, S. G., Wang, B., Beachy, P. A. and Baylin, S. B. (2003) Hedgehog signalling within airway epithelial progenitors and in small-cell lung cancer. *Nature* **422,** 313–317.

8. Romer, J. T., Kimura, H., Magdaleno, S., et al. (2004) Suppression of the Shh pathway using a small molecule inhibitor eliminates medulloblastoma in Ptc1 (+/–)p53(–/–) mice. *Cancer Cell* **6,** 229–240.

9. Sanchez, P., Hernandez, A. M., Stecca, B., et al. (2004) Inhibition of prostate cancer proliferation by interference with SONIC HEDGEHOG-GLI1 signaling. *Proc. Nat. Sci. Acad. USA* **101,** 12,561–12,566.

10. Thayer, S. P., di Magliano, M. P., Heiser, P. W., et al. (2003) Hedgehog is an early and late mediator of pancreatic cancer tumorigenesis. *Nature* **425,** 851–856.

11. Kinsella, T. M. and Nolan, G. P. (1996) Episomal vectors rapidly and stably produce high-titer recombinant retrovirus. *Hum. Gene Ther.* **7,** 1405–1413.

12. Heemskerk, M. H., Blom, B., Nolan, G., et al. (1997) Inhibition of T cell and promotion of natural killer cell development by the dominant negative helix loop helix factor Id3. *J. Exp. Med.* **186,** 1597–1602.

13. Wiznerowicz, M. and Trono, D. (2003) Conditional suppression of cellular genes: lentivirus vector-mediated drug-inducible RNA interference. *J. Virol.* **77,** 8957–8961.

14. Rubinson, D. A., Dillon, C. P., Kwiatkowski, A. V., et al. (2003) A lentivirus-based system to functionally silence genes in primary mammalian cells, stem cells and transgenic mice by RNA interference. *Nat. Genet.* **33,** 401–406.

15. Deng, H., Lin, Q., and Khavari, P. A. (1997) Sustainable cutaneous gene delivery. *Nat. Biotechnol.* **15,** 1388–1391.

16. Dull, T., Zufferey, R., Kelly, M., et al. (1998) A third-generation lentivirus vector with a conditional packaging system. *J. Virol.* **72,** 8463–8471.

17. Zufferey, R., Nagy, D., Mandel, R. J., Naldini, L., and Trono, D. (1997) Multiply attenuated lentiviral vector achieves efficient gene delivery in vivo. *Nat. Biotechnol.* **15,** 871–875.

18. Kinzler, K. W., Ruppert, J. M., Bigner, S. H., and Vogelstein, B. (1988) The GLI gene is a member of the Kruppel family of zinc finger proteins. *Nature* **332,** 371–374.

19. Tanimura, A., Dan, S., and Yoshida, M. (1998) Cloning of novel isoforms of the human Gli2 oncogene and their activities to enhance tax-dependent transcription of the human T-cell leukemia virus type 1 genome. *J. Virol.* **72,** 3958–3964.

20. Regl, G., Neill, G. W., Eichberger, T., et al. (2002) Human GLI2 and GLI1 are part of a positive feedback mechanism in Basal Cell Carcinoma. *Oncogene* **21,** 5529–5539.

21. Davis, H. E., Rosinski, M., Morgan, J. R., and Yarmush, M. L. (2004) Charged polymers modulate retrovirus transduction via membrane charge neutralization and virus aggregation. *Biophys. J.* **86,** 1234–1242.

22. Boukamp, P. Petrussevska, R. T., Breitkreutz, D., Hornung, J., Markham, A., and Fusenig, N. E. (1988) Normal keratinization in a spontaneously immortalized aneuploid human keratinocyte cell line. *J. Cell Biol.* **106,** 761–771.

7

Cell Surface Marker and Cell Cycle Analysis, Hedgehog Signaling, and Flow Cytometry

Kristina Detmer and Ronald E. Garner

Abstract

Detailed cytological analysis of cells undergoing differentiation often reveals clues to the regulation of multiple cell features. The Hedgehog (Hh) signaling cascade is a master regulator of cell fate during differentiation and is implicated in the development of some neoplasias. Hh signaling affects the expression of cell surface markers of differentiation. We have used the flow cytometer to evaluate the effect of blockage of the Hh signal on the expression of cell surface markers of erythroid differentiation in an in vitro system. In addition, the effect of Hh signaling on the distribution of cells in the phases of the cell cycle over the course of erythroid differentiation was assessed. Inhibition of the Hh signal retards progression of the erythroid developmental program. Included is a discussion of some of the basic parameters, limitations, and interpretations of flow cytometric analysis used for CD marker expression and cell cycle studies.

Key Words: Flow cytometry; Hedgehog; cyclopamine; cell cycle; fluorescent antibodies; CD marker; propidium iodide; differentiation.

1. Introduction

Hedgehog (Hh) signaling is a potent regulator of cell fate determination and differentiation in many developmental systems (*1*). Alteration in Hh signaling can produce macroscopic morphological changes in developing tissues. A goal of much research is to identify the mechanisms by which Hh signaling regulates morphological change. At the cellular level, Hh signaling affects the distribution of cell surface markers. Few genes have been definitively identified as direct targets of Hh signaling, but cell cycle regulator genes are among them (*1,2*).

Flow cytometry is a convenient and informative technique for analyzing the distribution of cell surface markers and/or evaluating the distribution of a population of cells into the phases of the cell cycle. It offers many advantages.

From: *Methods in Molecular Biology: Hedgehog Signaling Protocols*
Edited by: J. Horabin © Humana Press Inc., Totowa, NJ

In each sample analyzed, typically more than 10,000 cells are evaluated. There are an abundance of commercial fluorescent antibodies as well as cytometry-compatible fluorochromes that allow for the identification of cell differentiation markers and determination of cell viability and DNA content. Whether bound to the cell membrane or incorporated into the cellular DNA, the different fluorochromes allow simultaneous evaluation of different cellular components. Fluorescence is measured during passage of the cells, one at a time, through a flow cell in a cytometer. The fluorochromes are excited with laser light of defined wavelength. Light emitted by the excited fluorochromes is detected by photomultiplier tubes, which detect specific regions of the visible light spectrum. Fluorescence detection takes advantage of the Stokes Shift phenomenon in which the emission maximum of a fluorochrome is at a longer wavelength than the excitation maximum. Most single laser instruments use 488 nm as their excitation wavelength and are capable of detecting emitted light at three distinct wavelength regions simultaneously. With three channels for gathering data, several different parameters can be evaluated in a single sample, so long as a different fluorochrome is available for each parameter. However, since fluorochromes emit in a range and not at a single wavelength, there may be overlap between emission ranges that can interfere with clear evaluation of cells that have been labeled with more than one fluorescent probe. Discriminating between fluorochromes with overlapping emission spectra is accomplished with a function referred to as compensation. Most of the newer flow cytometers can be set to compensate automatically for overlapping emission spectra.

To capitalize on the accuracy and efficiency of flow cytometry, the investigator should take some time to understand the principles of flow cytometry, as there are limitations to any technique. A standard source of information in the field of flow cytometry, including basic scientific principles, use of dyes, operation of machines, and interpretation of results is Howard Shapiro's Practical Flow Cytometry *(3)*.

Properly designed flow cytometric experiments can be extremely powerful. Today's flow cytometry laboratory can provide applications covering nearly all aspects of cell biology. In the past 30 years, flow cytometry applications have been developed that assess cellular mitogenic activity, apoptotic condition, phosphorylation of second messengers, activation-state, and solute movement between intracellular and extracellular compartments. Whatever the chosen application, one should always remember that a flow cytometer neither has an intuitive mechanism for telling what is and is not a cell, nor can it interpret significance. The user does this, either by selecting the threshold values or by using gating to identify discrete cell populations. Furthermore, a visual microscopic evaluation should be used to confirm the value of the threshold as a relevant parameter. Repeated observations and consistency between homogenous samples of known value confirms the validity of the limits chosen by the investigator.

In experimental systems in which the Hh signal can be modulated, morphological changes reflect the end result of changing a developmental program. The timing of the appearance or disappearance of markers of a particular developmental stage can be altered, presumably caused directly or indirectly by the alteration of the Hh signal. Changes in the number of markers on the cells will be reflected in changes in fluorescence intensity in the population. The cytometer measures and records fluorescence intensity per cell as each cell passes through the detector. Thus, changes in the frequency with which a marker is found on one cell population versus another can be compared. Each of the subpopulations of cells that pass through the laser is counted as a subtotal of events (cells). Histogram statistics are calculated with the internal instrumental statistics package, which compares numbers of events (cell counts) within a single population defined by its X parameter (degree of fluorescence) with other discrete populations. The user defines the populations with limits demarcated on the histogram graph, and the dependent population statistics are calculated and viewed in a table displayed below the histogram. Percentage of total events (cells) and mean fluorescence are provided by the instrumental analysis program.

Further information can be obtained with the use of multiparameter analysis. Multiparameter analysis using side and forward scatter in combination with fluorescent markers can dissect out variations in a cell population as well as allowing visualization of distinct subpopulations. Side scatter is a measurement of cellular complexity and reflects differences in intracellular granulation and cell surface irregularities. Forward scatter correlates with the size of a cell. When measuring the expression of a cell surface marker with a fluorescent antibody detection system, forward or side scatter provides an independent parameter for cellular identification. Cellular size and shape can be used in conjunction with immunofluorescence to identify specific cellular subpopulations. For example, lymphocytes, monocytes, and granulocytes can be distinguished from each other by their combined forward and side scatter profiles *(4)*. Heterogeneous expression of the cell surface antigens CD4 and CD8 distinguish between T helper and cytotoxic T lymphocytes, which are indistinguishable on the basis of forward and side scatter. On the other hand, a highly fluorescent population of particles might be shown by forward scatter to be too small to represent living cells. The ability to set thresholds allows the user to assess, identify, and disregard artifacts from an experiment, but a flow cytometric analysis is affected by the general rule of garbage in, garbage out. That is, the cleaner the preparation, the stronger and more reliable the data.

We have examined the development of primary human bone marrow cells in semi-solid medium in the presence of cyclopamine, an inhibitor specific for Hh signaling. At intervals, cultures were harvested and the expression of erythroid lineage-specific and developmental stage-specific cell surface markers were examined by flow cytometry.

2. Materials

2.1. Cell Culture and Harvest

1. Cyclopamine (LC Laboratories, Boston, MA) is dissolved in tissue-culture grade DMSO (Sigma St. Louis, MO) and stored at −20°C. Stock solutions are 20 mM and working solutions are 2 mM (*see* **Note 1**).
2. Recombinant Sonic Hh (R&D systems, Minneapolis, MN) is dissolved at 50 µg/mL in sterile phosphate buffered saline (PBS) containing 0.1% (w/v) bovine serum albumen and stored in single use aliquots at −20°C. It is not used in the experiments described here, but is included as a useful source.
3. Human bone marrow CD34$^+$ cells are purchased from Cambrex Bio Science Walkersville, MD.
4. Methylcellulose medium containing hematopoietic cytokines (StemCell Technologies, Vancouver, British Columbia; *see* **Note 2**).
5. 16-gauge blunt needles.
6. 35-mm culture dishes.

2.2. Cell Staining

1. Newborn calf serum (NCS) (Atlanta Biologicals, Atlanta, GA).
2. PBS: In 1 L water dissolve 8 g NaCl, 0.2 g KCl, 1.44 g Na_2HPO_4, and 0.24 g K_2HPO_4. Adjust to pH 7.4 with HCl. For working solutions, NCS is added to PBS to a concentration of 2% (PBS-2%NC).
3. Propidium iodide (PI) (Sigma St. Louis, MO) 20 mg/mL aqueous stock solution. Store protected from light.
4. Fluorochrome-conjugated antibodies (BD Biosciences San Jose, CA). Stored at 4°C protected from light. Do not freeze; use sterile technique when removing aliquots. Commonly used fluorochromes are fluoroscein isothiocyanate (FITC) and phycoerythrin (PE).
5. 12 × 75 mm polystyrene Falcon round bottom tubes (Fisher Pittsburgh, PA) with or without a cell strainer cap, which incorporates a 35-µm nylon mesh. The cap is useful for systems in which the cells are "sticky".
6. DNase I 2000 U/mL (New England Biolabs Beverly, MA) supplied with 10× DNase I buffer, optional.

2.3. Alternate Staining Protocol for Limited Cells or Large Number of Samples

1. 96-well round bottom tissue culture plates (Fisher).
2. 1.4-mL round bottom storage tubes (minitubes) and miniracks (Matrix Technologics Corp.) The minitubes and racks are configured so as to allow easy transfer of material from the 96-well tissue culture plates with a multichannel pipettor.

2.4. Distribution of Cells in Phases of the Cell Cycle

1. RNase A Type II (Sigma) 50 mg/mL aqueous stock.
2. 100% ethanol at −20°C.
3. PI (Sigma) 20 mg/mL aqueous stock solution. Store protected from light.

2.5. Flow Cytometry

Sheath fluid (Fisher) non-fluorescent, particle-free isotonic solution for use in flow cytometry.

3. Methods

Typically, a minimum of 10,000 events is recorded for each flow cytometric experiment. The rate at which cells pass through the detection chamber is proportional to the concentration of cells in suspension. Consequently, for timely collection of data, minimum cell concentrations of 500,000/mL are desirable. In determining the number of cells to harvest, allowance should be made for cell losses during washing stages. It is essential that the final cell suspension be free from clumps of cells or other particulate matter as the tubing leading to the detection chamber is easily clogged.

The protocols below describe the analysis of primary hematopoietic cells cultured in semisolid medium. These protocols are also suitable for adherent or suspension cells grown in liquid culture and for primary cells harvested from experimental animals in which Hh signaling has been modulated by retroviral, transgenic, or knockout approaches. Discussion of modifications needed to adapt the protocol will be included as appropriate in the **Notes** section. Protocols are given for analysis of cell surface markers, and an example is shown in **Fig. 1**. As it is increasingly recognized that Hh signaling affects the control of the cell cycle *(2,5,6)*, a protocol for determining the distribution of cell cycle phases within a population is included, and sample results are shown in **Fig. 2**.

The premise of cell cycle analysis derives from measuring the increase in DNA content by PI fluorescence as the cells progress through the S-phase of the cycle. Since the PI is excluded from living cells, a chemical fixation step is needed to provide access to the nuclear DNA. Alcohol dehydration of the membrane or hypotonic sodium citrate (with or without the incorporation of low concentration of nonionic detergent) can be used to infuse PI into the cells. Since PI will complex with RNA and fluoresce, the RNA is removed by digestion with RNase. PI has an emission spectrum that is fairly broad and peaks at approx 545 nm; it is read in the FL2 channel.

Diploid cells in G_0 or G_1 are identified as the population with the lowest mean fluorescence. These cells enter S-phase, where DNA content increases; this is seen as a plateau in plots of fluorescence vs number of events that can vary in height and angle of descent based on the number of cells with a $2n+$ DNA content. The area under the whole $G_0/G_1-S-G_2/M$ curve should reflect the total population. Therefore, converting the total number to percentage reflects the mitogenic dynamics of the cell population. When the cells are subjected to anti-mitogenic treatments, the change in the percentage of cells

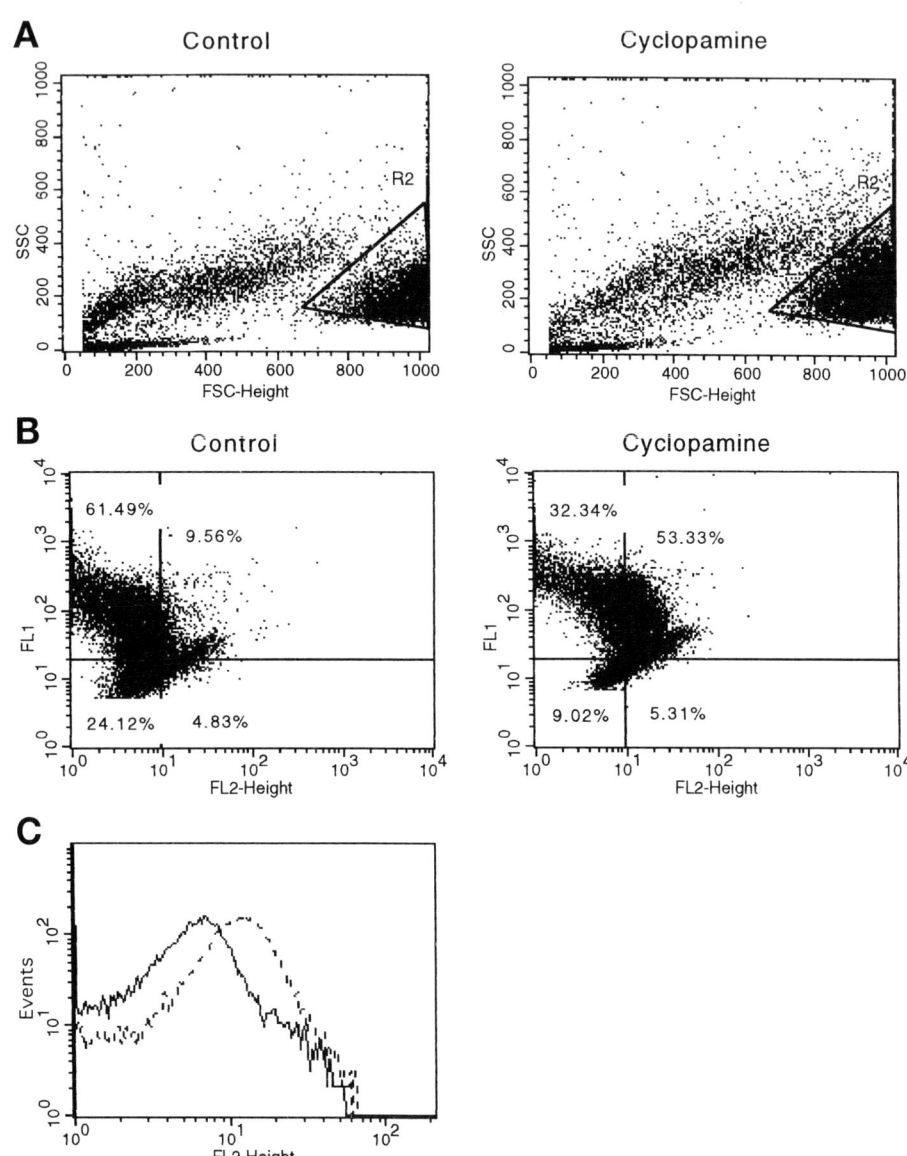

Fig. 1. Glycophorin A and CD36 expression on human bone marrow CD34$^+$ cells cultured in vitro in the presence or absence of 10 μM cyclopamine. Bone marrow CD34$^+$ cells were cultured in 1% methylcellulose medium in Iscove's MDM containing 30% fetal bovine serum, 1% bovine serum albumin, 10^{-4} M2-mercaptoethanol, 2 mM L-glutamine, 50 ng/mL recombinant human (rh) stem cell factor, 10 ng/mL rh granulocyte–monocyte colony stimulating factor (GM-CSF), 10 ng/mL rh interleukin-3, and 3 U/mL rh erythropoietin. After incubation for 5 d, the cells were harvested and labeled

in each mitogenic compartment may be used to determine the effectiveness and target of the cell cycle blockade. With appropriate selection of fluorochromes, cell cycle analysis can be coupled with the measurement of cell surface antigen expression to correlate expression of differentiation markers with proliferative state. This multiparameter approach is used extensively in lymphocyte activation to reveal the kinetics of cytokine receptor expression during lymphocyte activation *(7)*.

3.1. Preparation of Samples for Assay of Cell Surface Markers by Flow Cytometry

1. Add cyclopamine to the methylcellulose medium to a final concentration of 10 μM followed by vortex mixing (*see* **Note 3**).
2. Plate cells at a density of 25,000 cells/mL if they are to be cultured for no more than 6 d. If they are to be cultured for more than 6 d, plate cells at a density of 5000 cells/mL. The total volume of additions should be equal to one-tenth of the volume of methylcellulose medium to ensure a proper final viscosity. After all additions, mix the medium vigorously with a vortex mixer, and let stand for 5 min to allow bubbles to rise. Draw up into a syringe and dispense into culture dishes through a blunt 16-gauge needle. Incubate cultures for the desired time in a humidified incubator at 37°C, 5% CO_2.
3. To harvest, suspend cells in 10 volumes ice-cold PBS. All the following steps are carried out at 4°C. Centrifuge in a swinging bucket rotor at 200*g* for 7 min.
4. Repeat the wash with cold PBS-2%NC. Suspend cells in 1 mL PBS-2%NC (*see* **Note 4**).
5. Centrifuge as before and suspend at $1–5 \times 10^6$ cells/mL in PBS-2%NC (*see* **Note 5**). Add DNase I if desired (*see* **Note 6**).
6. Dispense ~0.5 mL cells into 12×75 mm tubes.

---

with FITC-conjugated anti-glycophorin A antibody and PE-conjugated anti-CD36 antibody. Data collection and analysis were carried out using CellQuest software. **(A)** Distribution of cells by forward scatter and side scatter. At least two well-defined cell populations can be distinguished. **(B)** Quadrant analysis of glycophorin A and CD36 expression on the R2 population identified by forward and side scatter. The cyclopamine-treated culture has a larger proportion of cells that are both glycophorin A and CD36 positive, upper right quadrant, than does the control culture. In the control culture, the majority of cells are glycophorin A positive and CD36 negative, upper left quadrant. **(C)** Distribution of CD36 fluorescence intensity in control (solid line) and cyclopamine-treated cultures (dashed line). The R2 population from cyclopamine-treated cultures shows greater fluorescence intensity than cells from control cultures. In the erythroid developmental program, CD36 expression precedes glycophorin A expression. Unlike glycophorin A expression, CD36 expression is lost as erythrocytes mature. The above results are consistent with a model in which inhibition of Hh signal retards the maturation of developing erythrocytes.

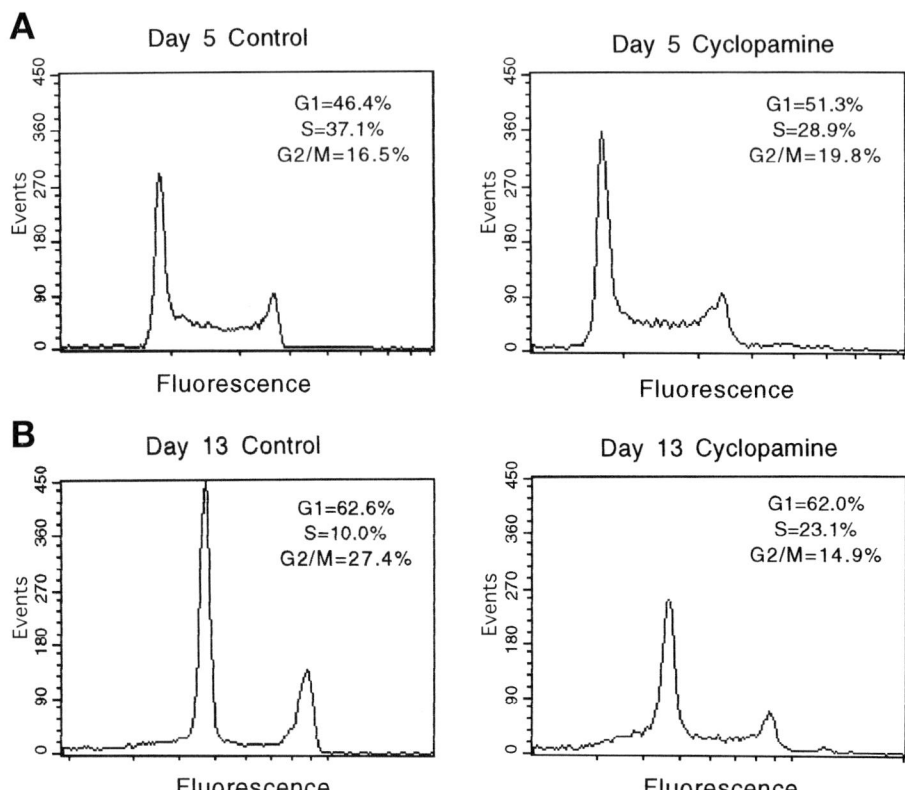

Fig. 2. Cell cycle phase distribution. Erythroid progenitor cells were cultured in the presence or absence of 10 μM cyclopamine for the indicated period before being stained with PI. The two peaks represent diploid and tetraploid cells; the intermediate fluorescence represents intermediate DNA content as the cells pass through S phase. Data collection and analysis was carried out using CellQuest software. (**A**) After culture for 5 d. (**B**) After culture for 13 d. On day 5, the cells are rapidly proliferating as indicated by the ratio of S-phase cells to G1-phase cells. By day 13, the proportion of cells in S-phase in untreated cells has dropped substantially, indicating that the proliferative phase of the developmental program is near its end.

7. Make a 2× stock solution of fluorochrome-conjugated antibody in PBS-2%NC. Add ~0.5 mL to cells, mix, and incubate on ice for 30 min. Cover tubes with aluminum foil or otherwise protect from light. Reserve a tube to which no antibody has been added for the negative control (*see* **Note 7**).
8. Centrifuge, wash with PBS-2%NC and resuspend in ~0.5 mL. PBS-2%NC. If cells are limited, care should be taken to avoid cell losses on the sides of the tube during washing. Hold cell suspension up to light and flick to see if particles are

detectable. Transfer to another 12 × 75 mm tube through a cell strainer cap if necessary (*see* **Note 8**).

3.2. Alternate Protocol for Staining Samples with a Limited Number of Cells or a Large Number of Samples

This protocol is particularly convenient when screening several samples with a panel of antibodies.

1. Harvest samples. If harvesting primary tissue, lyse red cells as described in **Note 5** and suspend in PBS-2%NC. Add DNase I if needed. All solutions and centrifugations should be carried out at 4°C.
2. Dispense on ice 40 µL aliquots into the wells of a 96-well round bottom tissue culture plate.
3. Make 2× stocks of each antibody in PBS-2%NC. Add 40 µL of 2× antibody to the appropriate samples. Wrap plate in aluminum foil.
4. Mix 10 min on a platform shaker, preferably on ice.
5. Incubate 30 min at 4°C.
6. Add 120 µL PBS-2%NC to each well (*see* **Note 9**).
7. Centrifuge on a platform swinging bucket rotor at 200*g* for 7 min.
8. Check to see that the cells are on the bottom of the wells. Remove the supernatant liquid by turning the plate upside down and giving a quick flick of the wrist.
9. Add 150 µL PBS-2%NC to each well.
10. Remove the contents of each well to a minitube in a minirack. The miniracks hold 96 tubes in a 12 × 8 array and are labeled in the same way as the 96-well plates.
11. Add an additional 100–150 µL PBS-2%NC to each minitube.
12. To collect data, place each minitube in a 12 × 75 mm polystyrene tube, and mix by vortexing. Collect data with a flow cytometer, *see* **Section 3.4**.

3.3. Preparation of Cells for Cell Cycle Analysis

1. Harvest cells and suspend in PBS-2%NC at 1×10^6 cells/mL. Mix 0.75 mL cell suspension with 0.25 mL chilled (−20°C), 100% ethanol added dropwise with gentle mixing. After incubation at room temperature for 15 min, centrifuge the cells as before, wash with PBS-2%NC and suspend in 1 mL PBS-2%NC.
2. Add 1 mg/mL RNase A Type II and 50 µg/mL of PI. Incubate 1 h before analysis (*see* **Note 10**).
3. Collect data by flow cytometry, *see* **Section 3.5**.

3.4. Flow Cytometry of FITC and PE Double-Labeled Cells

1. Inspect samples to be sure they are free of visible particles. Remember rule number 1: if you can see it, you cannot run it (*see* **Note 11**).
2. Cells are introduced through the sample injection port at a rate of 12 µL/min and travel through a flow cell within a moving stream of sheath fluid (*see* **Note 12**).
3. A 488 nm argon laser excites the fluorochromes. A photodiode detects forward light scatter (FSC). Separate photomultiplier tubes detect side scatter light (SSC),

light emitted in the 500–560 nm range (FL1), light emitted in the 545–625 nm range (FL2), and light emitted at >650 nm (FL3). The instrument settings must be adjusted before meaningful data can be collected. These adjustments must be determined empirically. However, FSC and SSC are typically set to collect data in the linear mode (Lin) and FL1, FL2, and FL3 are set to the logarithmic (Log) mode.

4. On the flow cytometer computer, call up a dot plot and set the x and y axes to FL1 and FL2, respectively. While unstained cells are passing through the cytometer, the threshold, compensation, and gain settings are adjusted so that when unstained cells pass through the detector they are seen in the square bounded by approximately 2×10^1 fluorescence units on each axis (*see* **Note 13**).

5. Collect and analyze data with the appropriate software. FITC-labeled antibodies are detected in the FL1 channel, and PE-labeled antibodies are read in the FL2 channel. It is desirable to collect at least 10,000 events.

3.5. Cell Cycle Analysis by Flow Cytometry

1. The fluorescence profiles of PI-labeled cells are collected in the FL2 channel.
2. An example of a cell cycle analysis is showing the distribution of cells in the G_1, S, and G_2 phases of the cell cycle is shown in **Fig. 2**. A modification of this method can be used to determine the DNA synthesis time (*see* **Note 14**).

4. Notes

1. Cyclopamine can also be dissolved in ethanol and many investigators use alcoholic solutions of cyclopamine in their experiments, but in our hands DMSO solutions give more reproducible results in the semisolid colony assay.
2. The hematopoietic cytokines contained in the methylcellulose medium is determined by the experimental question.
3. The concentration of cyclopamine or recombinant Hh required to achieve a biological result varies substantially among developmental systems.
4. If the investigator desires to collect data from live cells only, suspend the cells in 1-mL PBS-2%NC containing 5 µg/mL PI. PI stains nucleic acids of dead cells, but not live cells. Subsequent washings do not include PI, as PI does not dissociate once bound to nucleic acid. The fluorescent dead cells can be removed from the flow cytometric analysis (*see* **Note 13**).
5. Cells grown in suspension can simply be harvested by centrifugation and stained with PI. Cells from soft primary tissues such as spleen or liver can be harvested by mashing the tissue through a tissue strainer (Fisher) with the plunger of a 3-cm^3 syringe. The tissue strainer fits a 50-mL blue-capped Falcon tube for convenient collection. The mesh of the strainer is washed with PBS-2%NC to increase recovery of cells. Red blood cells are suspended in an aqueous solution of 0.8% ammonium chloride/0.1 mM EDTA on ice for 15 min. Dilute with 2 volumes of PBS-2%NC with or without 5 µg/mL PI. Centrifuge, suspend in PBS-2%NC and stain with antibodies.
6. Some cell types are prone to lysis with release of DNA. The DNA entangles cellular debris and non-lysed cells causing clumping, which will interfere with proper

antibody staining and will clog the flow cytometer. Clumping can be prevented by the addition of DNase I to a final concentration of 20 U/mL. If DNase I is used, magnesium must be added to the buffer. This can conveniently be done with the 10× buffer supplied with the enzyme by manufacturers.

7. The antibody dilution must be determined empirically. Useful dilutions of the manufacturer's stock can range from 1:10 to 1:3000.

8. The minimum volume of cell suspension in a 12×75 mm tube is ~0.3 mL.

9. A multichannel pipettor greatly facilitates this step.

10. The amount of PI bound to non-living cells and hence the fluorescence intensity is proportional to the nucleic acid content. As a cell progresses through the cell cycle, the DNA content doubles until the cell goes through mitosis. For PI staining to accurately reflect the DNA content of cells, RNA is removed by digestion with RNase.

11. Powdered gloves are a common source of particles. Non-powdered gloves should be used in the preparation of samples for flow cytometry. If particles are present, vortexing followed by filtering through a cell strainer or addition of DNase I and magnesium may be helpful.

12. The flow rate can be varied. Samples with a low concentration of cells can be sampled at a faster flow rate.

13. It is possible to look only at living cells if the cells have been washed with a solution containing PI during the labeling procedure (*see* **Note 4**). During the instrument set up, data on unstained cells are collected and shown on a dot plot with FSC on the *x*-axis and FL3 on the *y*-axis. Living cells will exclude PI, while dead cells will form a highly fluorescent population high on the *y*-axis. A "gate" can be drawn around the living cells and the computer set to collect only events that fall within that gate.

14. Cells are grown for several hours in the presence of bromodeoxyuridine (BrdU), and then labeled with FITC-conjugated anti-BrdU antibody and PI. The rate of passage of BrdU-labeled cells through the S-phase can be quantified by comparing their mean PI fluorescence relative to that of G_1 and G_2 cells *(8)*.

References

1. McMahon, A. P., Ingham, P. W., and Tabin, C. J. (2003) Developmental roles and clinical significance of Hedgehog signaling. *Curr. Top. Dev. Biol.* **53,** 1–114.

2. Teh, M. T., Wong, S. T., Neill, G. W., Ghali, L. R., Philpott, M. P., and Quinn, A. G. (2002) FOXM1 is a downstream target of Gli1 in basal cell carcinomas. *Cancer Res.* **62,** 4773–4780.

3. Shapiro, H. M. (2003) *Practical Flow Cytometry*, 4th Ed., Wiley-Liss, Hoboken, NJ.

4. Caldwell, C. W. (2001) Quality control and quality assurance in Immuno-phenotyping. In *Flow Cytometry in Clinical Diagnosis* (Keren, D. F., McCoy, J. P. J., and Carey, J. L. eds), American Society of Clinical pathologists Press, Chicago, IL, pp. 117–156.

5. Roy, S. and Ingham, P. W. (2002) Hedgehogs tryst with the cell cycle. *J. Cell Sci.* **115,** 4393–4397.

6. Neumann, C. J. (2005) Hedgehogs as negative regulators of the cell cycle. *Cell Cycle* **4**, 1139–1140.

7. Schmid, I., Cole, S. W., Zack, J. A., and Giorgi, J. V. (2000) Measurement of lymphocyte subset proliferation by three-color immunofluorescence and DNA flow cytometry. *J. Immunol. Methods* **235**, 121–131.

8. Begg, A. C., McNally, N. J., Shrieve, D. C., and Karcher, H. (1985) A method to measure the duration of DNA synthesis and the potential doubling time from a single sample. *Cytometry* **6**, 620–626.

8

Detecting Tagged Hedgehog with Intracellular and Extracellular Immunocytochemistry for Functional Analysis

Ainhoa Callejo, Luis Quijada, and Isabel Guerrero

Abstract

In this chapter, we explain different strategies to analyze the extracellular Hedgehog (Hh) morphogen distribution and Hh intracellular trafficking by immunohistochemistry techniques. For this purpose, it has been very useful to have a transgenic fly line that expresses a Hh-green fluorescent protein (GFP) fusion protein. These flies can be used to study the way Hh spreads through the anterior compartment where it signals, and analyze in detail how Hh is internalized by its receptor Patched. In addition, this Hh-GFP fusion made without lipid modifications (cholesterol or palmitic acid) can be used to investigate the function of these lipids on Hh in terms of spreading, internalization, and signaling abilities.

Key Words: Hedgehog protein (Hh); green fluorescent protein (GFP); patched (Ptc); wing imaginal disc cells; Gal 4 driver; UAS vector; extracellular labeling.

1. Introduction

The Hedgehog (Hh) pathway is crucial in many developmental programs that establish the body plan of an individual. The active Hh protein is synthesized as a precursor that undergoes autocatalytic cleavage. This peptide is additionally modified at its N- and C-termini by palmitoyl and cholesterol adducts, respectively. Within a developing organ, Hh is secreted by a discrete subsets of cells; a graded and short-range response to Hh signaling occurs in the cells receiving the signal. The receptor of Hh, Patched (Ptc), is a 12 pass-transmembrane protein. Ptc keeps the Hh pathway silenced in its unliganded state: in the absence of Hh, Ptc suppresses the activity of Smoothened (Smo), a seven pass-transmembrane protein and a positive modulator of the pathway. Hh binding to Ptc results in the release of Ptc-mediated inhibition of Smo and, the subsequent activation

From: *Methods in Molecular Biology: Hedgehog Signaling Protocols*
Edited by: J. Horabin © Humana Press Inc., Totowa, NJ

of the transcription factors: Cubitus interruptus protein in *Drosophila* and the orthologous Gli proteins in mammals. One peculiarity of the Hh signaling pathway is that *ptc* is upregulated in response to increasing amounts of Hh. Therefore, Hh controls both its own activity and its own spreading by maintaining high levels of the receptor. By this means, a morphogenetic gradient is formed in the Hh-receiving cells.

Much of this knowledge has been derived from experimental studies performed in imaginal discs of the larva of *Drosophila* (reviewed in Ref. *[1]*). The wing imaginal disc is a sac-like structure in the larva formed by epithelial cells. One surface of this structure comprises a monolayer of pseudostratified columnar epithelial cells with their apical membranes orientated towards the disc lumen. This specialized group of cells will give rise to the wing. The overlaying surface is a squamous epithelium, named the peripodial membrane that does not give rise to any cuticular structure in the adult fly. Two populations of cells, with different adhesion affinities, divide the epithelium of columnar cells into posterior (P) and anterior (A) compartments. Hh is synthesized in P cells, where it is able to diffuse long distances. However, the spreading of Hh is limited by the interaction with the Ptc receptor expressed in the A cells. When Hh reaches Ptc, a concentration gradient of Hh is formed at the boundary that separates the anterior and posterior (A/P) compartments. The journey of Hh through the epithelium is a highly regulated process and not a mere diffusion from its source. The panoply of genes involved in controlling the adequate spreading of Hh is large, and still increasing (**Fig. 1**).

The methods described in this chapter focus on the study of the distribution of Hh along the epithelium of the wing imaginal disc. We describe (1) the tagging of Hh with green fluorescent protein (GFP) and its cloning in a UAS-expression vector, (2) the expression of the tagged protein in transgenic flies with Gal4 drivers, (3) the immunocytochemistry of imaginal discs for confocal microscopy, and (4) in vivo labeling with soluble substrates.

2. Materials

1. Antibodies: Rabbit polyclonal anti-Hh (1:800 dilution *[2]*), mouse monoclonal anti-Ptc (Hybridoma Bank [Apa1], 1:50 dilution), rabbit polyclonal anti-GFP, and rabbit polyclonal anti-βGal or mouse monoclonal anti-βGal.
2. Soluble substrates, such as dextran-red fluorescent (3.7 mM Red-dextran, lysine fixable, MW 3000).
3. Schneider's M3 medium: commercial disposal. Stored at 4°C.
4. 10× PBS (pH 7.0): 2.8 M NaCl, 55 mM KCl, 0.15 M NaHPO$_4$, and 29 mM KH$_2$PO$_4$.
5. 4% (w/v) paraformaldehyde in 1× PBS. Add paraformaldehyde powder and mix, boiling only briefly until paraformaldehyde is in solution. Adjust pH to 7.5–8. Cool before using. It can be stored frozen.

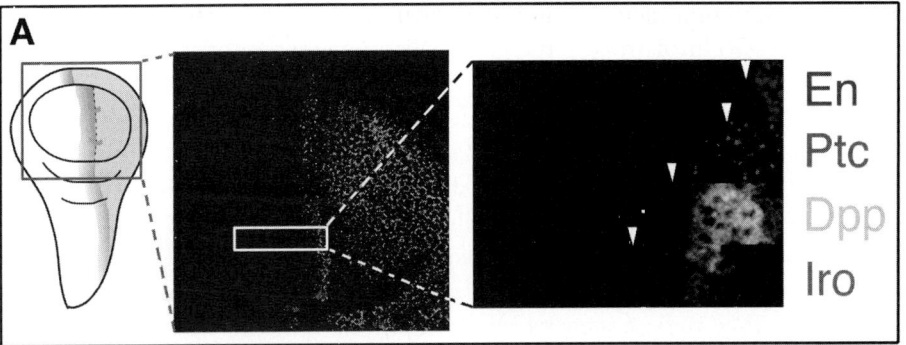

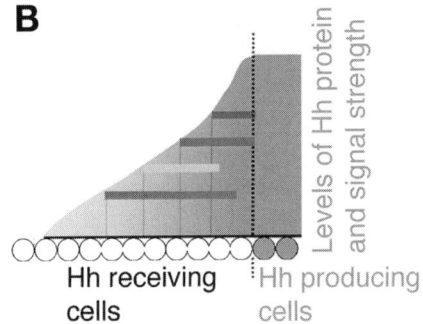

Fig. 1. Hh gradient in the wing imaginal disc. (**A**) Diagram of a wing imaginal disc with the posterior compartment cells expressing Hh (green) and the anterior compartment cells expressing Ptc (red) in response to Hh. The red frame indicates the territory represented in the adjacent panel, showing the real staining of the wing disc with Hh (green) and Ptc (red) antibodies. The right panel shows the expression of different Hh target genes in A compartment cells. The different thresholds of Hh signaling required for the activation of the different targets are indicated in the graph in (**B**).

6. Fly genotypes
   -UAS-Hh-GFP, UAS-HhN-GFP, and UAS-HhC85S-GFP *(3,4)*.
   -ap-Gal4/CyO *(5)*.
   -ubx>f + >Gal4, UAS-βgal *(6)*.
   -actin>CD2>Gal4 *(7)*.
   -Heat shock Flipase (HS Flp *[8]*).

3. Methods

The availability of antibodies for proteins in which we are interested is sometimes the limiting step in performing experiments. Production of specific antibodies is time consuming and there is no guarantee of good antigen–antibody recognition. Additionally, the resulting antibodies are sometimes not suitable for immunofluorescence staining. One way to overcome these problems is to generate a fusion construct to the GFP. This moiety allows for direct detection of the tagged protein.

One difficulty to fusing the GFP moiety in Hh is due to the fact that Hh is processed at both the N- and the C-terminus (**Fig. 2A**). Thus, the GFP open reading frame (ORF) must be included within the Hh ORF before its cleavage amino acid. Previous work tagging Hh with hemagglutinin (HA) showed that adding 3× HA between amino acid residues 255 (V) and 256 (H) resulted in a fusion protein that retained full Hh activity *(9,10)*. For the construction of HhGFP, GFP-coding sequences were amplified by PCR from a pEGFP-N1 vector and tagged in frame before the C-terminal auto-proteolysis cleavage site (…SH255V-GFP-H256GCF…). The coding region of Hh contains a unique PmlI restriction site coincidental with the codons for amino acid residues 255 and 256 that facilitates the cloning and avoids addition of extra amino acids encoded by the linkers used to amplify the GFP ORF by PCR. The GFP ORF was cloned in the PmlI restriction site of a pBS-Hh plasmid, in which the Hh cDNA was previously cloned. The orientation of the inserted GFP ORF fragment was checked by restriction analysis and sequencing. We also engineered mutant fusion proteins, such as HhN-GFP (without the cholesterol moiety), HhC85S-GFP (which lacks the palmitic acid adduct), and HhC85S-N-GFP (with no lipid modifications). To this aim, we induced point mutations via PCR, with primers that contained the needed changes. The spreading and signaling activity of these Hh mutant forms are different to wild-type Hh *(4)*.

In order to express a protein in a spatially and time-restricted manner in *Drosophila*, the Gal4/UAS system is an ideal approach *(11)*. For an extended review of the Gal4/UAS system, *see* Chapter 13 by Busson and Pret. We used the pUAS-T vector to clone the Hh-GFP ORF between the EcoRI and the NotI restriction sites (**Fig. 2B**). The pUAS-Hh-GFP plasmid was isolated and injected into fly embryos in order to obtain transgenic flies carrying the UAS-Hh-GFP construct. This chimera behaves, in terms of spreading and signaling, as the wild-type protein. Hh-GFP as well as all mutant forms are also normally processed. We used the same strategy to clone and obtain transgenic flies with Hh forms mutated to eliminate the lipid modifications (cholesterol and palmitic acid). The description of the methods to introduce DNA in *Drosophila* is beyond the scope of this chapter and the reader is referred to detailed protocols described elsewhere *(12)*.

3.1. Expression of the Tagged Protein in Transgenic Flies Using Gal4 Drivers

1. Balance transgenic flies for the engineered construct gene to avoid recombination and, at the same time, to map on which of the four chromosomes of *Drosophila*

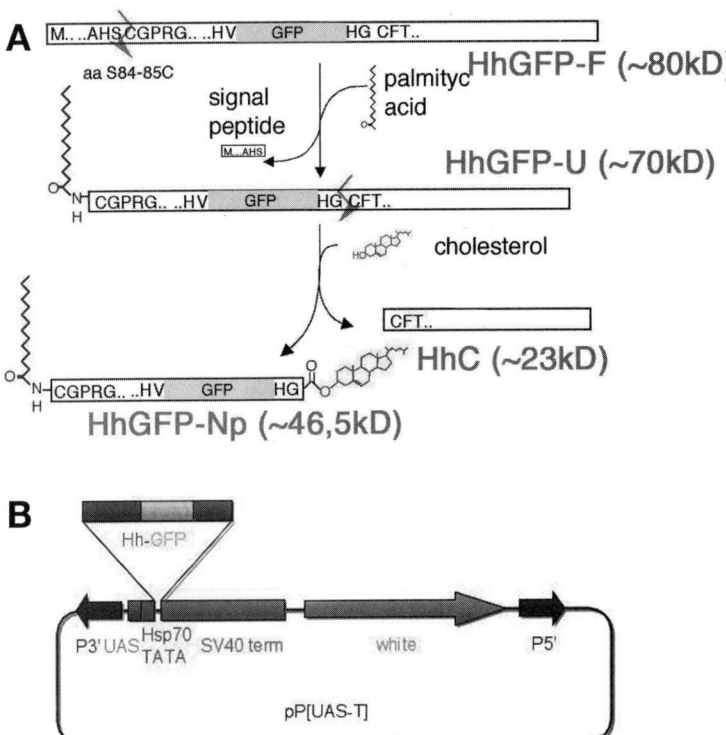

Fig. 2. Hh-GFP fusion protein. (**A**) Scheme of the predicted HhGFP fusion protein processing. GFP sequences are cloned in frame within Hh protein before C-terminal site of autocleavage. HhGFP-F, full-length protein; HhGFP-U, the unprocessed protein without the signal peptide; HhC, the C-terminal region of the processed protein; HhGFP-Np, the N-terminal region of the processed protein with the GFP fragment and the palmitic acid and cholesterol modifications. Therefore, the HhGFP chimera behaves as the wild-type Hh protein. (**B**) Cloning site of the Hh-GFP chimera in the pUAST expression vector.

the insertion has occurred. In our example (*see* **Note 1**), we used a fly stock with an insertion on the second chromosome, i.e., y w; UAS-Hh-GFP/CyO.

2. Cross the females with males carrying one of the various Gal4 drivers; in this example, we chose the *apterous*-Gal4 driver (*ap*-Gal4) based on its expression pattern which coincides with the whole of the dorsal compartment including both the anterior and the posterior domains (*see* **Note 2**).
   The crossing scheme is the following:

$$\text{(P)} \ ♀ \ yw; \ UAS\text{-}Hh\text{-}GFP/CyO \times ♂ \ yw/Y; \ ap\text{-}Gal4/CyO$$

$$\downarrow$$

$$♀♂ \ UAS\text{-}Hh\text{-}GFP/ap\text{-}Gal4 \text{ from the F1}$$

3. Dissect third instar larva in ice-cold 1× PBS. This should last no longer than 30 min.

4. Stain wing imaginal discs with specific antibodies that recognize protein products of several target genes of the Hh pathway (e.g., anti-Ptc).

3.2. Immunocytochemistry of Imaginal Discs for Confocal Microscopy

3.2.1. Intracellular Staining

1. Dissect and collect imaginal discs in an eppendorf tube with ice-cold PBS for no longer than 30 min. It is recommended to go immediately to the next step of the protocol to avoid protein degradation in the discs.

2. Fix in 4% paraformaldehyde in 1× PBS, for 30 min at RT, with gentle shaking.

3. Wash three times with PBS + 0.1–0.3% Triton (1 mL each wash, 10 min at RT, with gentle shaking) to get the disc tissue permeable to antibodies. Optional step: If you wish to stop here, wash several times with PBS (with or without Triton X-100 0.1–0.3%) and leave the discs in PBS at 4°C overnight. If you wish to continue, go directly to the next step.

4. Blocking: Incubate the discs 30–60 min with PBS + 0.2% Triton + 2% BSA (PBT/BSA) at RT, with gentle shaking.

5. Incubation with first antibody: Dilute the antibody in blocking solution PBT/BSA. Incubate with gentle shaking 2–3 h at RT or overnight at 4°C. Overnight incubation is preferred because the antibody is preserved better from degradation by the lower temperature.

6. Wash three times with PBS + 0.2% Triton.

7. Incubation with secondary antibody: Dilute the antibody in PBT. Incubate for 2 h at RT, with gentle shaking.

8. Wash three times with PBS + 0.2% Triton.

9. Rinse two times with PBS to remove all trace of Triton X-100.

10. Suspend the discs in mounting medium to preserve the fluorescence and decrease background, (e.g., Vectashield). Stored at 4°C.

11. Final analysis by confocal microscopy (*see* **Note 3**). The sample is useful for analyzing both Hh-cyan fluorescent protein (CFP) and Hh-yellow fluorescent protein (YFP), or Hh-CFP-Dally-like-YFP protein-protein interaction by fluorescence resonance energy transfer (FRET; *see* **Note 4**). It can also be used to study protein dynamics in living cells by the Fluorescence Recovery after photobleaching technique (FRAP; *see* **Note 5**).

3.2.2. Extracellular Staining with Anti-GFP Antibody

1. Dissect the larvae in ice-cold Schneider's M3 medium no longer than 30 min (*see* **Note 6**).

2. Incubation with first antibody: Transfer the discs to an eppendorf tube with fresh, ice-cold, M3 medium containing anti-GFP antibody (rabbit, 1:300 dilution). Incubate for 30 min, 4°C, with gentle shaking (*see* **Note 7**).

3. Wash the discs in ice-cold PBS, three times.

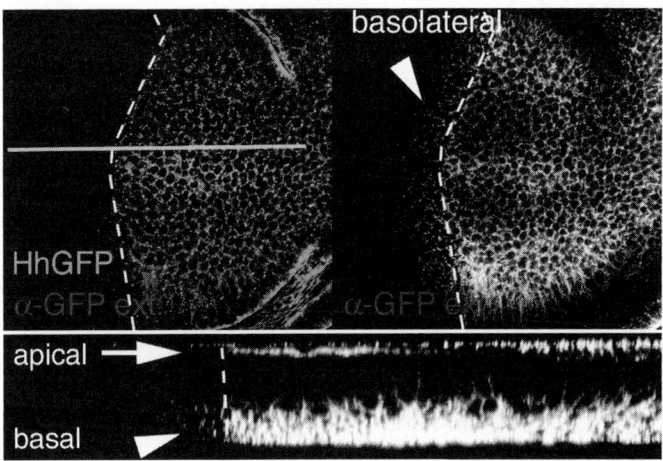

Fig. 3. Extracellular staining of Hh-GFP with anti-GFP antibody. Confocal basolateral section and a transverse view of a wing imaginal disc expressing Hh-GFP in the posterior compartment using a Hh-Gal4 driver. Observe the colocalization of Hh-GFP (green) and anti-GFP antibody (red in the first confocal section, white in the second confocal section).

4. First fixation: Fix the discs in PBS + 4% paraformaldehyde for 30 min at 4°C, with gentle shaking.
5. Second fixation: Fix the discs in PBS + 4% paraformaldehyde for 30 min at RT, with gentle shaking.
6. Continue from **step 3.2.1.3.** of the intracellular staining protocol. The sample is useful to reveal the colocalization between Hh and Ptc receptors in the extracellular environment (**Fig. 3**; *see* **Note 8**).

3.3. Staining with Soluble Substrates

1. Dissect third instar larva in ice-cold Schneider's M3 medium for no longer than 30 min.
2. Incubate imaginal discs in 3.7 m*M* Red-dextran (lysine fixable, MW 3000) diluted in M3 medium at 25°C (pulse).
3. Wash five times for 2 min in ice-cold M3 medium.
4. Incubate the discs for a chase period at 25°C in M3 medium prior to fixation in 4% paraformaldehyde:
   Pulse the discs for 5 min without a chase period to visualize the early endosomes of the endocytic compartment.
   Use a 5 min pulse and 60 min chase for late endosomes.
5. First fixation: Fix in 4% paraformaldehyde in 1× PBS for 40 min at 4°C.
6. Second fixation: Fix in 4% paraformaldehyde in 1× PBS 0.05% Triton X-100 for 20 min at RT.

7. Wash the discs and incubate with antibodies in PBS 0.05% Triton X-100 (*see* **Note 9**). The sample is useful for analyzing Hh endocytic trafficking after internalization (**Fig. 4**; *see* **Note 10**).

4. Notes

1. It is highly recommended to establish fly stocks with transgene insertions balanced over different chromosomes in order to facilitate mating strategies.

2. Using the *ap*-Gal4 driver we can ectopically express Hh in the dorsal compartment and observe Hh spreading in the anterior/ventral and posterior/ventral compartments (**Fig. 5**). It also allows a comparison of the spreading properties of wild-type Hh to Hh-N-GFP and Hh-C85S-GFP.

   Another way to express *Hh-GFP, Hhc85s-GFP*, and *HhN-GFP* ectopically in the imaginal disc is to generate random clones of these proteins (flip-out clones) by recombination (for details, *see* Chapter 12 by Bankers and Hooper). The ectopic clone is labeled by the lack of CD2 staining *(7)*, and is possible to study the activation of *decapentaplegic* as a target of Hh-GFP using the *dpp-LacZ* reporter and staining with anti-βgal antibody.

   To do this experiment, the cross scheme is the following:

   ♀*HS Flp; actin>CD2>Gal4; dppLacZ/CyO* × ♂ *yw/Y; UAS-Hh-GFP/CyO*
   ↓
   ♀♂*HS Flp; actin>CD2>Gal4; dppLacZ/UAS-Hh-GFP*

   The ectopic Hh expressing clone can be positively labeled by the expression of a LacZ reporter *(6)*. In this case, the cross scheme is the following:

   ♀ *ubx>>Gal4-βgal/CyO; HS Flp/TM6B* × ♂ *yw/Y; UAS-Hh-GFP/CyO*
   ↓
   ♀♂ *ubx>>Gal4-βgal/UAS-Hh-GFP; HS Flp*

   Incubate embryos at 24–48 h of development, at 37°C for 10 min to activate the *heat shock flipase* (*HS Flp*). With this technique, one can observe differences in signaling properties between wild-type Hh-GFP and non-lipidated mutant Hh forms (**Fig. 6**).

3. To obtain high-resolution confocal images from stained tissues, it is necessary to have a battery of highly specific antibodies, produced in different animals. Highly specific secondary antibodies allow the simultaneous staining of different antigens with combinations of different primary antibodies.

4. The FRET technique detects the energy transfer between an excited donor fluorophore and a nearby acceptor fluorophore. This energy transfer is dependent on the overlap of excitation spectrum of the acceptor with the emission spectrum of the donor, as well as the distance between the fluorophores. Energy transfer can only occur when a donor and acceptor are very close together (within nanometers) and their spectra sufficiently overlap. With this technique, it is possible to know if there is a physical interaction between two molecules. In this way, for example, we can observe whether Hh forms oligomers or whether there is a direct interaction between the glypican Dally-like and Hh in the extracellular matrix.

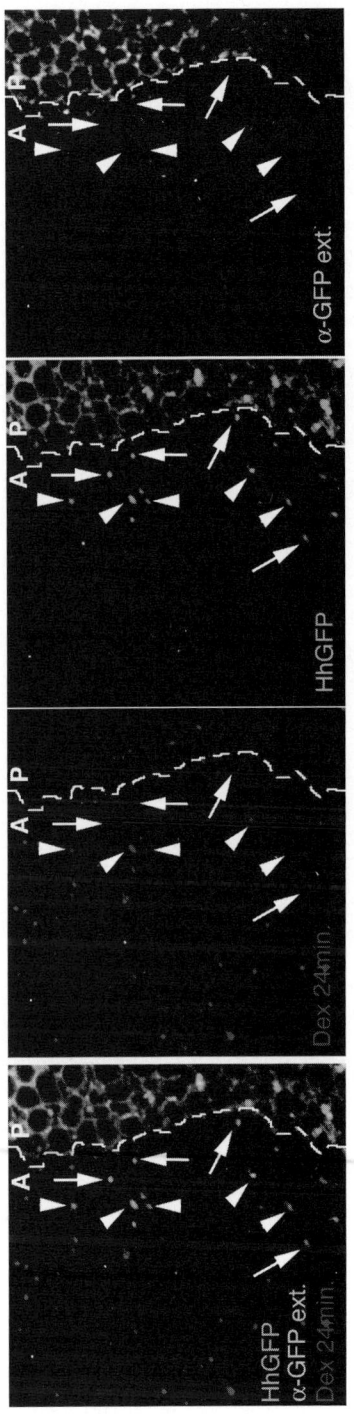

Fig. 4. Hh-GFP localizes in endocytic pathway. Hh-GFP/*hh-Gal4* wing discs were incubated first with dextran-red to label the endocytic compartment and subsequently with anti-GFP antibody at 4°C to label the extracellular Hh-GFP. Note that some Hh punctate structures (green), but not all, colocalize with dextrans (arrowheads) indicating that they are internalized vesicles. The punctate structures that do not colocalize with dextrans are also intracellular vesicles (arrows), because they are not labeled by extracellular staining approaches.

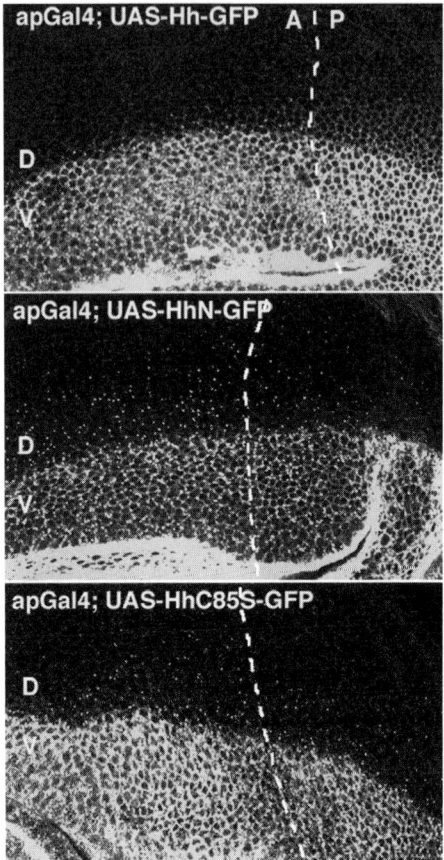
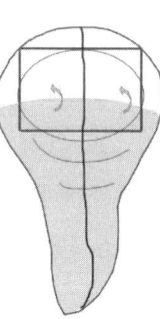

Fig. 5. Spreading properties of lipid-modified and -unmodified forms of Hh. UAS-HhGFP, HhN-GFP, and HhC85S-GFP proteins are expressed in the dorsal domain of the wing imaginal disc using the *ap-Gal4* driver, and spread to the ventral territory. Note that spreading properties of wild-type Hh and lipid unmodified Hh forms are different.

5. The FRAP technique consists in abolishing the fluorescence signal of a GFP-tagged protein. A cell field expressing a GFP chimera is briefly photobleached with a high-intensity laser, and the movement of unbleached fluorescent molecules into the bleached area is followed by low-intensity laser light. The FRAP technique can be used to study processes like vesicle transport, flow-based movement, viscosity of the environment of a protein and whether the protein is part of a much larger complex. It can also be used to analyze the dynamics of the Hh gradient by observing the time recovery of the Hh-GFP fluorescence after photobleaching.
Photoconversion and reversible photobleaching are phenomena that can seriously complicate photobleaching analysis. Photoconversion is the process by which a

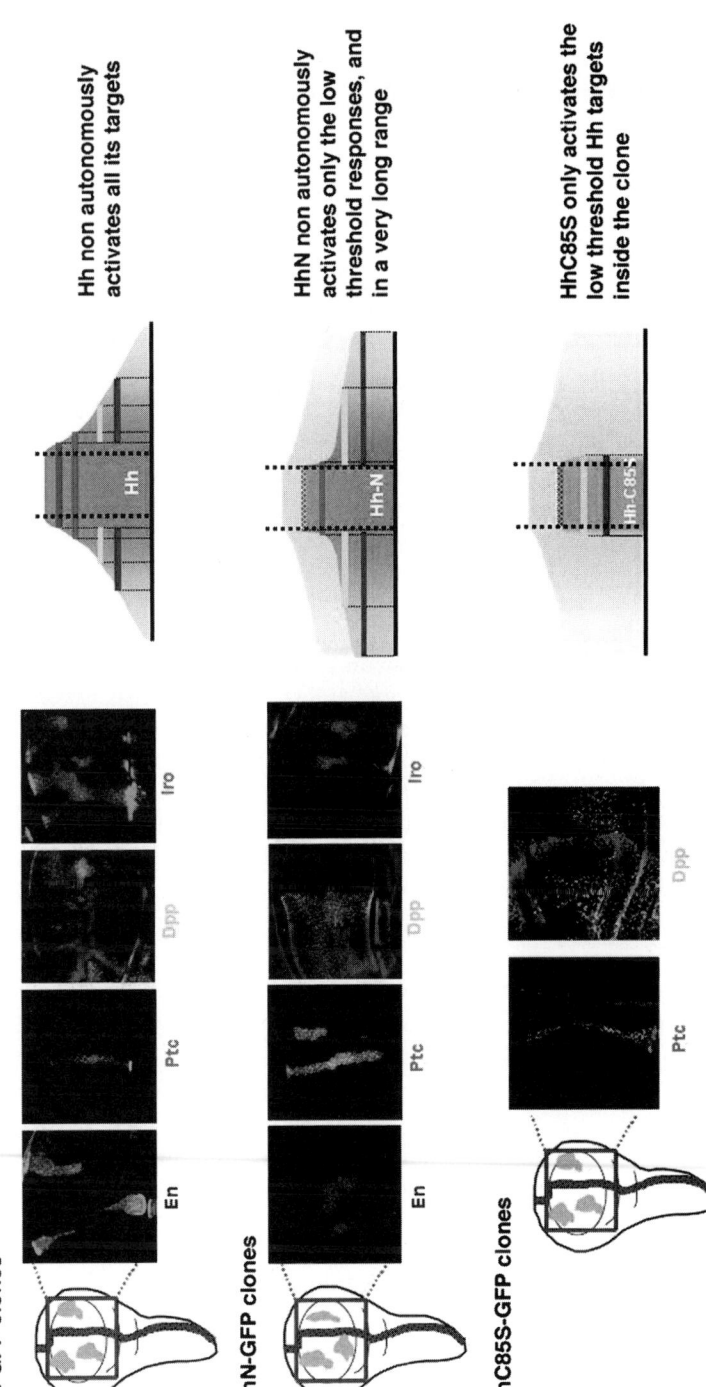

Fig. 6. Hh target activation in ectopic Hh-GFP, Hh-N-GFP, or Hh-C85S-GFP clones. Ectopic Hh clones induced in the anterior compartment of the wing imaginal disc, and analyzed for the signaling properties of the different forms of Hh-GFP. Note that HhN-GFP only activates non-autonomously and in a very long range, the low-threshold genes, such as dpp and iro. HhC85S-GFP has very poor signaling activity.

fluorophore is excited and becomes transiently or permanently altered in its fluorescence excitation and emission spectra. Reversible photobleaching occurs when a fluorophore's excitation state is changed by intense illumination, which appears to an observer as destruction of the fluorophore. The fluorophore reverts to its native excitation and emission spectra and becomes fluorescent again. Wild-type GFP readily undergoes photoconversion. Thus, FRAP experiments using wild-type GFP chimeras should be avoided. However, YFP or CFP chimeras can be used to avoid this problem. If the fluorophore bleaches too rapidly during acquisition, the excitation light intensity should be decreased and the gain on the detector increased to collect light more efficiently.

6. Schneider's M3 is an enriched medium that keeps alive and nourishes insect Schneider cells under growth conditions.

7. Antibodies were used at threefold higher concentration than for conventional staining.

8. To follow the Hh and Ptc binding, it is possible to label Hh and Ptc "in vivo" at low temperature to avoid internalization. An antibody against Ptc protein that recognizes its extracellular domain, and anti-GFP antibody to recognize Hh-GFP can be used. Furthermore, it is possible to follow the internalization of Hh-Ptc by incubating with the same antibodies at 25°C.

9. To preserve the Red-dextran staining in the imaginal disc, it is essential to reduce the detergent to a minimum in the incubation buffers.

10. The endocytic compartment can be labeled by incubation with Red-dextran in M3 medium at 25°C (pulse). To visualize early endosomes of the endocytic compartment, the discs are pulsed for 5 min and fixed without a chase period. To detect late endosomes, a 5 min pulse and 45 min chase are used. After fixation, the discs are washed and dissected for confocal microscopy to test whether Hh-GFP proteins colocalize with the red-fluorescence in the early or late endocytic compartment.

Acknowledgments

We are very grateful to Aphrodite Bilioni for comments on the manuscript. This study was financed by the Spanish MEC, grants BFU2005-04183 and by an institutional grant from the Fundación Ramón Areces.

References

1. Torroja, C., Gorfinkiel, N., and Guerrero, I. (2005) Mechanisms of Hedgehog gradient formation and interpretation. *J. Neurobiol.* **64**(4), 334–356.

2. Takei, Y., Ozawa, Y., Sato, M., Watanabe, A., and Tabata, T. (2004) Three Drosophila EXT genes shape morphogen gradients through synthesis of heparan sulfate proteoglycans. *Development* **131**, 73–82.

3. Torroja, C., Gorfinkiel, N., and Guerrero, I. (2004) Patched controls the Hedgehog gradient by endocytosis in a dynamin-dependent manner, but this internalization does not play a major role in signal transduction. *Development* **131**, 2395–2408.

4. Callejo, A., Torroja, C., Quijada, L., and Guerrero, I. (2006) Hedgehog lipid modifications are required for Hedgehog stabilization in the extracellular matrix. *Development* **133**, 471–483.

5. Calleja, M., Moreno, E., Pelaz, S., and Morata, G. (1996) Visualization of gene expression in living adult Drosophila. *Science* **274**, 252–255.
6. de Celis, J. F. and Bray, S. (1997) Feed-back mechanisms affecting boundary in the Drosophila wing. *Development* **124**, 3241–3251.
7. Pignoni, F. and Zipursky, S. L. (1997) Induction of Drosophila eye development by decapentaplegic. *Development* **124**, 271–278.
8. Struhl, G. and Basler, K. (1993) Organizing activity of wingless protein in Drosophila. *Cell* **72**, 527–540.
9. Burke, R., Nellen, D., Bellotto, M., et al. (1999) Dispatched, a novel sterol-sensing domain protein dedicated to the release of cholesterol-modified Hedgehog from signaling cells. *Cell* **99**, 803–815.
10. Pepinsky, R. B., Zeng, C., Wen, D., et al. (1998) Identification of a palmitic acid-modified form of human Sonic hedgehog. *J. Biol. Chem.* **273**, 14,037–14,045.
11. Brand, A. H. and Perrimon, N. (1993) Targeted gene expression as a means of altering cell fates and generating dominant phenotypes. *Development* **118**, 401–415.
12. Ashburner, M. (1989) *Drosophila. A Laboratory Handbook*, Cold Spring Harbor Laboratory Press, New York.

9

Confocal Analysis of Hedgehog Morphogenetic Gradient Coupled with Fluorescent *In Situ* Hybridization of Hedgehog Target Genes

Armel Gallet and Pascal P. Thérond

Abstract

Hedgehog (Hh) family members are secreted proteins that can act at short and long range to direct cell fate decisions during developmental processes. In both Drosophila and vertebrates, the morphogenetic gradient of Hh must be tightly regulated for correct patterning. The posttranslational modification of Hh by a cholesterol adduct participates in such regulation. We have shown that cholesterol modification is necessary for the controlled long-range activity of Drosophila Hh, as observed for its vertebrate counterpart Sonic Hh. The presence of cholesterol on Hh allows the observation of *l*arge apical *p*unctuate *s*tructures of Hh (Hh-LPSs) at a distance from the Hh source both in embryos and in imaginal discs. The Hh-LPSs apical distribution reflects the Hh gradient and is temporally regulated. Hh gradient modulation is directly related to the dynamic expression of the Hh target gene *serrate* (*ser*), shown by immunofluorescent detection of Hh coupled with fluorescent *in situ* hybridization of *ser*.

Key Words: Drosophila; morphogenetic gradient; Hedgehog-posttranslational modification; cholesterol; confocal immunofluorescence; *in situ* hybridization, *serrate*.

1. Introduction

It is unclear how cholesterol-modified Hedgehog (Hh), which is membrane-tethered through its cholesterol anchor, reaches distant cells. One possibility may involve large punctate structures (LPSs), formation of which is dependent on the cholesterol adduct on Hh *(1)*. We showed that in absence of the Hh receptor Patched, Hh-LPSs remain attached to the apical side of receiving cells *(2)*. In addition, assembly and movement of LPSs depend, respectively, on two genes, *dispatched* and *tout velu*, necessary for Hh long-range activity *(1,3,4)*.

From: *Methods in Molecular Biology: Hedgehog Signaling Protocols*
Edited by: J. Horabin © Humana Press Inc., Totowa, NJ

Altogether, these data allowed us to propose that Hh-LPSs provide a vehicle for Hh long-range activity and reflect a functional Hh gradient that spreads apically through epithelia.

In each abdominal segment of the embryo, Hh is expressed in two rows of cells under the control of Engrailed. Apical Hh-LPSs form a symmetric gradient several cells wide from the source. *ser* is expressed in the most distant cells from Hh cells *(5)*. Hh is the main repressor of *serrate (ser)* expression *(6)*; consequently, *ser* expression and the presence of Hh-LPS are mutually exclusive. Thus, *ser* represents a good marker to follow the limit of Hh long-range activity in the embryonic ventral ectoderm.

Development of a specific antibody that recognizes Hh provides a technical tool for the observation of Hh-LPSs. This allowed us to show that the slope of the apical Hh-LPSs gradient is dynamic during embryogenesis *(2)*. Confocal immunofluorescence and fluorescent *in situ* hybridization allows us to correlate the expression of the target gene *ser* and Hh-LPSs range in embryonic epithelium.

2. Materials

2.1. Embryos and Fix

1. Egg-laying plates: in a 500 mL beaker dissolve 3 g sugar and 7.5 g agar agar in 185 mL tap water, then add 65 mL apple or raisin juice. Boil the mixture using a microwave. Wait for few minutes until you can grab the beaker with your hands, then add 5 mL absolute ethanol + 2.5 mL glacial acetic acid. Pour the medium immediately into plates. Plates can be kept up to 10 d in a wet chamber at 4°C.
2. 4% bleach: diluted bleach to remove the chorion is made from supermarket bleach. The dilution is done with distilled water. Make fresh as required.
3. PBS–EGTA mix: 50 mL of 0.5 M EGTA, 50 mL of 10× PBS added to 265 mL distilled water. This mix can be stored at room temperature (RT) for several months.
4. Formaldehyde fixative (5%, *see* **Note 1**): Take 1.46 mL PBS–EGTA mix and add 530 μL stock solution of 37% formaldehyde (ACS Reagent, ref. F-1268, Sigma). Diluted formaldehyde is not stable, therefore, the solution should be made fresh as required. Mix well and add 2 mL heptane. The fixative solution should appear as two phases with an aqueous lower phase (containing the formaldehyde) and an upper heptane phase. Four milliliters of fixative mix are generally enough for an overnight lay from 80 to 100 flies.

2.2. Probe Preparation

1. RNAase free DiMethyl PryoCarbonate (DMPC, *see* **Note 2**) water: DMPC (ref. D-5520, Sigma) is a strong RNAase inhibitor. The distilled water used for RNA probe synthesis or in the hybridization buffer must be DMPC treated, 500 μL DMPC to 500 mL distilled water. Shake vigorously and wait for 12 h at 37°C, then autoclave the DMPC water.
2. DIG RNA-labeling Mix 10× (Roche).

3. tRNA (Sigma): stock in DMPC water at 40 mg/mL.
4. Stock of 2× carbonate buffer: dissolve 120 mM Na$_2$CO$_3$ and 80 mM NaHCO$_3$ in DMPC water. Adjust to pH 10.2.
5. Stock of stop solution: dissolve 0.2 M sodium acetate in DMPC water and adjust to pH 6.
6. Vacuum drier, such as SpeedVac for 1.5 mL Eppendorf tubes.

2.3. Hybridization

1. Deionized formamide: add two spoons of AG 501-X8 (D) resin (cat. no. 143-7525, Bio-Rad) to 250 mL formamide (Merck, ACS grade). Stir with magnetic stirrer until the resin color changes from blue to gold. Filter the formamide through a coffee filter and aliquot into 25 mL in 50 mL Falcon tubes. Store at −20°C for several years.
2. Ten milliliters of hybridization buffer (Hybe) are necessary for one hybridization reaction. Practically, we prepare 50 mL Hybe as it can be stored at −20°C if not entirely used. To 25 mL deionized formamide, add 12.5 mL of 20× SSC (made with DMPC water and autoclaved, see Ref. *[7]* for recipe), 50 mL heparin (Roche, stock at 100 mg/mL in DMPC water and stored at −20°C), 500 µL denatured Herring Sperm DNA (Roche, stock at 10 mg/mL), 50 µL of Tween-20, and complete with 11.9 mL DMPC water to reach a final volume of 50 mL.
3. DMPC PBT: 1× PBS made with DMPC water with 0.1% Tween-20 final concentration.
4. Alkaline phosphatase buffer: 100 mM Tris–HCl (pH 8.5) made fresh as required from a 1 M stock solution.
5. Vector Red Alkaline Phosphatase Kit I (ref. SK5100, Vector).

2.4. Immunostaining

1. PBTr: 1× PBS, 0.1% Triton X-100.
2. Blotto: PBTr, 10% bovine or goat serum, 0.01% azide. The serum must be first inactivated by heating for 30 min at 37°C and then filtered with a 45-µm filter using a syringe. You can make aliquots of ready to use serum and store them at −20°C.
3. Primary polyclonal antibodies: our rabbit anti-Hh must be pre-adsorbed on wild-type (WT) embryos to decrease the background. Re-hydrate WT embryos by sequential washing in 70, 50, and 30% ethanol. Rinse embryos twice with PBTr and wash 3× for 5 min in PBTr on a rocker. Rock the embryos in blotto for 2 h at RT, then incubate them with primary antibody diluted in the ratio of 1:10 in blotto for 20 min at RT. Keep the supernatant containing the pre-adsorbed antibody at 4°C. As blotto contains azide, the pre-adsorbed antibody can be kept at 4°C for several weeks.
4. Secondary antibodies: Streptavidin-Alexa[488] conjugated (Molecular Probes), anti-mouse Cy[5]-conjugated antibody (Jackson Laboratory), and anti-rabbit biotin-conjugated (Jackson Laboratory).
5. Mounting medium: 80% glycerol in PBS. Place under vacuum until bubbles no longer appear on the upper surface. This will take 12 to 24 h. The absence of microbubbles is of great importance as bubbles interfere with the confocal laser scanning.

3. Methods

3.1. Preparation of Embryos

1. Collect staged embryos (*see* **Note 3**) of the desired genotype in baskets and wash extensively with tap water.
2. Dechorionate embryos in basket with diluted bleach for 2 to 3 min and then rinse extensively with tap water.
3. Transfer embryos into the fixative mix and place on a rocker for 25 min. At this step, the use of small conical tubes (12 mL) is helpful for their devitellinization. Embryos are at the inter-phase.
4. Remove the aqueous phase and add 2 mL methanol. Remove the upper heptane phase and add again 2 mL heptane.
5. Shake vigorously or even vortex for few seconds. Devitellinized embryos sink to the bottom of the tube.
6. Remove all of the methanol and heptane. Rinse the embryos 2× with methanol and then 3× with absolute ethanol.
7. Store embryos in absolute ethanol in a 1.5-mL Eppendorf tube at −20°C (*see* **Note 4**).

3.2. Antisense DIG Labeled RNA Probe Synthesis

1. We use the "run-off technique" for mRNA probe synthesis which requires linear DNA. Either the PCR product of cDNAs (which requires the RNA polymerase initiation sequence be present in the PCR primers) or linearized vectors, such as pBluescript or pNB40, digested with an appropriate restriction enzyme (that cuts into the 5′ of the cDNA) can be used. Purify the DNA: phenol, phenol/chloroform, and chloroform extraction. Precipitate with sodium acetate (0.3 M final concentration) and absolute ethanol. Spin 15 min at 14,000g. Wash the pellet once with ethanol 75% in Diethyl pryocarbonate (DEPC) water. Discard the supernatant and dry the pellet.
2. Resuspend the DNA pellet in 4 μL DMPC water, add 1 μL of 10× DIG RNA labelling mix, 2 μL of 5× in vitro transcription buffer (provided with the RNA polymerase by Promega), 1 μL of 100 mM DTT, 1 μL RNasin (RNAase inhibitor from Promega or Takara), and 1 μL RNA polymerase (Promega; *see* **Note 5**).
3. Incubate for 2 h at 37°C, then add another 1 μL RNasin and 1 μL RNA polymerase, and let the reaction incubate an addition of 1 h at 37°C.
4. Add 15 μL DPMC water and 25 μL of 2× carbonate buffer. Incubate 20 min at 65°C (*see* **Note 6**).
5. Add 50 μL of stop solution and mix well.
6. Add 15 μL of 4 M LiCl (in DMPC water) and 2.5 μL tRNA (stock at 40 mg/mL). Mix.
7. Add 300 μL absolute ethanol and leave for 20 min at −20°C.
8. Spin 15 min at 14,000g at 4°C.
9. Discard the supernatant and rinse with 70% ethanol in DMPC water.
10. Spin 10 min at 4°C and discard carefully the supernatant.
11. Dry the pellet under vacuum and resuspend in 75 μL Hybe.

3.3. Whole Mount Embryo In Situ *Hybridization*

All the steps are done at RT except where specified. "Rinse" means wait until the embryos reach the bottom of the eppendorf tube (such steps take <1 min), "wash" means the embryos must shake on a rocker.

1. About 50 μL ethanol preserved embryos are necessary for one *in situ* hybridization. Take the fixed embryos from the freezer, remove the ethanol, first add 500 μL ethanol and then 500 μL mixed xylenes (ACS Reagent X-2377, Sigma). Rock for 30 min.
2. Remove the supernatant and rinse the embryos 5× with absolute ethanol.
3. Rinse 2× with methanol.
4. Rinse 3× with DMPC PBT.
5. Prefix the embryos in 5% formaldehyde in DMPC PBT for 25 min on a rocker. Fixation must not exceed 30 min otherwise the hybridization steps will be impaired.
6. Rinse 5× with DMPC PBT.
7. Incubate embryos in 4 μg/mL proteinase K in DMPC PBT (stock: 2 mg/mL stored at −20°C) for 4 min on a rocker (*see* **Note 7**).
8. Rinse 4× with DMPC PBT.
9. Post-fix embryos in 5% formaldehyde in DMPC PBT for 25 min on a rocker.
10. Rinse 5× with DMPC PBT.
11. Rinse the embryos once with a mix of 500 μL DMPC PBT/500 μL Hybe.
12. Rinse the embryos 3× with Hybe.
13. Prehybridization step: incubate the embryos for at least 1 h at 55°C in Hybe in an electronically regulated dry incubator without rocking (Bioblock Scientific). Prehybridization incubations ranging from 3 to 5 h are better.
14. To the 75 μL embryos (corresponding to 50 μL dehydrated embryos), add 75 μL Hybe plus 1.5 to 3 μL DIG RNA probe (*see* **Note 8**).
15. Hybridization step: incubate overnight at 55°C in a dry incubator (*see* **Note 9**).
16. Wash the embryos with Hybe 4× for 20 min each at 55°C in a dry incubator.
17. Wash the embryos once with a mix of 500 μL DMPC PBT/500 μL Hybe for 15 min with rocking.
18. Wash 4× for 15 min each with PBT (DMPC is not required after hybridization).
19. Incubate with anti-Dig antibodies (Roche) at the dilution of 1:1000 in PBT for 90 min on a rocker.
20. Wash 4× for 15 min each with PBT.
21. Rinse 3× with the alkaline phosphatase buffer.
22. Fluorescent staining must be performed as described in the Vector Red Alkaline Phosphatase Kit I and kept away from light. The kit takes advantage of a faint red stain—the coloration step will take 10 to 20 min. Stop the reaction by removing the staining solution and adding PBT.
23. Rinse 2× with PBT.
24. Wash 2× for 10 min each with PBT.
25. Keep the embryos in PBT at 4°C until the immunostaining process.

3.4. Immunostaining of Embryos

1. Rinse the embryos with PBTr after processing for *in situ* hybridization.
2. Wash 2× for 5 min each with PBTr.
3. Block for at least 2 h with Blotto (change the solution once) on a rocker. A blocking step of 4 to 5 h is best for subsequent Hh immunostaining.
4. Incubate overnight with primary pre-adsorbed antibodies at 4°C on a rocker. Our anti-Hh antibodies are used at a final dilution in the ratio of 1:200. If you perform a double stain (for example, anti-Ptc or anti-Wg and anti-Hh), both primary antibodies can be incubated simultaneously.
5. Wash 6× for 20 min each at RT with blotto.
6. As Hh is weakly expressed in embryos, we use an amplification step for Hh immunostaining. Incubate the embryos for 2 h at RT with an anti-rabbit antibody conjugated to biotin diluted in the ratio of 1:1000 in Blotto.
7. Wash 6× for 20 min each at RT with Blotto.
8. Incubate embryos overnight with fluorescent coupled antibodies at 4°C. For Hh detection, we use Alexa^{488} conjugated-streptavidin (from Molecular Probes) diluted in the ratio of 1:200 in Blotto. For Wg or Ptc detection, we use an anti-mouse Cy^5-conjugated antibody diluted in the ratio of 1:400 in Blotto. Both secondary antibodies can be mixed together. Avoid visible light which will bleach the fluorochromes.
9. Wash 6× for 20 min each at RT with Blotto.
10. Wash 4× for 10 min each at RT with PBTr.
11. Preparing the slide for mounting: as the ventral side of the embryos needs to be examined, it is important to direct their settling to facilitate confocal analysis. First clean the slide with a soft tissue. Then stick with uncolored nail polish, two cover slips (22 × 22 × 0.15 mm) 20-mm apart on the slide (**Fig. 1A**).
12. Mounting of embryos: remove the PBTr and add 50 μL PBS-glycerol. Mix with a yellow tip (pipetman 200 μL size). Cut the tip of a fresh yellow tip and "aspirate" up the embryos. Place them in the center of the microscope slide between the two stuck cover slips (**Fig. 1B**). Cover the embryos with a large cover slip (22 × 32 × 0.15 mm; **Fig. 1C**) over the two small covers slips on each side. The slide has to be kept at 4°C before scanning with the confocal microscope.

3.5. Image Capture and Analysis

1. Image capture: Fluorescent imaging was done on a Leica DMR TCS_NT confocal microscope with the 63× objective and numeric zoom 2. First, the ventral side of the embryos needs to face up. To do this, push the upper cover slip so that the ventral side of the embryos faces the top (**Fig. 1D**). As both apical spreading of Hh and transcriptional expression of Hh target genes need to be analyzed, the physical problem to overcome is both signals are not in the same plane: mRNAs are cytosolic, while Hh spreading occurs apically near the cell surface. Hence, the first confocal cross section must be done at the cell surface (corresponding to the domain of Hh spreading; **Fig. 2A**) and the second, 2-μm deeper than the first (just above or at the level of the nucleus; **Fig. 2B**). The "projection function" of the

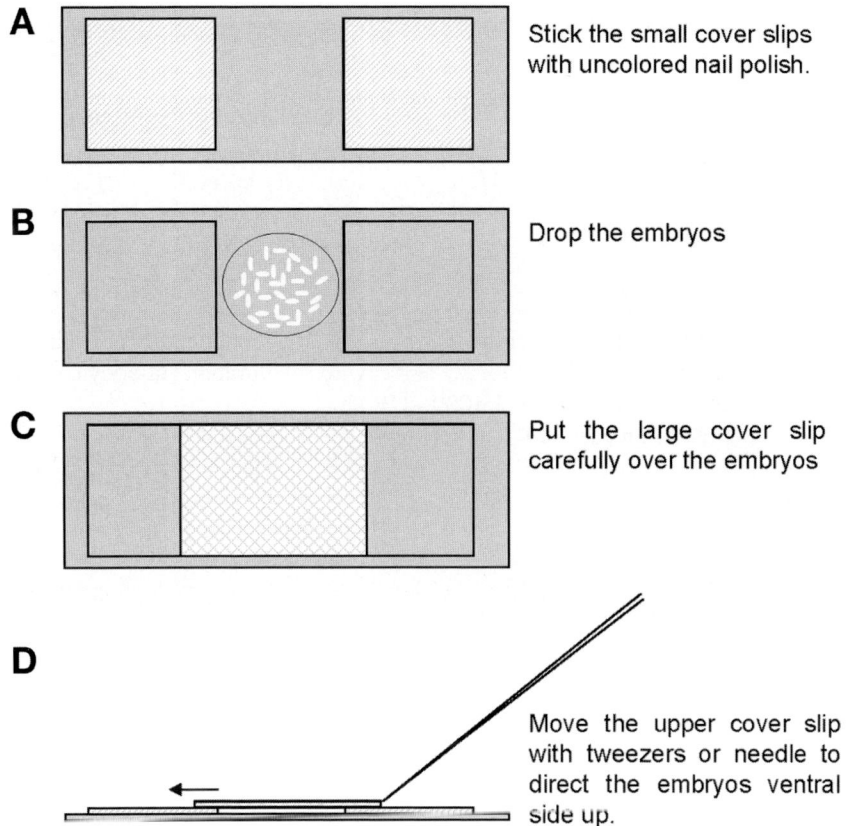

Fig. 1. Scheme showing the mounting technique for confocal analysis of embryos. (**A**) Stick two small coverslips on each side of the slide. (**B**) Drop the embryos in PBS–glycerol between the two cover slips. (**C**) Carefully put a large coverslip over the embryos. (**D**) Lateral view of (**C**). Using a needle, move the upper coverslip to orient the embryos.

Leica software was used to obtain the projected image (**Fig. 2C**). It is also possible to perform such projections using ImageJ software (http://rsb.info.nih.gov/ij/).

2. Plot analyses of Hh gradient: We use ImageJ software. After selecting the domain of analysis (rectangle), we perform the plot profile (in the "Analyze" menu of ImageJ; **Fig. 2D**). If you compare different labels on the same sample, do not forget to note the size (in pixels) of your rectangle of selection. Take note of the values indicated on the "*y*" axes which correspond to the gray value of the fluorescence intensity. Those values are arbitrary and can only be compared between embryos of the same sample but cannot be compared between two independent experiments owing to the variability in fluorescence intensity.

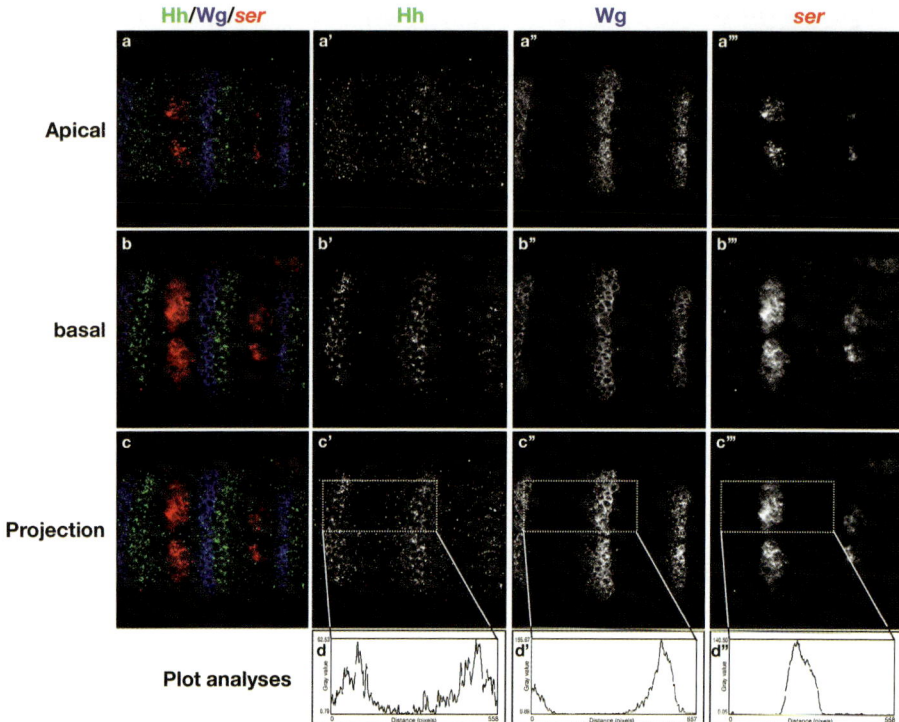

Fig. 2. Making the projection of both apical and basal confocal sections of embryos. (a–c) Early stage 11 embryos. (a) Apical section. (b) Basal section. (c) Projection of (a) and (b). (d) Plot analysis by ImageJ. (a–c) triple labeling for anti-Hh (green), anti-Wg (4D4; from DSHB) (blue) and *ser* mRNA (red). (a'-a''', b'-b''' and c'-c''') single channel of images shown in a–c, respectively. Hh-LPSs spreading occurs apically and symmetrically (a' and b'), while Wg spreading is asymmetric and more basal (a'' and b''). *ser* mRNA is strongly seen in the more basal section (a''' and b'''). Projection of both apical and basal views allows a relative comparison of the different stains (c). Plot analysis report the symmetric Hh spreading (d) the asymmetric Wg spreading (d') and the domain of *ser* expression relative to the Hh spreading (d'').

4. Notes

1. As most of the products for *in situ* mRNA hybridization on whole-mount embryos are toxic, you should work under a chemical hood.
2. DEPC (Sigma) can also be used; however, it is very toxic.
3. Hh is expressed and functional from stages 4 to 5 until the end of embryogenesis, hence to get a panel of developmental stages, let the flies lay for 15 to 16 h at 18°C, 10 to 12 h at 25°C or 8 to 9 h at 29°C.

4. Fixed embryos can be kept at −20°C for several years. It is interesting to note that the mRNA *in situ* hybridization works better on embryos that have been kept for at least 1 d at −20°C.

5. T7, T3 or SP6 RNA polymerases can be used, however better yields are obtained with T3 or SP6 polymerase.

6. This step is required to break the probe into small pieces of about 200 to 400 nucleotides. Twenty minutes of incubation are the average time to break a probe of 1500 to 2000 nucleotides. The time of incubation must be adjusted to the length of the probe.

7. Proteinase K treatment can be variable but must not exceed 8 min. If embryos are fragile (owing to the genotype, for example), then the proteinase K step can be eliminated. If this is the case, eliminate steps 8–10 and go directly to step 11. In such cases, the color reaction of step 22 will take longer.

8. The amount of the probe will depend on the efficiency of synthesis. Usually, 1.5 to 3 µL probe is enough for standard hybridization with *wg*, *ser*, or *rho* probes. However, in the case of weakly expressed mRNA, the volume of the probe can be up to 8 µL/75 µL Hybe.

9. The temperature of incubation depends on the strength of the base pairing of the probe to its target mRNA; 55°C is an average temperature. To avoid background and enhance specificity, the temperature can be raised up to 60°C.

References

1. Gallet, A., Rodriguez, R., Ruel, L., and Therond, P. P. (2003) Cholesterol modification of Hedgehog is required for trafficking and movement, revealing an asymmetric cellular response to Hedgehog. *Dev. Cell.* **4**, 191–204.

2. Gallet, A. and Therond, P. P. (2005) Temporal modulation of the Hedgehog morphogen gradient by a patched-dependent targeting to lysosomal compartment. *Dev. Biol.* **277**, 51–62.

3. Bellaiche, Y., The, I., and Perrimon, N. (1998) Tout-velu is a Drosophila homologue of the putative tumour suppressor EXT-1 and is needed for Hh diffusion. *Nature* **394**, 85–88.

4. Burke, R., Nellen, D., Bellotto, M., et al. (1999) Dispatched, a novel sterol-sensing domain protein dedicated to the release of cholesterol-modified Hedgehog from signaling cells. *Cell* **99**, 803–815.

5. Alexandre, C., Lecourtois, M., and Vincent, J. (1999) Wingless and Hedgehog pattern Drosophila denticle belts by regulating the production of short-range signals. *Development* **126**, 5689–5698.

6. Gallet, A., Ruel, L., Staccini-Lavenant, L., and Therond, P. P. (2006) Cholesterol modification is necessary for controlled planar long-range activity of Hedgehog in Drosophila epithelia. *Development* **133**, 407–418.

7. Maniatis, T., Fritsch, E. F., and Sambrook, J. (1982) *Molecular Cloning: A Laboratory Manual*, Cold Spring Harbor Lab., New York.

10

RNAi in the Hedgehog Signaling Pathway: pFRiPE, a Vector for Temporally and Spatially Controlled RNAi in *Drosophila*

Eric Marois and Suzanne Eaton

Abstract

RNA interference (RNAi) has become an irreplaceable tool for reverse genetics in plants and animals. The universality and specificity of this phenomenon allows silencing of virtually any chosen gene to examine its involvement in biological processes. Many strategies exist to reduce the expression of a particular gene using RNAi. Some rely on delivering directly to cells the ~21-nucleotide long interfering double-stranded RNA (dsRNA) species that are central mediators of the silencing process. Others rely on the transgenic expression of longer dsRNA molecules, leaving it to the cellular machinery to process these hairpins into short active dsRNA.

In this chapter, we describe a transgenic method to deplete a chosen protein from a specific *Drosophila* tissue following induction of long dsRNA. It was used to uncover the role of lipidic particles in Hedgehog signaling by silencing lipophorin in the fat body *(1)*, and we routinely use it to deplete specific proteins from wing imaginal disc subdomains *(2)*. The method, certainly not restricted to the study of Hedgehog signaling, allows fast and efficient construction of a plasmid incorporating various *Drosophila* genetic tools to allow heat-shock-induced expression of dsRNA at the desired time and in the desired tissue. For protocols involving injection of in vitro synthesized dsRNA in embryos to study Hedgehog signaling, see for example *(3)*. For genomic screens to identify Hedgehog pathway components in tissue culture cells by transfection of small interfering RNAs, *see* refs. *(4,5)*.

Key Words: Inducible RNAi; tissue-specific RNAi; *Drosophila* wing imaginal disc; Gal4/UAS; FLP/FRT; pFRiPE.

1. Introduction

In 2000–2002, multiple reports indicated that it is possible to deplete a protein of choice from a tissue of choice by RNA interference (RNAi) in flies *(6–14)*. All methods relied on the Gal4/UAS system *(15,16)* to tissue-specifically express

From: *Methods in Molecular Biology: Hedgehog Signaling Protocols*
Edited by: J. Horabin © Humana Press Inc., Totowa, NJ

long double-stranded RNA (dsRNA), usually from vectors expressing a gene fragment cloned in an inverted repeat configuration. Endogenous Dicer-2 is expected to cleave the resulting dsRNA into short interfering RNAs, which are packaged into the RISC complex that targets endogenous homologous mRNAs for degradation (for review on the RNAi process, *see* **ref.** *[17]*). Unlike in *Caenorhabditis elegans*, RNAi does not spread systemically in *Drosophila*, which allows the useful possibility to compare cells subjected to RNAi with normal cells in neighboring tissue or to knock down gene function in one organ only. Following these initial successful reports, many laboratories made inverted repeat constructs to knock down their proteins of interest. It quickly became notorious that cloning inverted repeats of a gene fragment in a head-to-head configuration is difficult, even if using recombination-deficient *E. coli* strains, such as Stratagene's SURE2 competent cells that can stabilize recalcitrant DNA structures. In our and other's experience, initial *E. coli* transformant colonies are easily obtained but usually loose the inverted repeat during subsequent growth. In the minority of cases in which *E. coli* retained the repeats, transgenic flies generated from such constructs did display efficient RNAi and were genetically stable over many generations (*[18]*; E.M. and S.E., unpublished). To circumvent the repeat instability problem in *E. coli*, a spacer sequence, for example an intron, can be inserted between the repeats to suppress their loss *(18)*. Some introns were even reported to improve the efficiency of particular RNAi constructs *(11,19)*. In our experience, introns will allow RNAi when placed between certain repeats but will suppress RNAi if placed between others. In several independent cases, a given inverted repeat was more efficient at triggering RNAi when devoid of spacer than when cloned in an intron-containing vector (Marois and Eaton, unpublished). This may be due to improper intron splicing in certain contexts. We speculate that unspliced introns may participate in secondary structures with the first synthesized repeat, inhibiting efficient pairing of the second repeat to the first. After testing four different spacer introns with variable success depending on the gene targeted, we wanted to devise a vector strategy offering both spacer-mediated repeat stabilization and the more reproducible RNAi efficacy of inverted repeats without an intervening spacer.

pFRiPE, the RNAi vector we present here, contains a spacer that is excisable in *Drosophila* at the genomic level, using a recombination reaction based on the flip recombinase (FLP)/flipase recognition target sequence (FRT) system *(20)*. While solving the problems associated with cloning head-to-head repeats, this allows the in vivo production of dsRNA without spacer. Cloning of the inverted repeats is facilitated by the use of the Gateway technology (Invitrogen Carlsbad, CA) which inserts two inverted repeats in a single in vitro recombination step *(21)*. Furthermore, excision of the spacer by FLP recombinase (flipase) can be controlled by heat shock, providing temporal control of dsRNA expression. Expression from pFRiPE (a pUAST derivative) is regulated by the GAL4/UAS system, conferring the advantages of RNAi induction in a spatially restricted fashion.

RNAi-mediated gene knockdown using pFRiPE provides several advantages when compared with generating null mutant clones. Firstly, gene knockdown can be induced within 48 h in large fields of cells without a requirement for cell division. In contrast, mutant cells may fail to divide, be eliminated or over the longer term produce compensation mechanisms that can blur result interpretation. Secondly, Gal4 control of RNAi allows reproducible targeting of an entire tissue compartment (for example, all Hedgehog-secreting cells), whereas cell clones generate largely unpredictable mosaics.

The organization of pFRiPE is outlined in **Fig. 1**. Inverted repeats are inserted by Gateway cassette replacement on either side of an excision cassette flanked by FRT recombination sites (**Fig. 1A**). Before excision, Gal4 transcribes an *HcRed* gene present within the excision cassette. The *HcRed* sequence is followed by a long transcription terminator to prevent premature RNAi activation (*see* **Note 1**). Upon spacer excision, an RNA inverted repeat is produced, of which the center of symmetry is a single remaining FRT site (**Fig. 1B**). Because the FRT site itself is an almost perfect palindrome, only three nucleotide mismatches are found at the center of symmetry (**Fig. 1C**).

In **Section 2.**, we will outline (i) the design of the gene fragment that will trigger RNAi, initially to be cloned in a Gateway entry vector; (ii) the preparation of recombination-ready pFRiPE vector; (iii) the recombination reaction that generates the desired construct by mixing the entry clone with pFRiPE; (iv) the selection of appropriate constructs for transgenic fly generation; and (v) some aspects of using pFRiPE transgenic flies.

2. Materials

1. Petri dishes containing LB medium with: kanamycin 30–50 µg/mL; ampicillin 100 µg/mL; ampicillin 100 µg/mL + chloramphenicol 40 µg/mL.
2. DB3.1 competent cells.
3. DH5α competent cells.
4. pENTR plasmid (Invitrogen; for example, pENTR1A, pENTR2B, or pENTR3C).
5. Restriction enzymes: *Bam*HI (or *Acc*65I/*Kpn*I), *Xho*I (or *Eco*RV, or *Not*I), *Eco*RI (or *Stu*I or *Kas*I), *Nhe*I (or *Aat*II).
6. Gateway LR enzyme mix (Invitrogen).
7. DNA purification columns (Qiagen Hilden, Germany).
8. TE: 10 m*M* Tris–HCl (pH 8) and 10 m*M* EDTA.
9. Oligonucleotides:
   - EM15   GCAGGCTCTTTAAAGGAACCAA
   - EM16   GCTGGGTCTAGATATCTCGAG

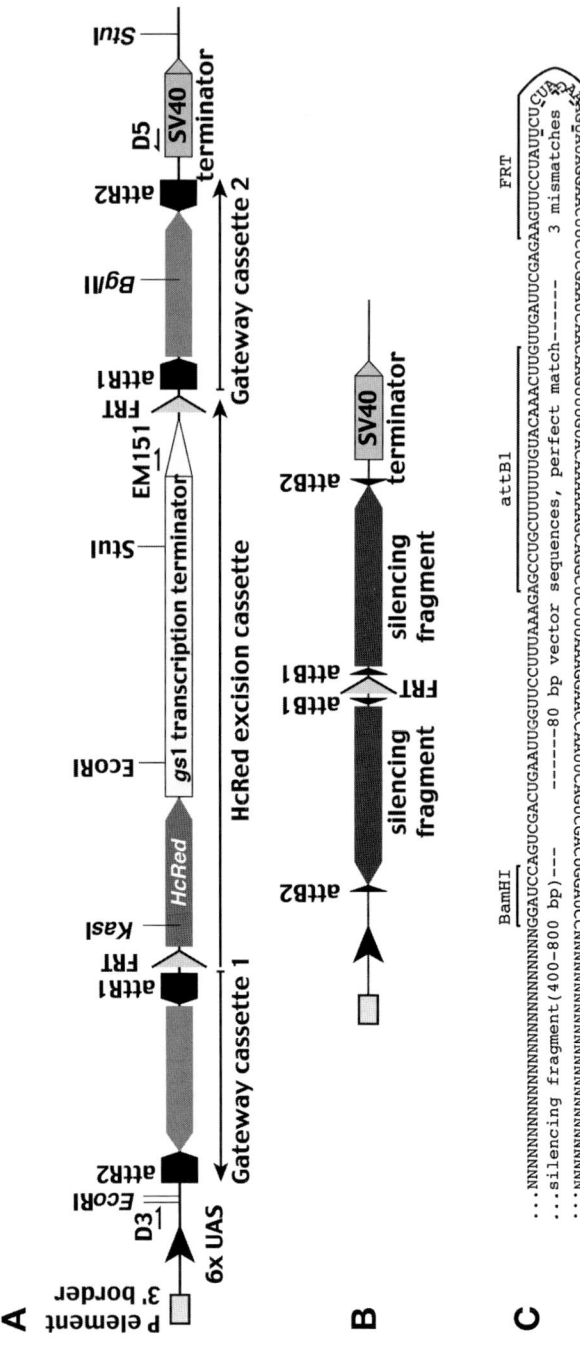

Fig. 1. Scheme of pFRiPE (scale is approximate). (**A**) pFRiPE before insertion of a silencing fragment. The new DNA sequences inserted into pUAST to construct pFRiPE are shown. Gateway cassettes 1 and 2 (flanked by attR1 and attR2 recombination sites) are in inverted orientation with respect to one another, resulting in corresponding orientation of the silencing fragment after the recombination reaction. *Bgl*II restriction site in Gateway cassette 2 serves for vector linearization prior to recombination. Tandem FRT sites provide flipase-mediated excisibility of the HcRed gene and the accompanying transcription terminator. D3, D5, and EM151 primer binding sites are shown. (**B**) RNAi construct generated from pFRiPE after in vitro LR recombination (which replaced the Gateway cassettes with silencing fragment) and after in vivo flipase-mediated HcRed cassette excision. A single remaining FRT site separates the repeats. (**C**) Nucleotide sequence of the center of symmetry of the inverted RNA repeat expressed from construct shown in **B**. pFRiPE sequence can be found at GenBank (Accession number DQ305103).

- D5   GGTAGTTTGTCCAATTATGTCACACC
- D3   CAACTGCAACTACTGAAATCTGCC
- EM151   GACAAGCGGCAATAAACGGGTA

3. Methods

3.1. Design of the Gene Fragment to Clone into pENTR

1. Run the entire mRNA sequence of the gene to be silenced in a program screening the sequence for potential genes that could be cross-silenced. BLAST is not sufficient, because every possible 21 nucleotide sequence from the target gene should be checked for potential targeting of other genes. We find the online program Deqor *(22)* (http://deqor.mpi-cbg.de/Deqor/deqor.html) very useful for this purpose (*see* **Note 2**). Upon running Deqor, gene regions devoid of 21 nucleotide stretches homologous to potential off-target genes can be selected. In addition, select a gene region with the following characteristics.
   - Size between 400 and 800 bp long (other sizes may function but were not tested);
   - Absence of either *Nhe*I or *Aat*II restriction sites (one of these enzymes will be used to linearize the pENTR plasmid prior to the Gateway recombination reaction, *see* **Note 3**);
   - Absence of the restriction sites incorporated in the oligonucleotides used to amplify this fragment for subsequent cloning (preferably *Xho*I and *Bam*HI, *see* **step 2** in this subsection);
   - Selected in any region of the mRNA (5′UTR, 3′UTR or coding sequence, depending on available possible controls, *see* **Section 3.5.**). If chosen to contain part of the coding sequence, care should be taken to avoid methionine-coding triplets in the fragment (*see* **Note 4**). If that is impossible, the fragment should be cloned as an antisense first repeat (*see* **Note 5**), i.e., as a *Bam*HI/*Xho*I fragment.

2. Design primers with *Xho*I and *Bam*HI sites at the extremities to amplify the chosen region. Add three extra nucleotides 5′ to the restriction site to ensure efficient restriction digestion of the PCR product. In case no appropriate gene region can be found devoid of *Xho*I and/or *Bam*HI sites, *Eco*RV or *Not*I can be used instead of *Xho*I; *Acc*65I (*Kpn*I) or *Sal*I instead of *Bam*HI. If the construct is desired "sense first", the PCR primers should be designed to amplify an *Xho*I/*Bam*HI fragment, i.e., 5′ primer contains the *Xho*I and 3′ primer contains the *Bam*HI site. If the construct is to be "antisense first", 5′ primer contains the *Bam*HI and 3′ primer contains the *Xho*I site.

3. PCR amplify the gene fragment from total cDNA (or from an individual cDNA clone if one is available) and clone the resulting PCR product into a pENTR plasmid using the chosen restriction sites. Cloning PCR products into pENTR1A, 2B or 3C is very efficient as long as the 3:1 insert:vector molar ratio is approximately respected. pENTR plasmids lose the *ccdB* gene upon successful ligation (*ccdB* is toxic to DH5α *E. coli*, but not to DB3.1, XL1-Blue or other F plasmid-containing *E. coli* strains) which ensures that virtually 100% of the obtained colonies are positive when transforming DH5α.

4. Sequence a few pENTR plasmids (or PCR products amplified from single pENTR *E. coli* colonies) to select one that does not contain PCR mutations. For both colony PCR screening and sequencing, use primers EM15 and EM16 (*see* **Section 2.**).

5. Miniprep the chosen pENTR plasmid using a standard alkaline lysis protocol (resuspend DNA in 50 µL TE).

3.2. Preparation of pFRiPE

1. Upon receipt of pFRiPE plasmid DNA, transform an aliquot into *E. coli* strain DB3.1 (other strains are not appropriate). Select on ampicillin 100 µg/mL + chloramphenicol 40 µg/mL. Chloramphenicol selection ensures that the two inverted Gateway cassettes each containing a chloramphenicol-resistance gene will not be lost by the bacteria. Chloramphenicol resistance will be lost in positive clones after the Gateway recombination reaction.

2. Purify pFRiPE by plasmid miniprep (resuspend in 50 µL TE buffer). It is important to apply chloramphenicol selection during *E. coli* growth.

3. Linearize 20 µL of the pFRiPE miniprep with *Bgl*II in a total volume of 100 µL. This step is necessary for the recombination reaction to occur efficiently.

4. Purify the linear DNA (for example with a Qiagen spin column). Measure DNA concentration and check the plasmid on an agarose gel (should be a band of 15,857 bp).

3.3. LR Reaction and Selection of Positive Final RNAi Constructs

1. Digest 6 µL of the pENTR construct miniprep with either *Nhe*I or *Aat*II (should not cut inside the silencing fragment) in a total volume of 20 µL. Heat inactivate the enzyme. Run 15 µL of the digest on an agarose gel to estimate DNA concentration and to check that the enzyme was active: a single linear band should be visible around 3 kb. If two bands are seen, an *Nhe*I or *Aat*II site was present in the gene fragment: the converse enzyme must be used. Save the remainder of the reaction for the next step.

2. Perform LR reaction in a total volume of 5 µL (1 µL Invitrogen LR enzyme mix, 1 µL LR buffer, up to 3 µL DNA) with about 75 ng *Bgl*II-linear pFRiPE and 45 ng *Nhe*I-linear ENTR plasmid (this approximates a 1:3 pFRiPE: pENTR molar ratio). Overnight incubation at room temperature yields best results. Caution: the reaction will fail if a large excess of ENTR plasmid is used.

   During this step, the attL1/2 recombination sites flanking the gene fragment in pENTR will directionally recombine with the attR1/2 sites flanking each Gateway *ccdB*-chloramphenicol resistance cassette in pFRiPE, resulting in fragment exchange and the formation of 25 nucleotide-long, attB1/2 sites. For more information about the Gateway system, see "Gateway cloning" on the Invitrogen web site (http://www.invitrogen.com/).

3. Add 1 µL proteinase K from the LR mix package, incubate 10 min at 37°C to digest the recombination enzymes.

4. Transform the whole reaction into chemically competent *E. coli* strain DH5α (*ccdB*-sensitive). Plate on ampicillin. Normally, several hundred colonies are obtained. If less than 100 colonies are obtained, something probably went wrong (too much pENTR plasmid used or one of the two plasmids was not linear).

5. Screen colonies by PCR with primer pair D5/EM151 (*see* **Fig. 1** for binding sites). Positive colonies yield a band of: size of the gene fragment + 431. If the LR reaction worked well, virtually all colonies are positive.

6. Miniprep several colonies (6 to 10 per construct: only 50% will be in the desired orientation, *see* **Note 6**). Check for correct orientation of the flip-out *HcRed* cassette using *Eco*RI or *Stu*I or *Kas*I, whichever of these enzymes do not cut inside the gene fragment of interest. The correct clones are those that yield a smaller *Eco*RI or *Stu*I insert or a larger *Kas*I insert. Discard minipreps that yield an insert of the wrong size (the flip-out cassette is reversed in those). The exact sizes of the correct inserts are:

   *Eco*RI: size of the gene fragment + 957

   *Kas*I: size of the gene fragment + 5303

   *Stu*I: size of the gene fragment + 1814

   Plasmid DNA from at least three correct clones (to reach sufficient DNA amount) can be pooled and cleaned using a Qiagen column followed by one chloroform extraction and injected into fly embryos to produce transgenic flies. If insufficient amounts of DNA are recovered, perform a larger-scale DNA purification from the correct *E. coli* colonies.

3.4. Sequencing the Final RNAi Constructs

Sequencing into either repeat is blocked due to single-molecule duplex formation during the annealing steps of the sequencing reaction, so that the sequence readout stops abruptly at the nucleotide where the repeat starts. This is already a sign that the plasmid does contain an inverted repeat. Sequencing can be made possible by restriction digestion of the DNA to be sequenced with an enzyme cutting both between the repeats (i.e., inside the flip-out cassette) and inside the vector backbone. This places the repeats on separate DNA fragments and prevents single-molecule duplex formation during sequencing. Enzymes that can be used for this purpose are again *Eco*RI, *Kas*I, or *Stu*I, as long as the chosen enzyme does not also cut inside the gene fragment. The restricted DNA must be purified before sequencing. A faster approach is to perform the colony PCR of **Step 3.3.5.** incorporating a third primer in the PCR: EM151, D5, and D3. Sequence the reaction product with EM151. At the end of the readable sequence of the cloned gene fragment, two sequences should overlap, representing the normal EM151/D5 PCR product plus the aberrant EM151/D3 PCR product. This sequence overlap proves that the gene fragment is present twice and in inverted orientations.

3.5. Using Transgenic pFRiPE Fly Lines and Checking for Decrease in Gene Expression Levels

Once transgenic pFRiPE fly lines are obtained (*see* **Note 7**), they are crossed with the desired Gal4 driver lines also containing the heat-shock *Flipase* transgene (*see* **Note 8**). We usually induce dsRNA expression 48 h before larvae reach the desired developmental stage (*see* **Note 9**), to allow sufficient time for protein depletion (but the optimal incubation time to reach sufficient depletion depends on each protein's turnover rate and should be determined empirically). RNAi is induced by incubating food vials for 1 h and 30 min in a warm water bath (37.2°C). This treatment leads to flipase-mediated FRT cassette excision in nearly 100% of cells. Occasionally, a small number of cells do not excise the cassette. Clones derived from these cells are identified by persistent HcRed fluorescence in the Gal4-expressing domain (*see* **Note 10**). RNAi does not occur in these clones. If desired, the flip-out cassette can be excised in the germline and new fly lines will be generated that express dsRNA under GAL4 control in the absence of heat shock (*see* **Note 11**). The latter procedure, combined with the use of temperature sensitive Gal80, can be extremely useful in experiments where sequential transgene induction is desired (*see* **Note 12**).

There is no perfect method to quantify the decrease in gene expression levels, because the amount will depend on the Gal4 driver used, will be restricted to the Gal4-expressing cells, and few methods are absolutely quantitative.

– *In situ* RNA hybridization (with a probe binding to the RNA message outside of the chosen silencing fragment) can provide an estimate of the reduction of mRNA levels in expressing tissue when compared with neighboring nonexpressing tissue. Reduction in RNA levels does not necessarily correlate with protein levels since some proteins may have a long half-life.

– RT-PCR, simpler to apply than in situ hybridization, has the same limitations and requires Gal4 expression in all the cells from which the RNA is prepared. If RT-PCR is performed, PCR primers should be designed outside the silencing fragment. For oligodT-primed cDNA, the RT-PCR primers should be chosen close to the 3′ end of the cDNA for good PCR efficiency.

– Antibody staining followed by confocal microscopy provides a more satisfying assessment, as protein rather than mRNA is visualized and the occurrence of possible noninduced cell clones can be directly observed.

– Western blotting can quantitatively assess the reduction in protein levels, in dissected organs where 100% of the cells express the Gal4 driver or in whole animals expressing dsRNA ubiquitously (for example, using tubulin-Gal4).

– If no antibodies to the protein of interest are available, but tagged constructs do exist (GFP or other tags), the efficiency of the RNAi construct can be assessed by Western blotting or confocal imaging to visualize reduction of tagged protein (provided the chosen silencing fragment is present in the tagged construct). In cases where the tagged protein is only available as a UAS construct, RNAi and tagged protein will be expressed in the same cells. However, it is still possible to obtain an internal control of RNAi efficiency by subjecting the flies to a mild heat shock. Only some cells will excise the HcRed cassette and activate RNAi. Resulting excision clones should show decreased levels of the tagged protein when compared with unexcised surrounding tissue in the Gal4-expressing domain.

– Ultimately, the phenotype of adult flies provides an indication of the efficiency of gene knock-down, but gene redundancy can obscure well-functioning RNAi constructs.

Once a pFRiPE construct triggering a phenotype has been obtained, controls need to be performed to ensure that this phenotype results from knock-down of the intended gene rather than off-target gene silencing or other artifacts. Here are possible approaches:

– Generate different pFRiPE constructs against the same gene. If different constructs derived from different regions of the cDNA show identical phenotypes, the probability of off-target effects becomes very low.
– Rescue the RNAi effect by transgenic expression of the same protein. This approach will work best if the rescuing construct does not share sequences with the silencing fragment (e.g., silencing fragment was chosen in the untranslated regions of the mRNA and these were not incorporated within the rescuing construct). Other types of rescue may be considered; for example metabolic (if knocking down an enzyme synthesizing an essential compound, this compound could be provided in the food) or with transgenic expression of a gene of conserved function from a different organism, as long as DNA homology to the silencing construct is low enough to avoid RNAi.
– To rule out artifacts due to potential transcriptional silencing of genes adjacent to the target gene or other integration site effects, several independent insertion lines should be tested for each pFRiPE construct. To rule out potential artifacts due to expression of peptides from the first repeat, flies that were not heat-shocked but carry the pFRiPE construct and Gal4 driver should be used as negative controls.

How well does "friping" work? To date we have constructed 19 pFRiPE RNAi constructs. For 12 of these constructs, molecular data (in situ hybridization, antibody staining, and loss of GFP fusion fluorescence) indicate that seven triggered efficient RNAi, while five were inefficient. Three additional constructs produced a strong phenotype indicative of RNAi, but specificity has not yet been confirmed. In the four remaining cases, reduction in RNA and protein levels was not tested; therefore, the absence of phenotype is due either to construct inefficiency, to gene redundancy or to gene dispensability. In the case of the *arrow* gene, a first pFRiPE construct clearly did not function. We made a second construct choosing a different region from the gene, which did trigger very efficient RNAi. Failures are therefore not necessarily because a gene was "immune" to RNAi, but because the chosen inverted repeat did not trigger RNAi efficiently. We have not detected a correlation between inherent characteristics of long inverted repeats and RNAi efficiency. If friping only a few genes, it is therefore recommended to try several fragments to maximize chances of success.

4. Notes

1. In an earlier generation of our RNAi vectors, the FRT-HcRed-FRT cassette was cloned immediately after the pUAST UAS sequence and only contained the short SV40 transcription terminator (amplified along with the *HcRed* gene from pHcRed1-N1 [Clontech]). The SV40 terminator was sufficient to abrogate expression from downstream-cloned protein-coding genes, but was insufficient for preventing RNAi from downstream-cloned, intron-separated inverted repeats. Therefore, in subsequent vectors including pFRiPE, we added a much longer terminator (~2.5 kb) from the glutamine-synthetase (*gs1*) gene 3′ region. The length of genomic sequence was meant to ensure that RNA polymerase would have sufficient time to fall off from template DNA before reaching the second inverted repeat. The *gs1* terminator appears to perform as intended in our current experiments.

2. The Deqor program was designed primarily to select 21 nucleotide siRNA for in vitro synthesis and injection into embryos. When running the program, one can ignore the siRNA output and use the rest of the displayed information to choose a sequence devoid of cross-silencers. When designing a silencing fragment to perform RNAi in *Drosophila*, the *Drosophila* transcriptome should be employed as the database for fragment scanning (check the corresponding box).

3. The Gateway LR reaction is only efficient (in an inverted repeat context) if pFRiPE and pENTR are both linear before mixing. pFRiPE must be linearized inside one of the two Gateway cassettes (so that fragment exchange by recombination recircularizes the plasmid). This is done with *Bgl*II. One can purify a large amount of *Bgl*II-linearized pFRiPE in advance, for use in all recombination reactions to be performed. pENTR is linearized with *Nhe*I or *Aat*II (whichever is not present in the gene fragment of interest).

4. If ATG triplets encoding a methionine in the original cDNA are present, a peptide from the gene of interest might be expressed prior to RNAi induction if the first repeat is cloned in the sense orientation. This should be avoided, as side-effects due to expressing fragments of endogenous proteins may exist. Therefore, the silencing fragment should be selected in a region of the gene that does not encode any methionine, or be cloned "antisense first". ATG triplets may however be present in the fragment, as long as they are not in frame with respect to the original coding sequence (though this may result in benign nonsense peptide expression). An unrelated consequence of the presence of ATG triplets within the first repeat is a possible decrease in the expression level of the HcRed gene contained in the excision cassette, resulting in weaker or absent red fluorescence in GAL4-expressing tissue even before RNAi induction. Indeed, ATG triplets upstream of the methionine codon of HcRed might be interpreted as translational starts by the biosynthetic machinery. If these ATGs are in frame with respect to *HcRed*, this can result in a peptide fusion to HcRed (which may preserve red fluorescence). If ATGs are not in frame with respect to HcRed, less frequent translation of HcRed (decreasing red fluorescence) might ensue. HcRed fluorescence is not related to the subsequent efficiency of RNAi, but can be useful to visualize possible cell clones that failed to excise the cassette.

5. If cloning the first repeat in the antisense orientation, one might be concerned that antisense-mediated gene silencing might occur prior to desired RNAi induction. We have not observed this so far with our constructs, probably because antisense gene silencing is inherently very inefficient. Antisense RNA molecules have to anneal to endogenous RNA cognates from which they are physically separated, whereas inverted repeats readily self anneal.

6. One drawback of the Gateway LR reaction (in the inverted repeat configuration) is that 50% of the plasmid clones resulting from the double-LR reaction display a flip-out HcRed cassette that is reversed with respect to UAS orientation. This happens when the attL1 site in one ENTR plasmid recombines with the attR1 site of Gateway cassette A, but the attL2 site in the same ENTR molecule recombines with the attR2 site of Gateway cassette B instead of A. Then a second ENTR plasmid recombines with the remaining crossed attR1 and attR2 sites and inverts the cassette. Therefore, after the LR reaction, *E. coli* colonies should be screened for correct orientation of the flip-out cassette. An RNAi construct with a reverted flip-out cassette might still function but (a) HcRed will not be expressed and (b) the reversed transcription terminator is unlikely to function properly, giving rise to a small risk of observing RNAi before heat shock. PCR cannot be used to determine cassette orientation (using a primer inside the flip-out cassette and a pUAST primer, such as the EM151/D5′ combination), because this yields the expected product even for clones in the wrong orientation (this phenomenon is due to primer extension into a repeat during each PCR cycle, the resulting product then serving as a primer for the other repeat in the next PCR cycle. Amazingly, the yield of these aberrant PCRs is as high as the normal reaction). Instead, a few clones must be mini-prepped and their orientation checked by restriction digestion.

7. We recommend selecting transgenic lines containing single genomic insertions of pFRiPE. If multiple insertions are present (especially on the same chromosome), one cannot exclude potential chromosomal deletions/translocations mediated by recombination between FRT sites at different insertion loci. Practically, several independent insertion lines with paler eyes should be established. Dark-eyed transgenic flies should be avoided as these are likely to be multiple insertions, or the multiple insertions separated by recombination.

8. In our experiments, we have used flies containing the P{hsFLP}22 insertion on the X chromosome in combination with various Gal4 drivers. With this insertion, excision only occurred after heat shock in most tissues (for example, wing imaginal discs). However, a subset of fat body cells (~50%) excised the FRT cassette without heat shock. Therefore, RNAi will occur before heat shock in these cells if expression is driven by a GAL4 driver active in the fat body. Other hs-FLP insertion lines might provide better regulated heat inducibility in this tissue, but have not been tested.

9. We have not studied RNAi in adult flies, but "friping" may be possible in the adult stage as well. For efficient heat-shock induction of the RNAi, we suggest heat-shocking food vials containing pupae on the walls of the tube 24–48 h prior to eclosion, or adults in the absence of food since flies tend to drown in the food during heat shock.

10. Most HcRed fluorescence produced before cassette excision will decay within 24 h of excision. Therefore, only unexcised clones will still fluoresce red when excited with laser light between 545 and 633 nm (excitation peak: 596 nm). Thus, antibody stains may be performed with secondary antibodies emitting wavelengths that overlap with the HcRed emission spectrum (e.g., Cy3, Cy5). If antibody staining is required prior to HcRed cassette excision, a green fluorescent secondary antibody should be used to avoid HcRed bleed through into the antibody signal. However, in many pFRiPE constructs, red fluorescence is faint enough even before cassette excision not to bleed through into the Cy3 or Cy5 channel, provided that primary antibodies yielding strong signals are used.

11. Some pFRiPE constructs (or transgenic construct integration sites) appear to yield more frequent unexcised clones than others. For example, a pFRiPE construct against *arrow* typically yields — two to six clones in the dorsal compartment of the imaginal disc wing pouch (driven with *apterous*-Gal4), which is still a minor proportion of the expressing cells.

    If the HcRed cassette in an existing pFRiPE RNAi construct appears to be excised inefficiently, derivative fly lines can be generated that do not contain the cassette any more. RNAi is then independent of heat shock-mediated excision, but can be made inducible by introducing temperature-sensitive Gal80 *(23)* in the Gal4 driver line.

    Generate excision lines by excising the *HcRed* cassette in the germ line. Heat-shock food vials containing larvae expressing both hs-FLP and the pFRiPE construct, twice in a two- or three-day interval. Cross the emerging flies to a balancer stock and establish several independent stocks containing the pFRiPE construct (now presumably missing the *HcRed* cassette). Test each for the presence/absence of the *HcRed* cassette (by PCR or cross with a Gal4 driver).

12. In a Gal80^{ts} + Gal4 driver genetic background, combining a pFRiPE derivative excision line with a different pUAST transgene containing a flip-out cassette (for example, the FRT-HcRed-FRT cassette) is an excellent method to achieve sequential transgene inductions: RNAi is first induced by inactivating Gal80 (flies are shifted from 18 to 29°C), later the second transgene is induced by heat shock-mediated *HcRed* cassette excision. This can serve to study the effect of a particular perturbation in the absence of a given protein. For an example, *see* ref. *(2)*.

Acknowledgments

This work was supported by an A. von Humboldt fellowship to E. M. and by a grant from the Deutsche Forschungsgemeinschaft (EA 4/2-1).

References

1. Panáková, D., Sprong, H., Marois, E., Thiele, C., and Eaton, S. (2005) Lipoprotein particles are required for Hedgehog and Wingless signaling. *Nature* **435,** 58–65.

2. Marois, E., Mahmoud, A., and Eaton, S. (2006) The endocytic pathway and formation of the Wingless morphogen gradient. *Development* **133,** 307–317.

3. Desbordes, S. C. and Sanson, B. (2003) The glypican Dally-like is required for Hedgehog signaling in the embryonic epidermis of Drosophila. *Development* **130**, 6245–6255.

4. Lum, L., Yao, S., Mozer, B., et al. (2003) Identification of Hedgehog pathway components by RNAi in Drosophila cultured cells. *Science* **299**, 2039–2045.

5. Nybakken, K., Vokes, S. A., Lin, T. Y., McMahon, A. P., and Perrimon, N. (2005) A genome-wide RNA interference screen in Drosophila melanogaster cells for new components of the Hh signaling pathway. *Nature Genet.* **37**, 1323–1332.

6. Kennerdell, J. R. and Carthew, R. W. (2000) Heritable gene silencing in Drosophila using double-stranded RNA. *Nature Biotechnol.* **18**, 896–898.

7. Martinek, S. and Young, M. W. (2000) Specific genetic interference with behavioral rythms in Drosophila by expression of inverted repeats. *Genetics* **156**, 1717–1725.

8. Fortier, E. and Belote, J. M. (2000) Temperature-dependent gene silencing by an expressed inverted repeat in Drosophila. *Genesis* **26**, 240–244.

9. Billuart, P., Winter, C. G., Maresh, A., Zhao, X., and Luo, L. (2001) Regulating axon branch stability: the role of p190 RhoGAP in repressing a retraction signal pathway. *Cell* **107**, 195–207.

10. Piccin, A., Salameh, A., Benna, C., et al. (2001) Efficient and heritable functional knock-out of an adult phenotype in Drosophila using a GAL4-driven hairpin RNA incorporating a heterologous spacer. *Nucleic Acids Res.* **29**, e55.

11. Reichhart, J. M., Ligoxygakis, P., Naitza, S., Woerfel, G., Imler, J. L., and Gubb, D. (2002) Splice-activated UAS hairpin vector gives complete RNAi knockout of single or double target transcripts in Drosophila melanogaster. *Genesis* **34**, 160–164.

12. Enerly, E., Larsson, J., and Lambertsson, A. (2002) Reverse genetics in Drosophila: From sequence to phenotype using UAS-RNAi transgenic flies. *Genesis* **34**, 152–155.

13. Kalidas, S. S. D. (2002) Novel genomic cDNA hybrids produce effective RNA interference in adult Drosophila. *Neuron* **33**, 177–184.

14. Giordano, E., Rendina, R., Peluso, I., and Furia, M. (2002) RNAi triggered by symmetrically transcribed transgenes in Drosophila melanogaster. *Genetics* **160**, 637–648.

15. Brand, A. H. and Perrimon, N. (1993) Targeted gene expression as a means of altering cell fates and generating dominant phenotypes. *Development* **118**, 401–415.

16. Duffy, J. B. (2002) GAL4 system in Drosophila: a fly geneticist's Swiss army knife. *Genesis* **34**, 1–15.

17. Filipowicz, W. (2005) RNAi: the nuts and bolts of the RISC machine. *Cell* **122**, 17–20.

18. Lee, Y. S. and Carthew, R. W. (2003) Making a better RNAi vector for Drosophila: use of intron spacers. *Methods* **30**, 322–329.

19. Smith, N. A., Singh, S. P., Wang, M. B., Stoutjesdijk, P. A., Green, A. G., and Waterhouse, P. M. (2000) Total silencing by intron-spliced hairpin RNAs. *Nature* **407**, 319–320.

20. Golic, K. (1991) Site-specific recombination between homologous chromosomes in Drosophila. *Science* **252**, 958–961.

21. Wesley, S. V., Helliwell, C. A., Smith, N. A., et al. (2001) Construct design for efficient, effective and high-throughput gene silencing in plants. *The Plant J.* **27,** 581–590.

22. Henschel, A., Buchholz, F., and Habermann, B. (2004) DEQOR: a web-based tool for the design and quality control of siRNAs. *Nucleic Acids Res.* **32,** W113–W120.

23. McGuire, S. E., Le, P. T., Osborn, A. J., Matsumoto, K., and Davis, R. L. (2003) Spatiotemporal rescue of memory dysfunction in Drosophila. *Science* **302,** 1765–1768.

11

Germline Clone Analysis for Maternally Acting *Drosophila* Hedgehog Components

Erica M. Selva and Beth E. Stronach

Abstract

Many of the genes of *Drosophila melanogaster* have their transcripts deposited in developing oocytes. These maternally loaded gene products enable an otherwise homozygous mutant embryo to survive beyond the first stage of development for which the gene product is required. Zygotic mutations that disrupt the Hedgehog signal transduction pathway typically yield a segment polarity 'lawn of denticles' cuticle phenotype. However, an embryo homozygous mutant for a gene can achieve normal embryonic segmentation precluding classification of the gene as a component of the Hh pathway, if wild-type transcripts from the mother are present. This chapter discusses the theory and importance of analyzing germline clone embryos for maternally acting genes involved in Hh signal transduction, and describes in detail the method to generate mutant germline clone embryos.

Key Words: Germline clone embryos; germline stem cells; Hh signal transduction; recombination; FLP recombinase; maternal transcripts; *Drosophila*.

1. Introduction

During oogenesis, transcripts of approx 30% of the estimated 13,000 genes present in the *Drosophila melanogaster* genome are deposited in the developing oocyte to produce a mature egg competent for fertilization *(1)*. The oocyte and the 15 nurse cells that nurture its development are derived by mitosis from a common diploid germline stem cell, and share a cytoplasm throughout oogenesis. It is only when the developing oocyte passes through the oviduct, where it is fertilized, that it completes meiosis, which will determine whether the egg receives the wild type or mutant chromosome. This can have significant consequences for recessive lethal mutations that encode maternally deposited products, as regardless of the embryo's genotype, for those genes that have maternally

From: *Methods in Molecular Biology: Hedgehog Signaling Protocols*
Edited by: J. Horabin © Humana Press Inc., Totowa, NJ

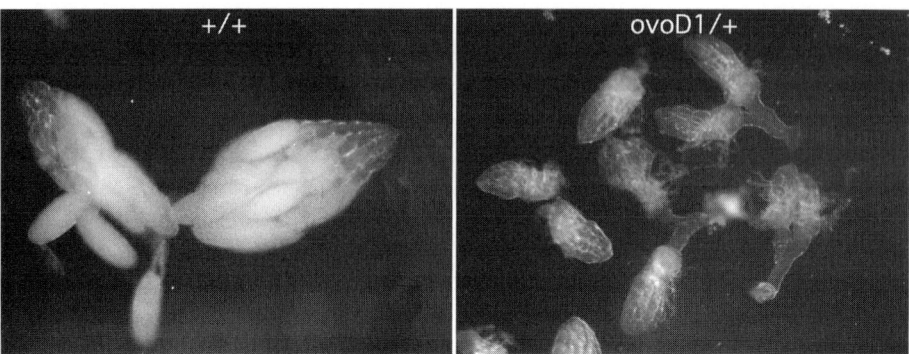

Fig. 1. ovo^{D1} arrest of oogenesis. Dissected ovaries from adult females wild type (left panel) and heterozygous for ovo^{D1} (right panel) visualized in brightfield.

deposited transcripts, there is an abundance of wild-type copies of the gene product in the newly fertilized oocyte. The presence of these wild-type transcripts in an embryo that is otherwise homozygous mutant at the genomic level, can permit it to survive through embryogenesis into later stages of development and, in some cases, to become a late pupa or pharate adult. Hence, the presence of maternally deposited transcripts can mask the first stage(s) in development that a gene is required.

The creation of germline mosaics has proven to be an invaluable tool for determining the earliest mutant phenotypes for recessive lethal mutations that also have maternal transcripts. Initially, these experiments were inefficiently performed using X-rays to induce homologous recombination and an X-linked dominant female sterile (DFS) mutation. The use of heat shock driven site-specific FLP recombinase from yeast, combined with the DFS mutation ovo^{D1}, has made generation of germline mosaics much more efficient making it a useful tool to analyze the zygotic lethal phenotype of genes with maternally deposited transcripts *(2–4)*.

The presence of a single copy of ovo^{D1} in females blocks oogenesis at an early stage preventing her ability to develop eggs (**Fig. 1**). The Perrimon laboratory has engineered lines that have ovo^{D1} distal to centromere proximal FRT sites, on the X and each arm of the second and third chromosomes. These lines are housed at the Bloomington stock center (**Table 1**). Recombination between FRT sites is induced by the heat shock driven expression of FLP present elsewhere in the genome. When FLP is expressed in the germline, stem cells that are in the G2-phase of the cell cycle with fully replicated DNA, can undergo recombination between their nonsister chromatids. Subsequent segregation of these chromosomes has the potential to yield three types of recombinant products.

Table 1
Germline Clone Stocks and Uses

Genotype	Use for germline clones	Comments[1]
$pr^1\ pwn^1\ P\{hsFLP\}$ $38/CyO;\ Ki^1\ kar^1$ ry^{506}	Flp source on 2 for clones on X	# 5258
$P\{hsFLP\}12,$ $y^1\ w^*;\ noc^{Sco}/CyO$	Flp source on X for clones on 2	# 1929
$P\{hsFLP\}1,\ y^1\ w^{1118};$ $Dr^{Mio}/TM3,\ ry^*\ Sb^1$	Flp source on X for clones on 3	# 7
$w^*\ ovo^{D1}\ v^{24}\ P\{FRT$ $(w^{hs})\}101/C(1)DX,$ $y^1\ f^1/Y;\ P\{hsFLP\}38$	Flp source on 2 and X-FRT site ready for clones on X	# 1813, FRT located at 14AB. C(1)DX is an attached X compound chromosome that ensures the ovoD males and C(1)DX females are the only progeny in the stock
$ovo^{D2}\ v^{24}\ P\{FRT(w^{hs})\}$ $9\text{-}2/C(1)DX,\ y^1\ f^1/Y;$ $P\{hsFLP\}38$	Flp source on 2 and X-FRT site ready for clones on X	# 1843, FRT located at 18A, ovo[D2] is a weak ovoD allele.These germline clone females lay many eggs, only a fraction are germline clone embryos most are collapsed ovo[D2] eggs. Avoid using if possible.
$P\{ovoD1\text{-}18\}2La$ $P\{ovoD1\text{-}18\}2Lb$ $P\{neoFRT\}40A/Dp$ $(?;2)bw^D,\ S^1\ wg^{Sp\text{-}1}$ $Ms(2)M^1\ bw^D/CyO$	2L-FRT correct males from the stock must be crossed with appropriate virgin females to generate males in the next generation with both FLP and FRT.	# 2121, FRT located at 40A. This stock is maintained with three chromosomes, a DFS ovoD, a dominant male sterile, S^1 $wg^{Sp\text{-}1}\ Ms(2)M1\ bw^D$, and the balancer, CyO. Hence only a fraction of the males and females in the population are fertile and competent to maintain the stock
$P\{FRT(w^{hs})\}G13$ $P\{ovoD1\text{-}18\}2R/$ $Dp(?;2)bw^D,\ S^1\ wg^{Sp\text{-}1}$ $Ms(2)M^1\ bwD/CyO$	2R-FRT correct males from the stock must be crossed with appropriate virgin females to generate males in the next generation with both FLP and FRT.	# 2125, FRT located at 42B, stock is maintained as above for # 2121.

(Continued)

Table 1 (*Continued*)

Genotype	Use for germline clones	Comments[1]
w*; P{ovoD1-18}3L P{FRT(w^hs)}2A/st^1 βTub85D^D ss^1 e^s/TM3, Sb^1	3L-FRT correct males from the stock must be crossed with appropriate virgin females to generate males in the next generation with both FLP and FRT	# 2139, FRT located at 79D-F. This stock is maintained with 3 chromosomes a dominant female sterile, ovoD, a dominant male sterile, st^1 βTub85D^D SS^1 e^s, and the balance TM3. Hence, only a fraction of the males and females in the population are competent to maintain the stock.
w*; P{neoFRT}82B P{ovoD1-18}3R/st^1 βTub85D^D ss^1 e^s/TM3, Sb^1	3R-FRT correct males from the stock must be crossed with appropriate virgin females to generate males in the next generation with both FLP and FRT	# 2149, FRT located at 82B, stock is maintained as above for 2139.

[1]All stocks can be obtained from the Bloomington Stock Center (http://flystocks.bio.indiana. edu). The numbers refer to the current stock number at Bloomington. Note these stock numbers are subject to change.

These recombinant stem cells can have either one or two copies of ovo^{D1} to yield sterile females that are unable to complete oogenesis. Only those females with germ cells that lack ovo^{D1}, but now have two copies of the mutant under study, will be capable of completing oogenesis and producing oocytes (**Fig. 2**). Hence, there is no wild-type genomic copy of the gene present in the only stem cells capable of completing oogenesis and as a result, maternal deposition of the wild-type transcript cannot occur. Fertilization of these mutant germline clone oocytes by fathers heterozygous for the mutation, allows for the analysis of the phenotype at the first stage in development when the gene is required. In regard to Hedgehog (Hh) signal transduction, use of the germline clone technique was instrumental in revealing that heparin sulfate proteoglycans are essential for Hh signal transduction *(5,6)*.

The first function of Hh signaling during embryonic development is to determine cell fates within each embryonic segment in collaboration with Wingless (Wg) signaling (*see* reviews *[7,8]*). In the early embryo, *hh* is expressed in stripes of epidermal cells in response to the homeobox protein, Engrailed (En). The secreted Hh ligand signals to cells immediately adjacent and just anterior

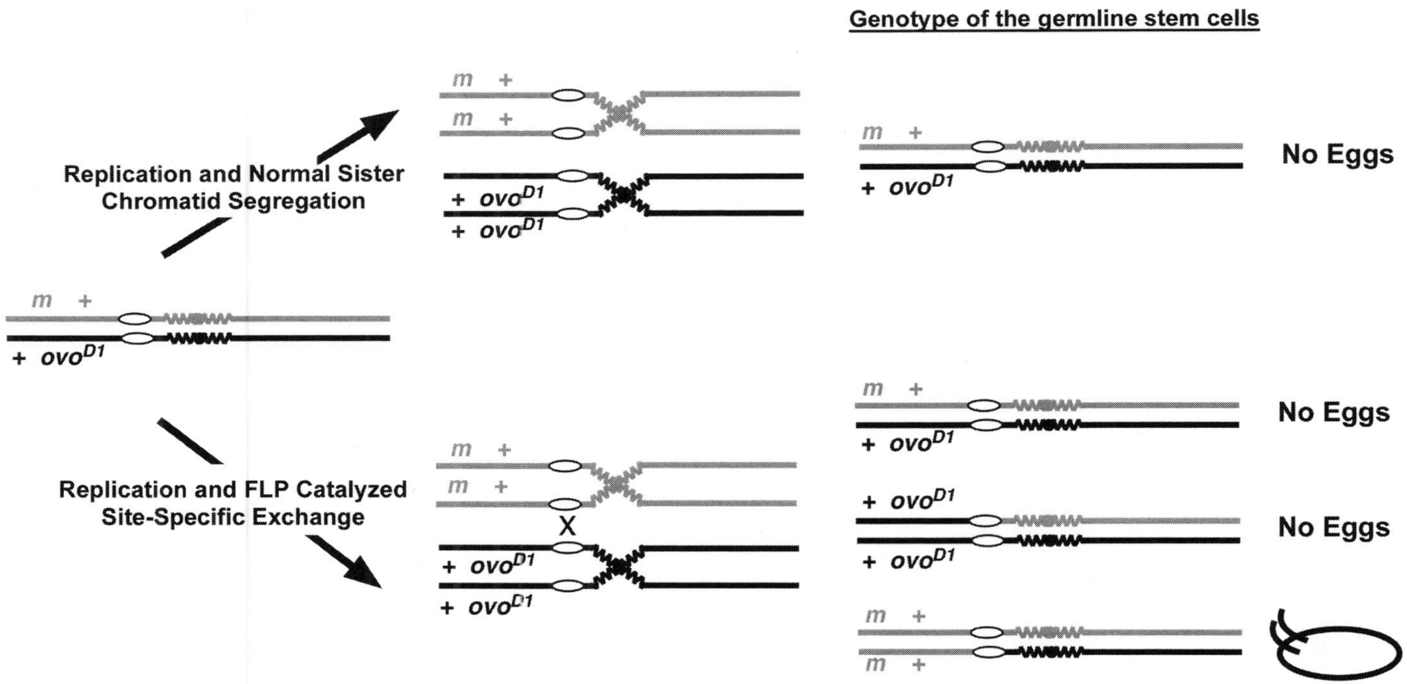

Fig. 2. Method to generate females that have homozygous mutant germline stem cells. See text for description.

133

to the *hh* expressing cells, to promote the expression of *wg*. In turn, the Wg ligand is secreted and signals to posterior epidermal cells to maintain the expression of *en* which drives *hh* expression. The juxtaposition of Hh and Wg signaling cells is crucial for the establishment of both the order and polarity of cell fates within each segmental unit. Combinations of pair-rule genes initiate the expression of both *hh* and *wg*; however, the maintenance of their expression becomes mutually dependent around stage 9 of embryonic development. In *hh* mutant embryos, epidermal Wg expression disappears due to the lack of induction of the Hh-signaling pathway and in *wg* mutant embryos, Hh expression fades from the epidermis *(9–12)*. Thus, loss of either *hh* or *wg* function in the embryo results in the absence of naked embryonic cuticle and, therefore, yields a "lawn of denticles" phenotype (**Fig. 3**). This provides a simple and sensitive method to identify mutations that might affect the Hh- or Wg-signaling pathways. Indeed, identification of the heparin sulfate glycosyltransferase, *tout velu* (*ttv*), was initially identified based on its germline clone "lawn of denticles" phenotype, which ultimately helped to define the importance of heparin sulfate proteoglycans in Hh signal transduction *(5,6)*.

This chapter describes the following: (1) How to generate females with homozygous mutant germline stem cells, (2) the isolation and processing of germline clone embryos that are targeted for specific phenotypic applications, and (3) how to analyze the germline clone embryos to determine if the mutation under study might play a role in the Hh signal transduction pathway.

2. Materials

1. Germline clone stocks (**Table 1**).
2. The mutation under study recombined onto an FRT chromosome.
3. Growth media (for recipes consult the Bloomington Stock Center, http://flystocks. bio.indiana.edu).
4. Plastic or glass vials with growth media.
5. Plastic or glass bottles with growth media (optional).
6. 37°C water bath or air incubator.
7. 18°C incubator (optional).
8. Plastic Petri dishes (60, 100, or 150 mm).
9. Embryo collection cages (Genesee Scientific, https://www.geneseesci.com or Harvard BioLabs Machine Shop, http://www.mcb.harvard.edu/bioshop).
10. Apple juice agar: For 1 L, autoclave 25 g agar in 750 mL H_2O then add 250 mL apple juice with 25 g dissolved sucrose. Swirl to mix. When cooled to ~60°C add 10 mL of 100% ethanol with 1.5 g dissolved tegosept antifungal (Fisher). Swirl to mix. Pour into Petri dishes or pipet into vials. Allow to solidify and cool to room temperature. Store at 4°C.
11. Yeast paste: Bakers yeast dissolved in H_2O to make a thick paste. Prepare in small amounts and use for short periods of time ~2 wk. Store at 4°C.
12. Spatula.
13. 5 or 10 mL syringe.

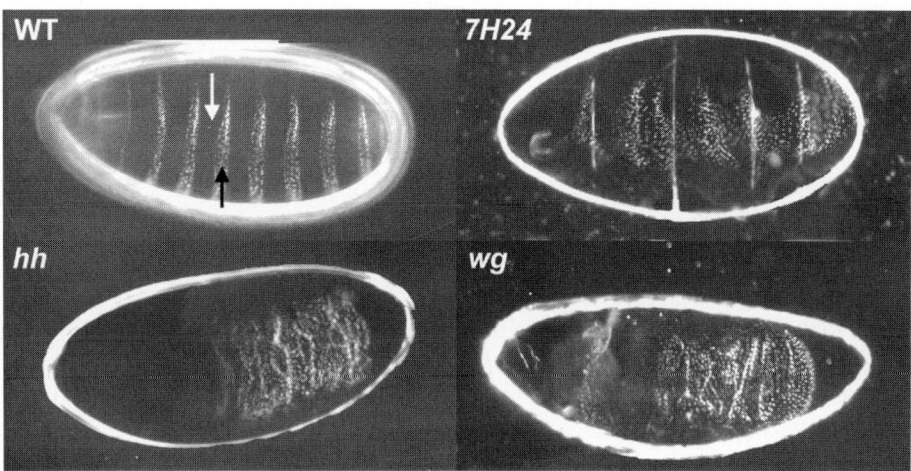

Fig. 3. Cuticle phenotypes for Hh/Wg pathway mutants. Shown are the denticle band pattern for wild-type embryos/fisrt instar larvae (WT, upper left, white arrow shows naked cuticle and black denticle bands) and the "lawn of denticles" phenotype for embryos that are homozygous for the zygotic mutations hh^1 (*hh*, lower left) and wg^{cx4} (*wg*, lower right). The germline clone cuticle phenotype for the ethylmethane sulfonate mutation, *7H24*, suspected of being involved in the Hh signal transduction pathway (*7H24*, upper right). Images taken in darkfield; vitelline membranes not removed.

14. Embryos collection baskets (Genesee Scientific, mesh basket or hand made).
15. Nitex nylon mesh, 120 µm (Genesee Scientific, https://geneseesci.com).
16. Screw capped tubes (5, 15, or 50 mL) to make collection baskets.
17. Dissecting forceps or needles (Fine Science Tools, http://www.finescience.com).
18. Wash bottle with H_2O.
19. Soft bristle paintbrush.
20. 120 mL plastic beakers.
21. 50% bleach solution.
22. Pasteur or transfer pipets.
23. Glass microscope slides.
24. Glass cover slips 22 mm^2.
25. Hoyer's mounting media: In a fume hood dissolve 30 g gum arabic in 50 mL H_2O (heat to 60°C), add 200 g chloral hydrate slowly and dissolve completely. Add 20 g glycerol. Centrifuge at 10,000g and then filter the supernatant through glass wool (*13*).
26. Hoyer's lactic acid: Add lactic acid to Hoyer's 1:1.
27. Slide Warmer.
28. Dissecting microscope.
29. Compound microscope with phase contrast and dark-field optics.
30. CO_2 to anesthetize adult flies.

3. Methods

3.1. Generating Males to Make Germline Clone Embryos

The first step in generating females capable of making germline clones is to obtain males that possess the relevant chromosome arm with an FRT close to the centromere, the DFS, ovo^{D1}, distal to the FRT and a source of FLP recombinase under the control of a heat shock promoter (*hsFLP*). Site-specific FLP recombinase recognizes FRT sites and catalyzes recombination between these sites on homologous chromosomes.

X chromosome: The presence of the C(1)DX chromosome allows the ovo^{D1}, FRT males with FLP recombinase present on the second chromosome to be kept as a self–maintaining stock. The males needed for Cross 2 (below) are the stock males, proceed to Cross 2 using stock males.

The autosomes: ovo^{D1} males with FLP recombinase are generated as the progeny from a cross between virgin females that are homozygous *hsFLP* on the X and balanced on the second or third chromosome. For the purposes of example, all crosses will be shown for making germline clones on the left arm of chromosome 2 (Cross 1). Complete genotypes for these stocks are given in **Table 1**.

Cross 1:

♀♀ *hsFLP12 y¹ w*; Sco/CyO* × ♂♂*ovoD1 FRT40A/CyO*

£

♂♂ *hsFLP12 y¹ w*/Y; ovoD1 FRT40A/CyO*

3.1.1. Recognition of Correct ovoD Males from Stocks

1. Males with ovo^{D1} FRT on the left and right arms of the second chromosome are identified as curly-winged flies with red eyes (**Table 1**). Male flies with brown eyes, curly or not, have the dominant male sterile chromosome. If working with both the left and right arm of the second chromosome, be sure not to confuse the stocks, as they are phenotypically identical.

2. Males with ovo^{D1} FRT on the left and right arms of the third chromosome are distinguished as stubble-bristled flies with orange eyes (**Table 1**). Because the FRT on the left arm of chromosome 3 is marked with the mini-*white* gene, the eye color of these males is slightly darker than that of ovo^{D1} males for the right arm of chromosome 3.

3.2. Generating Females with a Homozygous Mutant Germline

1. Generate females that are heterozygous for the mutation under study and the ovo^{D1} DFS chromosome by crossing ovo^{D1} males from Cross 1 (orange eyed and balanced) to females that have the mutation balanced and present distal to an FRT site on that chromosome arm (Cross 2).

2. Induce recombination between FRT sites on these chromosomes by heat shock to drive the expression of *FLP* recombinase (*see* **Note 1**).

3. To achieve the best results, the timing and duration of heat shocks should be determined empirically for each mutation. General guidelines are given below.

Cross 2:

    ♂♂ *hsFLP12 y¹ w*/Y; ovoD1 FRT40A/CyO* × ♀♀ *m¹ FRT40A/CyO*
             £ heat shock third instar lar

    ♀♀ *hsFLP12 y¹ w*/+; m¹ FRT40A/ovoD1 FRT40A* (genotype of most somatic cells, identified by the absence of *Cy*).

    ♀♀ *hsFLP12 y¹ w*/+;* m¹ *FRT40A/* m¹*1 FRT40A* (genotype of germline stem cells competent to complete oogenesis).

3.2.1. Size and Timing of the Cross

In order to obtain an overall good yield of germline clone embryos, it is imperative at this stage that you have sufficient flies present in the cross so that the larvae churn up the food (Cross 2). If insufficient eggs/larvae are present at this stage, the larvae are likely to die during the heat shock.

1. For vials, 30 virgin females that are balanced for the mutation under study and 10–15 *ovo*^{D1} males are sufficient to yield enough larvae after 48 h (day 2) of egg laying. If you only wish to analyze the cuticle phenotype of your mutation, the cross can be kept in the vials for up to 4 d; when performing a genetic screen for Hh mutants, for example, this is particularly useful.

2. If you need to generate a large number of germline clone embryos, bottles can also be used. For bottles at least three times, the number of flies is required. Use approx 90 virgin females balanced for the mutation under study and 30–40 *ovo*^{D1} males. The number of females is most important so fewer males can be used in Cross 2 if their numbers are limiting.

3.2.2. Timing and Duration of the Heat Shock

1. For vials, heat shock on both days 5 and 6 for 1 h at 37°C for crosses that have been raised at 25°C. As room temperature can vary among laboratories and crosses can be maintained at a lower temperature if desired, day 5 corresponds to the first emergence of wandering third instar larva from the growth media.

2. For bottles, heat shock for 3 h at 37°C when wandering third instar larvae first emerge. The extended time of the heat shock is needed to get complete heating of the growth media (*see* **Note 2**).

3. Heat shocks can be given in either a 37°C water bath or air incubator. However, it is helpful to add a drop or two of H$_2$O to vials or bottles when performing heat shocks in an air incubator, as the growth media tends to dry out, which can result in larval death during the heat shock (*see* **Note 3**).

3.2.3. Collection of Virgin Females with a Homozygous Mutant Germline

The heat shocked progeny that arise from Cross 2 will have the desired females with homozygous mutant germlines. These females must be collected as virgins so the males they mate with can be controlled (**Section 3.3.**). For all chromosomes, the correct females from this cross can be identified by the absence of the dominant marker on the balancer chromosome. When the progeny begin to eclose, they can be maintained at 18°C to extend the time the females remain as virgins. At 18°C, if you clear the vials/bottles carefully, females can be collected in the morning and evening and assumed to be virgins. If the parents in Cross 2 remain in the vials/bottles for only 2 d, significant numbers of virgin females can only be collected from these vials/bottles for 3–5 d (*see* **Note 4**).

3.3. Selection of Males to Cross to the Germline Clone Females

In this step, heterozygous males are crossed to the germline clone females. The males that are selected for Cross 3 vary depending on what you plan to do with the germline clone embryos. There are a range of applications that include: examination of the cuticle phenotypes for mutant characterization or a genetic screen to identify new Hh-signaling mutants, examination of molecular markers in developing embryos, or tests for complementation of the germline clone phenotype.

Cross 3:

$$♀♀ \; hsFLP12 \; y^1 \; w^*/+; \; ovoD1 \; FRT40A/m^1 \; FRT40A \times ♂♂ \; m^2 \; FRT40A/CyO$$

£

collect embryos for analysis.

3.3.1. Embryos for Cuticle Phenotype

1. Mate germline clone female virgins to males from the stock that was used to generate them. If you have a second allele of the mutation, it is preferable that these males are used (Cross 3). This will reduce the penetrance of background phenotypes caused by second site mutations carried on the FRT mutant chromosome.
2. Determine if your mutation can be rescued paternally by performing a careful count of the numbers of hatching larva. If 50% of the larvae hatch, this indicates that the wild-type paternal copy contributed by the balancer chromosome rescues completely. If no larvae hatch, this suggests that the function of the gene under study is required before zygotic transcription (stage 4) and cannot be rescued by the wild-type paternal copy (*see* **Note 5**).
3. Collect unhatched embryos/larvae for preparation of cuticles.

3.3.2. Embryos for Examination of Molecular Markers

To examine molecular markers in developing germline clone embryos, it is necessary to cross the germline clone females to males with a balancer chromosome with a reporter that distinguishes embryos that receive the balancer from those that received the mutant chromosome. For Hh signaling, it would be best to choose a reporter that is expressed at a stage of embryonic development prior to when Hh signaling occurs and is maintained throughout most of embryogenesis. These include so-called blue balancers with *lacZ* under the control of a pair-rule promoter, such as *fushi tarazu* (*ftz*) or green balancers that express green fluorescent protein under the control of the *twist* promoter. Consult the Bloomington Stock Center (http://flystocks.bio.indiana.edu) for balancers that have these reporters and other useful alternatives.

3.3.3. Embryos for Complementation Analysis

If you need to map the mutation under study, you can use the males you are crossing to your germline clone females to perform complementation analysis. To narrow down the chromosomal location of your mutant gene, there is a vast collection of deficiencies that can be used to look for complementation of the germline clone phenotype. The Bloomington Stock Center houses a large collection of deficiencies. This analysis can be extended to specific mutations that fall into this region, if desired. For complementation analysis, embryonic cuticle preparations would be best to assess rescue of the Hh segment polarity phenotype (**Fig. 3**).

3.4. Collection of the Germline Clone Embryos

All collections are performed on a hard agar media made with apple or grape juice, supplemented with some yeast paste after the agar has solidified. The agar prevents the females from pushing the eggs down into the growth media and the yeast makes it a desirable place for the females to lay their eggs.

3.4.1. Preparing Collection Media

The scale of the collection dictates the size of collection cages that should be used.

1. For small crosses with 10–20 germline clone females, the collection agar can be pipeted into regular growth vials.
2. For larger collections in the range of 100 germline clone females, prepare 60-mm Petri dishes with the collection agar. For this scale, the flies must be housed in a collection cage. These can either be made with plastic beakers that have the bottom replaced with nylon mesh or purchased. For most applications, 60-mm Petri dishes are sufficiently large for any germline clone embryo collection. However, if needed, 100- or 150-mm Petri dishes can be prepared and appropriate sized collection cages purchased.

3.4.2. Setting Up the Collection

Apply yeast paste to the surface of the collection media prior to placing the flies in the collection cages. For Petri dishes, this can be done with a spatula. For vials, it is easiest to put the yeast paste in a 5- or 10-mL syringe and apply a small amount on the collection media. After the germline clone females and the desired males have been crossed together in the collection cages, it will take few days before the cross begins to produce a large quantity of embryos.

3.4.3. Collection for Cuticle Preparations

Few embryos are needed for the cuticle preparations shown in **Fig. 3** and because you are examining the terminal lethal stage of embryogenesis, the cross can remain in the vials for 1 or 2 d to increase the number of embryos.

1. Transfer the cross to a new collection vial.
2. If you have sufficient embryos on the surface of the agar, replace the yeast paste since many of the paternally rescued larvae will migrate into it. This will decrease the number of larvae to contend with in the cuticle preparation.
3. Age the embryos for 24 h to allow all the embryos in the vial to complete embryogenesis and deposit cuticle.
4. After aging, remove the replaced yeast paste again to further reduce larvae.
5. The cuticles can be processed immediately or placed at 4°C for 1 or 2 d before processing.

3.4.4. Collection for Antibody Staining and In situ Hybridization

Once the females begin laying well, embryos should be collected on a regular rotation. This is most easily done by performing overnight (0–16 h) and all day (0–8 h) collections. Alternatively, one long collection of 20–22 h can be performed to look at all stages of embryogenesis. Longer collections should be avoided, as they tend to have many larvae. For the analysis of Hh signaling, the 8-h collections tend to have a high percentage of embryos at the appropriate stages of embryonic development.

3.4.5. Harvesting and Dechorionating Embryos

1. After the collection is complete, remove dead flies or pieces of flies from the collection agar with dissecting forceps.
2. Transfer embryos to collection baskets (*see* **Note 6**). Remove embryos from the plate using H_2O and a soft bristle paintbrush to gently dislodge them from the collection agar. Yeast paste that remains on the collection agar should be completely dissolved. Decant the liquid containing the embryos into a collection basket. Embryos that remain stuck to the agar can be removed with a wash bottle.
3. Wash embryos extensively with H_2O in the collection baskets to remove residual yeast. This can be done with a wash bottle or at the sink with a slow flow of deionized H_2O from a tap.
4. Remove the chorion by placing the collection basket into a Petri dish or small beaker with a 50% bleach solution.
5. Incubate for 3–5 min (*see* **Note 7**). During the incubation, the embryos should be gently swirled in the collection baskets and rinsed with the bleach solution using a pasteur or transfer pipet.
6. Wash embryos extensively to remove the bleach.

3.5. Analyzing Germline Clone Embryos

Detailed protocols for antibody staining *(13,14)* and *in situ* hybridization *(15,16)* of embryos have been developed and published elsewhere. Below are guidelines for these procedures and a description of the reagents that will facilitate the analysis of Hh signaling in germline clone embryos.

3.5.1. CUTICLE PREPARATIONS

Following dechorionation, the embryos are mounted directly in Hoyer's media with lactic acid. Terminal stage germline clone embryos have deposited cuticle and are generally brown. Therefore, some brown embryos should be visible in the collection. After successful dechorionation, embryos tend to stick together in clumps.

1. Blot the mesh to remove excess water.
2. Using a dissecting microscope, pick up the embryo clumps off the nylon mesh with dissecting forceps, trying to avoid live larvae.
3. Transfer directly to a glass slide that has 50–60 µL of Hoyer's media pipeted on it (*see* **Notes 8** and **9**).
4. Gently swirl with the tip of the forceps to get an even distribution of embryos. Be careful not to create bubbles in the Hoyer's media.
5. Place a 22-mm^2 cover slip on top of the Hoyer's and incubate overnight on a 55°C slide warmer with three pennies on top to keep the distribution of Hoyer's under the cover slip uniform.
6. Inspect cuticles under a compound microscope with darkfield or phase contrast optics.

3.5.2. Cuticle Preparations for Publication

The vitelline membranes can be removed (white ellipse surrounding the embryos in **Fig. 3**). This takes more time and lowers the yield.

1. Blot as in **Section 3.5.1**.
2. With forceps, place the clumps of dechorinated embryos in a 1.5 mL eppendorf tube with 0.75 mL heptane and 0.75 mL methanol.
3. Shake the mixture vigorously two to three times for 20 s each to remove the vitelline membrane of the embryos. Embryos in which the vitelline membrane has been removed will sink to the bottom of the tube in the lower methanol phase.
4. Aspirate off the organic solvents and wash two times with methanol.
5. Remove as much methanol as possible, then add 60 µL Hoyer's directly to the embryos in the tube and gently pipet up and down to mix. Try to avoid making bubbles in the Hoyer's.
6. Pipet onto a glass slide and proceed as in steps 3.5.1.5. and 3.5.1.6.

3.5 3. Antibody Staining and In situ Hybridization

For both antibody staining and *in situ* hybridization, embryos must first be fixed. We generally follow the method of Patel for embryo fixation *(14)*. However, several other fixation protocols and devitellinization options are described in detail in **ref.** *(13)*.

There are several useful monoclonal antibodies available at the Developmental Studies Hybridoma Bank at the University of Iowa that work well on embryos fixed as in Patel *(14)*. These are anti-Wg 4D4 and anti-En 4D9. Due to the feedback loop that exists between Wg and Hh during embryonic segmentation, it can be difficult to determine if a mutation under study disrupts Hh signaling or Wg signaling or both. The downstream markers, *patched (ptc), serrate (ser), and rhomboid (rho)*, are differentially expressed in response to Hh and Wg signaling in the embryo. Wg protein represses *rho* and *ser* expressions, while Hh activates *rho* and *ptc* *(17)*.

Recently, it has been shown by *in situ* hybridization to *rho* that its embryonic expression pattern is clearly distinct in Hh vs. Wg mutant embryos *(18)* and hence, provides a sensitive marker in the embryo to determine pathway specificity of the mutation under study.

4. Notes

1. The optimum developmental stage to induce recombination was empirically determined *(2)* to be during later stages of larval development. Since the heat shock cannot specifically target germline stem cells, this timing also somewhat restricts the amount of different tissues that undergo recombination, as most of the proliferative phases of imaginal disc development are completed at this stage. However, depending on the genotype, it may still be possible to visualize somatic clones in these adult females as recombination is not limited to the germline stem cells.

2. It is important to keep one vial that is NOT heat shocked as a control. Females collected from this vial should not lay any eggs, confirming that the DFS is blocking oogenesis and the genotype of the flies in the cross is correct.

3. It is also wise to set a timer for the heat shocks, because if left beyond the appropriate time, complete or significant larval death will occur depending on how long the larvae were exposed to the elevated temperature.

4. A note of caution when working on the second chromosome, the dominant marker *Cy* on the balancer is somewhat temperature sensitive and may not be as curly at 18°C. Hence, balanced progeny can be difficult to distinguish from the germline clone females strictly based on wing phenotype.

5. If your mutation is not paternally rescuable, to distinguish in cuticle preparations those embryos receiving the wild-type paternal copy contributed by the balancer from those that have received the mutant copy, you must recombine an embryonic marker, such as *trachealess (trh,* left arm of chromosome 3) onto the mutant chromosome. In this case, the same stock must be used for both the germline clone females and the males to which they are crossed.

6. Homemade collection baskets can be constructed in a variety of different ways. They can be made of screw cap tubes of different sizes with ends removed and holes made in the lids. A nylon mesh filter is then placed over the ends of the tubes

and secured with the cap. We use 20 mL glass scintillation vials that have the bottoms removed and purchase caps that have holes in them. For these, the nylon mesh can be cut to an exact size to fit in and be secured with the cap.

7. Freshly prepared bleach solution works best.
8. Hoyer's solution is very viscous to pipet. Since the final volume is not critical, cut the end of the pipet tip.
9. This volume of Hoyer's should be sufficient for up to 100 embryos, if more embryos need to be mounted use two slides or a larger volume of Hoyer's and cover slip.

References

1. Arbeitman, M. N., Furlong, E. E., Imam, F., et al. (2002) Gene expression during the life cycle of *Drosophila melanogaster*. *Science* **297**, 2270–2275.
2. Chou, T. B. and Perrimon, N. (1992) Use of a yeast site-specific recombinase to produce female germline chimeras in Drosophila. *Genetics* **131**, 643–653.
3. Chou, T. and Perrimon, N. (1996) The autosomal FLP-DFS technique for generating germline mosaics in *Drosophila melanogaster*. *Genetics* **144**, 1673–1679.
4. Chou, T. B., Noll, E., and Perrimon, N. (1993) Autosomal P[ovoD1] dominant female-sterile insertions in Drosophila and their use in generating germ-line chimeras. *Development* **119**, 1359–1369.
5. The, I., Bellaiche, Y., and Perrimon, N. (1999) Hedgehog movement is regulated through tout velu-dependent synthesis of a heparan sulfate proteoglycan. *Mol. Cell.* **4**, 633–639.
6. Han, C., Belenkaya, T. Y., Khodoun, M., Tauchi, M., Lin, X., and Lin, X. (2004) Distinct and collaborative roles of Drosophila EXT family proteins in morphogen signaling and gradient formation. *Development* **131**, 1563–1575.
7. Ingham, P. W. and Martinez Arias, A. (1992) Boundaries and fields in early embryos. *Cell* **68**, 221–235.
8. Perrimon, N. (1994) The genetic basis of patterned baldness in Drosophila. *Cell* **76**, 781–784.
9. DiNardo, S., Sher, E., Heemskerk-Jongens, J., Kassis, J. A., and O'Farrell, P. H. (1988) Two-tiered regulation of spatially patterned engrailed gene expression during Drosophila embryogenesis. *Nature* **332**, 604–609.
10. Martinez-Arias, A., Baker, N., and Ingham, P. W. (1988) Role of segment polarity genes in the definition and maintenance of cell states in Drosophila embryo. *Development* **103**, 157–170.
11. Heemskerk, J., DiNardo, S., Kostriken, R., and O'Farrell, P. H. (1991) Multiple modes of engrailed regulation in the progression towards cell fate determination. *Nature* **352**, 404–410.
12. Bejsovec, A. and Martinez Arias, A. (1991) Roles of wingless in patterning the larval epidermis of Drosophila. *Development* **113**, 471–485.
13. Rothwell, F. and Sullivan, W. (2000) Fluorescent analysis of Drosophila embryos. In *Drosophila Protocols*, Cold Spring Harbor Laboratory Press, New York, pp. 141–157.
14. Patel, N. H. (1994) Imaging neuronal subsets and other cell types in whole-mount Drosophila embryos and larvae using antibody probes. *Methods Cell Biol.* **44**, 445–487.

15. Parthasarathy, N., Lecuyer, E., and Krause, H. M. (2005) Optimized protocols for fluorescent *in situ* hybridization in Drosophila tissues. In *Methods in Molecular Biology: Protocols in Confocal Microscopy* (Paddock, S., ed.) Humana Press Inc.
16. Tautz, D. and Pfeifle, C. (1989) A non-radioactive *in situ* hybridization method for the localization of specific RNAs in Drosophila embryos reveals translational control of the segmentation gene hunchback. *Chromosoma* **98,** 81–85.
17. Hatini, V. and DiNardo, S. (2001) Divide and conquer: pattern formation in Drosophila embryonic epidermis. *Trends Genet.* **17,** 574–579.
18. Franch-Marro, X., Marchand, O., Piddini, E., Ricardo, S., Alexandre C., and Vincent J. P. (2005) Glypicans shunt the Wingless signal between local signaling and further transport. *Development* **132,** 659–666.

12

Clonal Analysis of Hedgehog Signaling in *Drosophila* Somatic Tissues

Christine M. Bankers and Joan E. Hooper

Abstract

To fully understand how animals develop, it is often necessary to remove the function of a particular gene in a specific cell type or subset of cells. In *Drosophila melanogaster*, mosaic animals have been widely utilized to study cell fate, growth and patterning, and restriction of cell fate. This chapter describes using FLP recombinase to generate mosaic *Drosophila*, discussing the chromosomes and cross scheme, how to induce the clones, how to properly identify the appropriate progeny, and how to prepare and analyze the tissues, clones, and phenotypes. It then presents three examples, applying this technique to study Hedgehog signaling. The first example describes moderate-sized *costal* clones in imaginal discs, using green fluorescent protein (GFP) as a marker and *dppLacZ* and Engrailed expression as phenotypic reporters. The second describes filling the adult eye with *roadkill* mutant clones, using *white* as a marker and scoring morphology. The third describes clonal misexpression of a truncated form of Smoothened, using GFP and *yellow* as markers.

Key Words: *Drosophila*; mitotic recombination; FLP recombinase; clonal analysis; Hedgehog signaling.

1. Introduction

1.1. Background and Theory

To fully understand how animals develop, it is often necessary to remove the function of a particular gene in a specific cell type or subset of cells. By creating mosaics, cell autonomy, side-by-side growth rate comparisons, and gene expression levels can all be studied while circumventing lethality to the animal. In *Drosophila melanogaster*, mosaic animals have been widely utilized to study cell fate, growth and patterning, and restriction of cell fate (reviewed in Refs. *[1,2]*). *Drosophila* is particularly useful as a model system because of the wide range of genetic tools available, the relatively low cost of maintaining the animals,

From: *Methods in Molecular Biology: Hedgehog Signaling Protocols*
Edited by: J. Horabin © Humana Press Inc., Totowa, NJ

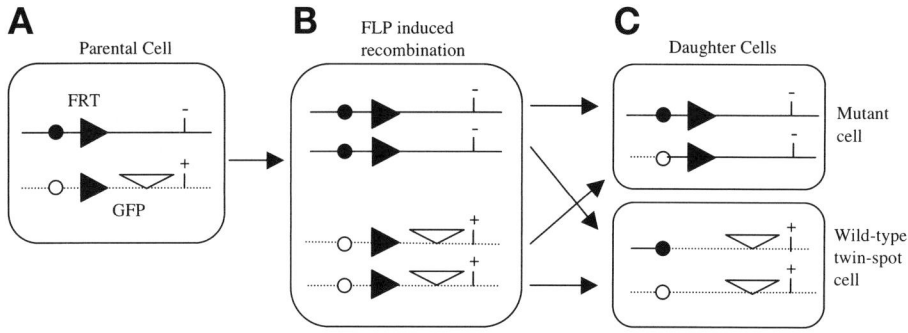

Fig. 1. Generating mutant clones using FLP-mediated mitotic recombination. **(A)** In the heterozygous parental cell, one homolog carries the wild-type form of the gene (+) and the cell marker GFP (∇). The other homolog carries a mutant form of the gene (−). Both carry the identical FRT insertion (black arrows). **(B)** FLP recombinase induces recombination between FRT sites. **(C)** Segregation of the recombinant chromosomes produces two daughter cells: the cell with two copies of the mutant allele has lost the GFP marker, while its twin sister has the wild-type alleles and two copies of GFP. Subsequent cell divisions will result in clones originating from each of these daughter cells.

and the fast generation time. In this chapter, we will discuss some considerations for successful mosaic analysis, including tissue to be used, cell and phenotypic markers, and size and frequency of clones.

In *Drosophila*, mitotic recombination is the preferred method for creating loss-of-function mosaics (**Fig. 1**). Recombination between homologous chromosomes can be induced by chromosome damage (e.g., X-rays) or by site-specific recombinases (*see* following paragraph). Segregation of the crossover products during mitosis then generates homozygous daughter cells in a heterozygous background; a homozygous mutant cell and its wild-type twin. In mitotically active tissue, this will lead to clones of mutant cells, accompanied by wild-type clones (twin-spots). The distribution of the resulting clones is random; their size depends on the mitotic rate and timing of clone induction. Since mitotic recombination can be induced simultaneously in many individuals, this approach allows generation of large numbers of clones with minimal effort.

Early clonal analyses relied on X-rays to induce mitotic recombination. However, X-ray doses that generate reasonable frequencies of clones (in 1–10% of animals) also cause cell death and even lethality to the animal. The advent of transgenics has brought a more user-friendly method, based on the yeast site-specific FLP recombinase *(3,4)*. FLP recombinase catalyzes efficient exchange between FRT sites; the placement and orientation of FRT sites will determine the recombination products. Application of this system to mitotic recombination

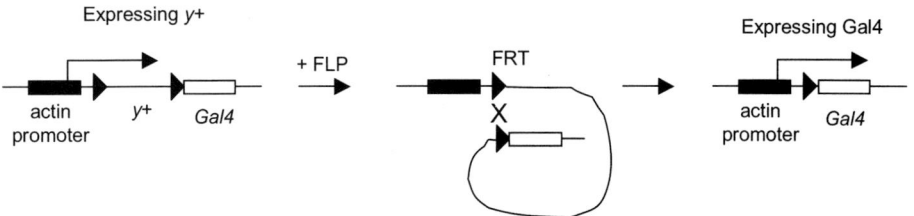

Fig. 2. Clonal misexpression using the FLP-out variation of the FLP/FRT system. In the parental cell, a constitutive promoter (e.g., a fragment of the *actin5C* promoter) drives the expression of the marker gene (e.g., y^+), which is followed by a transcription termination sequence (not shown). The marker gene with its termination sequence is flanked by two FRT elements in the same orientation. FLP recombinase drives mitotic recombination between the two FRT sites, thus excising the marker gene. This puts the downstream coding region (e.g., *Gal4*) under control of the actin promoter. Gal4 is therefore expressed in all y^- cells. Early FLP-out constructs incorporated CD2 as a marker gene, relying on detection of the CD2 antigen to detect clones *(26)*. An improved version uses y^+ as the marker gene and adds a UAS-*GFP* as a linked transgene *(14)*. The GFP allows identification of FLP-out clones in internal tissues, while the y^+ allows identification of the clones in adult cuticle.

is straightforward (**Fig. 1**). Each of the five major chromosome arms (X, 2L, 2R, 3L, and 3R) is available with an FRT element inserted close to the centromere. A chromosome arm is constructed carrying a mutant allele distal to an FRT site; a homolog is constructed with the same FRT site, a wild-type allele, and a cell marker. Flies are then generated carrying both homologs, along with an inducible FLP recombinase. Expression of FLP recombinase then catalyzes efficient exchange between FRT sites. The recombination products segregate to daughter cells, giving rise to homozygous clones in a heterozygous background. When the recombinase is under control of heat shock promoter, clone size and frequency can be controlled by varying the temperature, duration, and timing of heat shock.

The advent of transgenics has also brought the possibility of misexpressing transgenes. While the Gal4/UAS system (*[5]*; *see* Chapter 13 of this book) allows exquisite temporal and tissue control of misexpression, clonal misexpression is sometimes desirable. A tandem arrangement of FRT sites offers an elegant method for doing so (**Fig. 2**).

1.2. Considerations for Experimental Design

While the FLP–FRT system allows efficient and controlled generation of clones, there are many variations to its application in *Drosophila*. The following issues must be considered to ensure a successful experiment.

1. *In what tissue will you analyze your clones?* Imaginal discs are used most often because they are mitotically active, easy to process, and essentially two dimensional. The latter quality makes it possible to analyze the discs as whole-mounts and to capture all of the relevant information in a single photograph. Most importantly, the wealth of cell markers and phenotypic reporters make imaginal discs excellent for analyzing Hedgehog (Hh)-pathway responses. Wing discs are most often used for information about the magnitude of the Hh response. The eye disc offers a distinct alternative, with a gradient of differentiation across the disc and an intermingling of Hh-expressing photoreceptors with Hh-responsive "mystery cells" behind the morphogenetic furrow.

   The adult eye can be analyzed in living animals, as clones and their twin-spots can easily be marked with *white*$^+$ (*w*$^+$). While eye morphology is very sensitive to aberrant differentiation and patterning, external eye morphology must be documented by scanning electron microscopy (SEM) and internal morphology requires plastic-embedding and sectioning. Adult wing morphology is easily documented and is a sensitive reporter for Hh-dependent patterning. However, clone markers are limited (e.g., *multiple wing hair* is available only for 3L and *yellow* (*y*) is difficult to see and photograph). Other cuticle is even more difficult to mark and/or visualize.

2. *How large would the ideal clone be?* Small clones may be best for analyzing growth rates, medium clones are ideal for analyzing changes in target gene expression, and large clones are ideal for effects on morphology. Timing of clone induction is the principal way to control clone size: earlier induction yields larger clones. To fill a compartment with mutant tissue clones, it is possible to incorporate a *Minute (6)* or other recessive cell lethal mutation *(7)* onto the wild-type homolog. Analysis of clones in a minute background is complicated by nonautonomous effects on growth *(8)*; therefore, cell lethals are most often used to fill a tissue with mutant clones (*see* Experiment 3, this chapter).

3. *Will a wild-type twin-spot aid interpretation? What control clones will you need?* Wild-type twin-spots are essential internal controls if there is a possibility for growth defects or cell death. In addition, control clones that differ from the mutant clones only by the presence of the mutant allele usually need to be generated in parallel crosses.

4. *Which clone markers and phenotypic markers should be used?* In general, the clone markers are dictated by the tissue choice: *w* in the eye, *y* in other cuticle, green fluorescent protein (GFP) or some other foreign antigen in imaginal discs. While the markers are generally expressed in wild-type and heterozygous cells, this is not appropriate in tissues where the mutant cells may be dispersed (e.g., brain). For an alternative method, that is more appropriate for marking individual cells (*see* **Note 1**). In general, phenotypic markers in imaginal discs are antibodies to the products of Hh-target genes (e.g., Patched (Ptc), Cubitus interruptus (Ci), Engrailed (En), Collier (Col)) or *LacZ* reporter transgenes (e.g., *ptc-LacZ* or *dpp-LacZ*) whose expression is detected by antibodies against β-galactosidase (β-Gal). These *LacZ* reporters are especially useful where the gene product is a

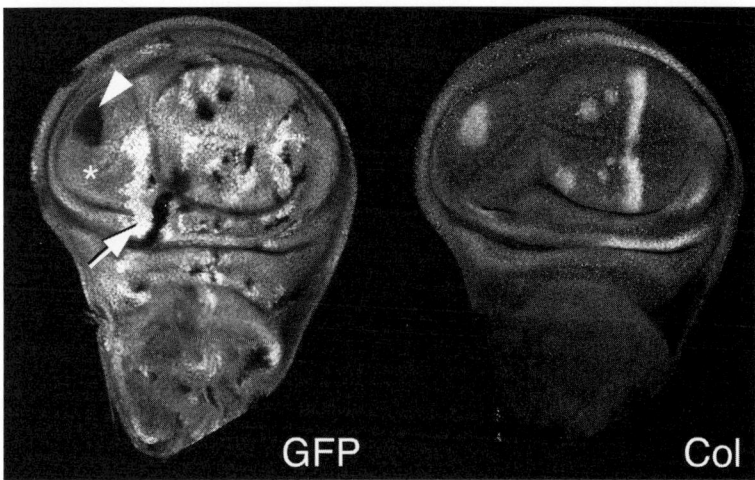

Fig. 3. Using GFP to negatively mark clones in imaginal discs. Clones mutant for *cos* were generated in wing imaginal discs. **(A)** Cells that lack Cos also lack GFP and appear black (arrowhead); their wild-type twin-spots express two copies of GFP and appear white (arrow). Heterozygous cells that have not undergone mitotic recombination express only one copy of GFP and appear grey (asterisk). **(B)** The six clones mutant for *cos* that lie in the anterior wing pouch express ectopic Collier, while the clones in the posterior compartment and notum do not. Anterior is to the left and the notum is down.

secreted protein (e.g., Hh). β-Gal is often used to mark clones, but it cannot be used both to mark clones and ascertain your phenotype (*see* **Note 4**). We prefer GFP as a clone marker in imaginal discs, as it can be visualized without antibodies and it will simultaneously mark clones and their twin spots (**Fig. 3**). In a typical experiment, we simultaneously visualize GFP (to negatively mark clones), Ci (via a rat monoclonal), and En (via a mouse monoclonal), using three-channel detection, which is commonly available on fluorescent microscopes.

1.3. The FRT Chromosomes

The FRT chromosomes are the crux of these experiments. While many useful chromosomes are available (e.g., **Table 1**), you may need to build your own to obtain the ideal combinations of FRT elements, cell markers, and mutations. This is particularly true if you need to add *LacZ* reporters, to create double mutants, or to test a transgene for its ability to rescue a loss-of-function phenotype (*see* **Notes 2** and **3** for elaborations). Remember that different FRT insertions vary in their efficiency of recombination and that you cannot combine different FRT insertions (e.g., FRT42B and FRT42D) in an experiment.

Table 1
Available Genotypes to Induce Clones of Hedgehog-Signaling Components

Gene	Chromosome location	Available chromosomes/genotype	Ref.
ci	102A1-3	*y w hsFLP122; FRT42D, P[ci+], hsp70GFP; ci^{94}*	*(21)*
smo	21B7	*y w hsFLP122;* $\left(\dfrac{FRT40A\,smo^2}{FRT40A} \right)$	*(22)*
ptc	44D5-E1	*y w hsFLP122;* $\left(\dfrac{FRT42D\,ptc^{IIW}}{FRT42D} \right)$	*(22)*
cos$^{\#}$	43B1-2	*y w hsFLP122;* $\left(\dfrac{FRT42D\,armLacZ}{FRT42D\,cos^{W1}} \right)$	*(23)*
Pka-C1	30C5	*y w hsFLP122;* $\left(\dfrac{FRT40A\,myc}{FRT40A\,pka^{dco}} \right)$	*(24)*
fu^{a}	17D1	*fu^{94} FRT18A; hsFLP86E/+*	*(25)*

$^{\#}$*costa (cos)* is referred to by its original name, *costal2 (cos2)* in many publications. *cos*, is located at 43B1-2 and encodes a divergent kinesin. It should not be confused with *Costal-1* (recently renamed *PKA-R2*), which is located at 46D1-4, and encodes a regulatory subunit for PKA.

^{a}*fu* mutant animals are viable through the third instar, so clones are not necessary to analyze mutant tissue.

1.4. The FLP Chromosomes

In choosing a FLP chromosome, there are two important considerations. Firstly, the FLP recombinase must be expressed in the relevant tissue, with the right timing to create the desired size and distribution of clones. FLP under control of a heat shock-inducible promoter (*hsFLP*) allows ubiquitous expression of the FLP recombinase; the timing is determined by when the animals are heat shocked, and the level of recombinase is determined by the temperature and duration of heat shock. For instance, heat shock during the first instar (24–48 h after egg laying) gives large clones, while heat shock during the second instar (48–72 h) gives smaller clones (for different *hsFLP* lines—*see* **Note 4**). Tissue-specific FLP expression is possible (e.g., an *eyeless* (*ey*) promoter fragment for driving expression in the eye), with UAS-*FLP* offering the entire arsenal of tissue-specific Gal4 lines *(5)* to drive FLP expression. Secondly, the FLP should be on a different chromosome from the FRT elements. This simplifies cross schemes and troubleshooting.

1.5. The Crossing Scheme

Clonal analysis is a numbers game, where dozens of experimental animals may need to be processed to find the perfect clone. Thus, the crossing scheme should maximize the yield of the correct genotype and incorporate strategies for their identification. We routinely use flies homozygous for both a FLP chromosome and an FRT chromosome (e.g., *y w hsFLP122; FRT82B, ubiGFP*) as mothers. This increases the percentage of progeny that may contain clones from 12.5% to 50%. In addition, use *TM6B, Tb*, or *CyO, y+* to balance any FRT chromosomes carrying recessive lethal alleles. By picking the *Tubby+* (or *yellow*) larvae, only the experimental progeny are processed (*see* **Note 5** for identification of larval markers). These two steps save enormous effort and are generally worth the extra generation(s), they may entail for their construction.

1.6. Troubleshooting

Before embarking on the full experiment, it is prudent to test the critical steps: viability and detection of progeny that should contain clones, induction, size and frequency of clones, and detection of clone and phenotypic markers. *See* **Note 6** for troubleshooting FLP/FRT combinations and clone induction, and Chapter 8 (this book) on optimizing antibody detection.

The remainder of this chapter will present three specific examples, discussing the chromosomes and crossing scheme, how to induce the clones, how to properly identify the appropriate progeny, and how to prepare and analyze the tissues, clones, and phenotypes. The first example describes moderate-sized *cos−* clones in imaginal discs, using GFP as a marker and *dpp-LacZ* and engrailed (En) expression as phenotypic reporters. The second describes filling the adult eye with *roadkill* (*rdx*) mutant clones, using *white* (*w*) as a marker and scoring morphology. The third describes clonal misexpression of a truncated form of smoothened (Smo), using GFP and *yellow* as markers.

2. Materials

2.1. Antibodies

The Developmental Hybridoma Studies Bank (http://www.uiowa.edu/~dshbwww/) has many relevant antibodies (e.g., En, Smo, Cos2, Fu) available for a nominal fee. Other antibodies against proteins in the Hh pathway include Ci (holmgren@casbah.acns.nwu.edu *[9]*), Ptc (iguerrero@cbm.uam.es *[10]*), Collier (alain.ghysen@univ-montp2.fr *[11]*), and Araucan (jfdecelis@cbm.uam.es *[12]*). Antibodies against Myc, β-Gal, GFP, or other cell markers are available from a variety of commercial sources. Species-specific fluorescent secondary antibodies are also widely available. When combining detection of different antigens (e.g., En, using a mouse monoclonal, and Ci, using a rat monoclonal), it is critical to use secondary antibodies that have been cross-adsorbed to remove reactivity against the relevant species (e.g., the secondary antibody-directed against mouse IgG must be cross-adsorbed against rat IgG).

2.2. FLP, FRT, and Balancer Chromosomes

The Bloomington Stock Center (http://flystocks.bio.indiana.edu/) carries a wide variety of FLP and FRT chromosomes with many combinations of markers and mutations. Many more are available from the researchers who built them. The list in **Table 1** is a starting point for available chromosomes involving Hh signaling components. The Bloomington Stock Center also carries useful balancer chromosomes: *CyO, y+* for the second chromosome, *TM6B, Tb* or *TM6B, y+* for the third chromosome, and *T(2;3) SM6a; TM6B, Tb* (a translocation between SM6a and TM6B, which segregates the second and third chromosomes together).

2.3. Lab. Equipment

1. 25°C incubator.
2. 38°C water bath.
3. Dissecting microscope.
4. Dissecting tools: three-well dissecting dishes, fine forceps, iridotomy spring scissors (Harvard Apparatus), tuberculin syringes (1 mL with 25-gauge needle).
5. Table top rotator for 1.5-mL tubes.
6. Slides, coverslips (#1), and weights (approx. 50 g).
7. Confocal microscope, for analyzing imaginal discs; SEM, for analyzing adult eyes.
8. Cryostat, if analyzing adult eyes.

2.4. Buffers and Reagents

1. Phosphate buffered saline (PBS) with 0.1% Tween 20 (PBT; 1× PBS, 0.1% Tween 20, and 0.2% BSA). Store at 4°C up to 1 week.
2. Fix (4% formaldehyde, 1× PBS, 50 mM EGTA, pH 8.0).
3. Block (PBT, 2% serum). Store at 4°C. (The species from which the serum is derived is not important when using highly specific primary and secondary antibodies. We routinely add sodium azide (0.02%) to assure that there will be no microbial growth.)
4. Ethanol: 25, 50, 75, and 100%.

3. Methods

3.1. Loss-of-Function Clones Marked by GFP

The first experiment generates moderate-sized *cos⁻* clones in imaginal discs (**Fig. 3**). Mutant clones are marked negatively by GFP, with *dpp-LacZ* and En as reporters for phenotype.

3.1.1. Making the Larvae

The experimental cross is:

$$\text{♂ } y\,w;\; \frac{FRT42D, cos^{31}}{CyO, y+} \times \text{♀ } y\,w\,hsFLP;\, dpp\text{-}LacZ,\, FRT42D,\, ubiGFP$$

The control cross is:

$$\text{♂ } y\,w;\; \frac{FRT42D}{CyO, y+} \times \text{♀ } y\,w\,hsFLP;\, dpp\text{-}LacZ,\, FRT42D,\, ubiGFP$$

Place 10 males and 10 females in yeasted vials of food, turn onto fresh food every day, and heat shock every vial on its third day. This will ensure a constant supply of larvae ready to dissect. Day 1, collect eggs in vials at 25°C. Day 2, turn parents onto new vials. Day 3, immerse the first vial 1–2 in. into a 38°C water bath for 2 h, then return the vial to the 25°C incubator. Day 6, larvae will be third instar larval, ready to dissect (*see* **Note 7**).

3.1.2. Dissecting the Larvae

1. Pick about 20 larvae from the food into water in the first well of the three-well dish. The ideal age is just prior to wandering third instar—large but not yet crawling out of the food. (The discs from wandering third instar larvae have deep folds and are therefore difficult to photograph.).
2. Rinse several times with water, to remove food particles. Fill the remaining two wells with cold PBS.
3. Under the dissecting microscope, sort the larvae into the PBS: y in one well and y^+ in the other. In larvae, y makes the mouthparts and the denticles brown rather than black. This is most easily seen against a light background at moderate magnification with oblique lighting. Educate your eye using larvae from the two parental lines (y and y^+).
4. Discard the y^+ larvae, as only the y larvae will have clones.
5. Dissect the larva by gently holding its tail with forceps. Cleanly snip off the head (approximately one-third of the larvae) and dispose of the tail end.
6. Turn the head inside out (like a sock) by steadying the cut end of the head against the forceps and pushing against the mouth with a needle.
7. Remove excessive fat and guts associated with the inverted heads using the needles. Dissected heads can sit in cold PBS for up to 10 min before fixing without affecting quality.

3.1.3. Fixing the Larvae

1. Transfer the dissected heads to 1.5-mL microcentrifuge tubes containing 0.5–1.0 mL of fix.
2. Incubate at room temperature for 20 min. Pipette off the fix and dispose of properly.
3. Wash the heads by adding 1-mL methanol, letting the heads settle, and pipetting off the methanol.

4. Repeat twice with methanol, then add 95% ethanol. Heads can be stored in ethanol at –20°C for many months. It is prudent to stockpile these heads for later use.

3.1.4. Antibody Labeling of Imaginal Discs

Both primary and secondary antibodies should be diluted in block and can be reused many times if stored at 4°C. Plan on at least 0.5-mL diluted antibody per 1.5-mL tube of heads. Optimize all antibodies separately prior to combining for double labeling (*see* Chapter 8). Check species-specificity of secondary antibodies by leaving out each primary antibody and check crossover of detection by leaving out each secondary antibody. Antibodies should be preadsorbed to reduce nonspecific labeling. This is done by incubating your diluted antibody with fixed embryos or fixed control tissue for 30–60 min while rotating.

1. Remove ethanol from larval heads and wash three times with cold PBT.
2. Rotate heads in block for 1 h at room temperature.
3. Remove block and add preadsorbed primary antibodies to heads and rotate overnight at 4°C. We use mouse monoclonal En 4D9 at a dilution in the ratio of 1:10 and rabbit antibody to β-Gal (Cappel) at a dilution in the ratio of 1:1000.
4. Remove and save primary antibody.
5. Wash heads with cold PBT three times every 20 min for 1 h. Rotate between washes.
6. Add block to heads and incubate at room temperature for 30 min while rotating samples.
7. Remove block, add preadsorbed secondary antibody to heads and rotate at room temperature in the dark for at least 2 h. We use Cy5 anti-mouse (Jackson Immuno-Research, cat. no. 715-175-151) and Alexa546 anti-rabbit (Molecular Probes, cat. no. A11010), both at a dilution in the ratio of 1:500. Remember to keep all fluorescent antibodies in the dark during all incubations and during storage.
8. Wash heads with cold PBT three times every 20 min for 1 h, rotating in the dark, with the last wash going overnight at 4°C.

3.1.5. Mounting and Analyzing Imaginal Discs

1. Using a transfer pipette, transfer the heads in PBT to a microscope slide and carefully blot away any excess PBT.
2. Replace the PBT with just enough 50% glycerol to cover the heads.
3. Locate the imaginal discs (eye-antennal, wing, and/or leg) under a dissecting microscope with oblique transmitted light. Gently pull them away from the rest of the tissue using needles. Never touch the disc with the sharp edge of the needle; use only the rounded side. Make a separate grouping of the imaginal discs and dispose of the remaining tissue. Be sure to remove any scraps of the tracheal trunks, as they will interfere with subsequent flattening.
4. Remove excess glycerol, while leaving just enough to cover the imaginal discs.
5. Place a thin line of the antifade-mounting media (e.g., PermaFluor from Thermoshandon) along the center of a coverslip and gently lay the coverslip down over the discs.

Fig. 4. Using a recessive cell lethal to fill the adult eye with mutant cells. The recessive cell lethal mutation, *RpS17*, is on the same chromosome arm as the wild-type allele of the gene of interest, the FRT site, and the cell marker, w^+. The homolog carries the FRT site and your mutation. *eyless:FLP* drives recombination in most cells in the eye. Cells that remain heterozygous are w^+ (pigmentation), cells homozygous for the mutant allele are w (white), and cells homozygous for *RpS17* (and your wild-type allele) have died.

6. Place a weight on the coverslip to flatten the discs and to remove bubbles.
7. Analyze your discs for clones and reporter expression on a confocal microscope once the mounting media has set.

3.2. Loss-of-Function Clones Marked by White

The next experiment fills most of the adult eye with *roadkill* (*rdx*) mutant clones using *white* (*w*) as a marker and examining eye morphology using thin sections and SEM (**Fig. 4**).

3.2.1. Making the Flies

This crossing scheme uses *eyFLP* to drive expression of FLP continuously in the larval eye. A *cell lethal* (*cl*) has been incorporated onto the wild-type chromosome proximal to the FRT element, so that the wild-type twin-spots die. The resulting eyes are mostly mutant cells marked with *w*, interspersed with heterozygous w^+ cells.

The experimental cross is:

$$\female \; y \, w \, eyFLP; \; \frac{FRT82B,cl(3R),w^+(3R)}{TM6B,Tb} \times \male \; \frac{FRT82B,rdx^5 red \, e}{TM6B,Tb}$$

The control cross is:

$$\female \; y \, w \, eyFLP; \; \frac{FRT82B,cl(3R),w+(3R)}{TM6B,Tb} \times \male \; \frac{FRT82B,red \, e}{TM6B,Tb}$$

The eyes of the emerging Tb^+ adults should be almost entirely white.

3.2.2. Analyzing Adult Eye Clones

The eye clones are photographed under a dissecting scope, with diffuse reflected light (**Fig. 4**). Due to the curvature of the eye, it is impossible to get the entire eye in focus at higher magnifications, but it is possible to document the patches of color that show the mutant tissue, using lower magnification. If the detailed external morphology will be documented by SEM, then the flies should be dehydrated in a graded ethanol series: 12 h each in 25, 50, 75, and 100% ethanol. If the internal, cellular morphology will be analyzed, then the flies should be prepared for thin sectioning *(13)*. While it may not be necessary to analyze every eye on all of the progeny in this experiment, it is extremely useful to stockpile flies for future use. Select the progeny with good eye clones, fix and/or dehydrate them using the appropriate steps, and store them in 100% ethanol until needed.

3.3. Clonal Misexpression of a Transgene

The final experiment uses the FLP-out method (**Fig. 2**) to clonally misexpress a truncated form of Smo (UAS-*SmoΔC*). AYG refers to the FLP-out construct pictured in **Fig. 2**, with an actin promoter, a y^+ marker, and *Gal4 (14)*. With UAS-*GFP* on the same chromosome, all cells that express SmoΔC are marked by GFP as well as by *yellow*.

The experimental cross is :

$$\female \; y \, w; \; AYG, \, UAS\text{-}GFP \times \male \; y \, w \, hsFLP; \; UAS\text{-}Smo\Delta C$$

The control cross is:

$$\female \; y \, w; \; AYG, \, UAS\text{-}GFP \times \male \; y \, w \, hsFLP; \; UAS\text{-}Smo$$

1. Setup the crosses and heat shock as described in **Section 3.1.1.**, though the heat shock should only be for 15 min when isolated clones are desired. Longer heat shocks will result in more FLP-out events, and a 2-h heat shock can drive the

FLP-out in virtually every cell. Since the FLP-out event permanently switches on Gal4 expression, this variation could be used to turn on ubiquitous expression at any stage.

2. Imaginal discs are prepared and processed as described in **Section 3.1.** Note that clones will also be marked with *y*, making it possible to locate and analyze them in adult cuticle.

3. Since all progeny can carry clones, there is no need to sort larvae into progeny classes. As in the previous experiments, it is essential to do a small-scale test run to ensure proper induction of clones and antibody labeling conditions.

4. When all conditions have been optimized, proceed with the large-scale experiment and stockpile excess heads in ethanol and store at −20°C.

4. Notes

1. The mosaic analysis with a repressible cell marker (MARCM) system can be used to mark only the mutant cells *(15,16)*. The cell marker is under control of a UAS promoter, which is activated by Gal4 and repressed by Gal80. A transgene that ubiquitously expresses Gal80 is inserted distal to the FRT site on the wild-type homolog. Mitotic recombination simultaneously eliminates the wild-type allele and the repressor (Gal80), so that only cells homozygous for the mutation derepress the marker.

2. Double mutant clones involving two genes on the same chromosome arm (e.g., *smo* and *pka*) are straightforward. When the genes are on different chromosome arms, a useful approach is to use a rescuing transgene. For instance, *smo* is on 2L and *cos* is on 2R. A *smo*⁺ transgene inserted into 2R will link *cos*⁺ and *smo*⁺. In a *smo*⁻ background, this allows analysis of *smo*⁻, *cos*⁻ double mutant clones *(17)*.

$$\female \ y \ w \ hsFLP; \ smo^3, \ al, \ FRT4 \times 2B, \ P[w^+, \ smo^+, \ hs:GFP]5/CyO, \ y^+$$

$$\times \ \male \ y \ w; \ smo^1, \ dp, \ FRT42B \ cos^{31}/CyO, \ y^+$$

3. While gene replacement is difficult in *Drosophila*, it is much easier to overexpress a mutant transgene in clones lacking activity of the endogenous gene. Here, we describe this "poor man's gene replacement" applied to determine the activity of a truncated Smo protein in the absence of the endogenous *smo* *(18)*. UAS-Smo∆C expression is driven by wing-specific Gal4 expression using the line MS1096. At the same time, *smo*⁻ clones are induced by *hsFLP* and marked by loss of GFP. The keys are the recombinant chromosome carrying both *hsFLP* and *MS1096*, and constructing a line homozygous for both the GFP-FRT chromosome and the UAS -Smo∆C chromosome.

$$\female \ y \ w \ hsFLP, \ MS1096; \ \frac{smo^3, FRT40A}{CyO, y^+}$$

$$\times \ \male \ y \ w; \ ubiGFP2L, \ FRT40A; \ UAS\text{-}Smo\Delta C$$

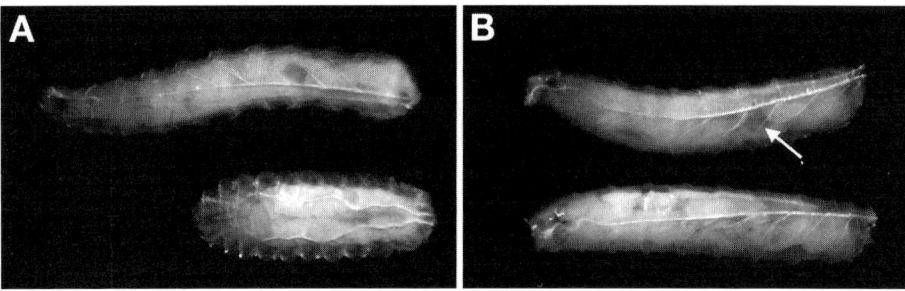

Fig. 5. The Tb phenotype and differentiating male and female larvae. **(A)** Wild type, *Tb+* larva (top) and *Tb* larva (bottom). *Tb* (Tubby) larvae are short and thickset when compared with wild-type larvae. This phenotype also affects the pupa and adult abdomen. **(B)** Male (top) and female (bottom) larvae. Male larvae can be distinguished by the gonads (arrow). The top larva in (A) is also a male, with its gonad visible.

    smo⁻ clones will be negatively marked by GFP in *yellow* larvae and UAS-*SmoΔC* will be expressed throughout the wing pouch.

4. *hsFLP* transgenes differ markedly in their efficiency of inducing recombination. *hsFLP1* **(3)** is relatively low, *hsFLP122* (also known as *hsp70FLP;* **[19]**) is high, and *hsFLP38* (on the second) and *hsFLP86E* (on the third) are intermediate **(20)**.

5. Male and female third instar larvae can be distinguished by their gonads, visible in the fifth abdominal segment in a dorsolateral position under a dissecting microscope. The larval testes are large, clear ovals, just lateral to the dorsal tracheal trunks. They are best seen against a dark background, but may require just the right angle. It is helpful to roll them around while sorting them. *Tb*⁺, or Tubby⁺, is quite easy to differentiate from *Tb*. *Tb* larvae are shorter and appear a bit thicker when compared with *Tb*⁺ larvae, which are longer and leaner (*see* **Fig. 5**).

6. While troubleshooting and optimizing conditions to induce clones, it may be necessary to adjust things slightly or even begin over. If there are no clones present, then carefully review every step. Are the FRT chromosomes compatible, with identical FRT sites? Are you using the right progeny class? Next, setup troubleshooting crosses to identify the faulty chromosome. We like to use *eyFLP* in troubleshooting crosses to check FRT chromosomes, because it eliminates the need to heat-shock larvae. *White* is a good clone marker for troubleshooting because clones are visible in adult eyes without tissue processing. If you find clones are present, but in very low numbers, slightly increasing the water bath temperature and the duration of the heat shock should increase clone frequency. If, however, the issue is the size of clones, the age of the progeny when the heat shock is performed can be changed. If you heat-shock embryos, there will be a small number of large clones but more lethality. If you wait for 48–72 h after egg laying, there will be many small clones.

7. If you do not work during weekends, culturing the vials at room temperature will synchronize the larvae with your work in the week. Alternatively, set up a larger culture on Wednesday and dissect the following on Monday or Tuesday.

References

1. Postlethwait, J. H. (1976) Clonal analysis of Drosophila cuticle patterns. In *The Genetics and Biology of Drosophila* (Ashburner, M. and Wright, T. R. F., eds), Vol. 2c, Academic Press, New York, pp. 359–441.
2. Ashburner, M. (1989) *Drosophila: A Laboratory Handbook*, Cold Spring Harbor Laboratory Press, Cold Spring Harbor, NY.
3. Golic, K. G. and Lindquist, S. (1989) The FLP recombinase of yeast catalyzes site-specific recombination in the Drosophila genome. *Cell* **59,** 499–509.
4. Xu, T. and Rubin, G. M. (1993) Analysis of genetic mosaics in developing and adult Drosophila tissues. *Development* **117,** 1223–1237.
5. Brand, A. H. and Perrimon, N. (1993) Targeted gene expression as a means of altering cell fates and generating dominant phenotypes. *Development* **118,** 401–415.
6. Garcia-Bellido, A. and Dapena, J. (1974) Induction, detection and characterization of cell differentiation mutants in Drosophila. *Mol. Gen. Genet.* **128,** 117–130.
7. Stowers, R. S. and Schwarz, T. L. (1999) A genetic method for generating Drosophila eyes composed exclusively of mitotic clones of a single genotype. *Genetics* **152,** 1631–1639.
8. Moreno, E., Basler, K., and Morata, G. (2002) Cells compete for decapentaplegic survival factor to prevent apoptosis in Drosophila wing development. *Nature* **416,** 755–759.
9. Motzny, C. K. and Holmgren, R. (1995) The Drosophila cubitus interruptus protein and its role in the wingless and hedgehog signal transduction pathways. *Mech. Dev.* **52,** 137–150.
10. Capdevila, J. and Guerrero, I. (1994) Targeted expression of the signaling molecule decapentaplegic induces pattern duplications and growth alterations in Drosophila wings. *EMBO J.* **13,** 4459–4468.
11. Crozatier, M. and Vincent, A. (1999) Requirement for the Drosophila COE transcription factor Collier in formation of an embryonic muscle: transcriptional response to notch signaling. *Development* **126,** 1495–1504.
12. Diez del Corral, R., Aroca, P., G mez-Skarmeta, J.L., Cavodeassi, F., and Modolell, J. (1999) The Iroquois homeodomain proteins are required to specify body wall identity in Drosophila. *Genes Dev.* **13,** 1754–1761.
13. Bentrop, J., Schwab, K., Pak, W. L., and Paulsen, R. (1997) Site-directed mutagenesis of highly conserved amino acids in the first cytoplasmic loop of Drosophila Rh1 opsin blocks rhodopsin synthesis in the nascent state. *EMBO J.* **16,** 1600–1609.
14. Ito, K., Awano, W., Suzuki, K., Hiromi, Y., and Yamamoto, D. (1997) The Drosophila mushroom body is a quadruple structure of clonal units each of which contains a virtually identical set of neurones and glial cells. *Development* **124,** 761–771.
15. Lee, T. and Luo, L. (2001) Mosaic analysis with a repressible cell marker (MARCM) for Drosophila neural development. *Trends Neurosci.* **24,** 251–254.

16. Lee, T. and Luo, L. (1999) Mosaic analysis with a repressible cell marker for studies of gene function in neuronal morphogenesis. *Neuron* **22,** 451–461.

17. Methot, N. and Basler, K. (2000) Suppressor of fused opposes hedgehog signal transduction by impeding nuclear accumulation of the activator form of Cubitus interruptus. *Development* **127,** 4001–4010.

18. Hooper, J. E. (2003) Smoothened translates Hedgehog levels into distinct responses. *Development* **130,** 3951–3963.

19. Struhl, G. and Basler, K. (1993) Organizing activity of wingless protein in Drosophila. *Cell* **72,** 527–540.

20. Chou, T. B. and Perrimon, N. (1992) Use of a yeast site-specific recombinase to produce female germline chimeras in Drosophila. *Genetics* **131,** 643–653.

21. Methot, N. and Basler, K. (1999) Hedgehog controls limb development by regulating the activities of distinct transcriptional activator and repressor forms of Cubitus interruptus. *Cell* **96,** 819–831.

22. Chen, Y. and Struhl, G. (1996) In vivo evidence that Patched and Smoothened constitute distinct binding and transducing components of a Hedgehog receptor complex. *Cell* **87,** 553–563.

23. Sisson, J. C., Ho, K. S., Suyama, K., and Scott, M. P. (1997) Costal2, a novel kinesin-related protein in the Hedgehog signaling pathway. *Cell* **90,** 235–245.

24. Li, W., Ohlmeyer, J. T., Lane, M. E., and Kalderon, D. (1995) Function of protein kinase A in hedgehog signal transduction and Drosophila imaginal disc development. *Cell* **80,** 553–562.

25. Lefers, M. A., Wang, Q. T., and Holmgren, R. A. (2001) Genetic dissection of the Drosophila Cubitus interruptus signaling complex. *Dev. Biol.* **236,** 411–420.

26. Jiang, J. and Struhl, G. (1995) Protein kinase A and Hedgehog signaling in Drosophila limb development. *Cell* **80,** 563–572.

13

GAL4/UAS Targeted Gene Expression for Studying *Drosophila* Hedgehog Signaling

Denise Busson and Anne-Marie Pret

Abstract

The GAL4/upstream activating sequence (UAS) system is one of the most powerful tools for targeted gene expression. It is based on the properties of the yeast GAL4 transcription factor which activates transcription of its target genes by binding to UAS *cis*-regulatory sites. In *Drosophila*, the two components are carried in separate lines allowing for numerous combinatorial possibilities. The driver lines provide tissue-specific GAL4 expression and the responder lines carry the coding sequence for the gene of interest under the control of UAS sites. In this chapter, the basic GAL4/UAS system and its extensions, namely those allowing precise temporal control and reversible expression, are described. In addition, a list of GAL4 and UAS lines and schematic maps of GAL4 and UAS vectors useful in the study of Hedgehog (Hh) signaling is given. Finally, uses of the GAL4/UAS system to resolve some of the questions addressed in the study of the Hh pathway are presented.

Key Words: GAL4; UAS; GAL80; FLP/FRT; GAL4 driver lines; UAS responder lines; targeted gene expression; spatio-temporal control; reversible expression; gain–of–function.

1. Introduction

In 1993, Brand and Perrimon published a landmark article describing the GAL4/upstream activating sequence (UAS) system for targeted expression in *Drosophila*, which constitutes one of the most powerful tools for studying gene function *(1)*. This bipartite system is based on the properties of the yeast *Saccharomyces cerevisiae* GAL4 transcription factor, which activates transcription of its target genes by binding to specific *cis*-regulatory sites called UAS *(2,3)*. In *Drosophila*, the two components of the system are carried in separate parental lines, the GAL4 driver line in which the *gal4* gene is expressed in a tissue-specific pattern and the UAS responder line in which the gene of interest

From: *Methods in Molecular Biology: Hedgehog Signaling Protocols*
Edited by: J. Horabin © Humana Press Inc., Totowa, NJ

is under UAS control. Mating of the UAS responder flies with the GAL4 driver flies results in progeny bearing the two components, in which the UAS-*geneX* is expressed in a transcriptional pattern that reflects that of the driver (*see* Refs. *[4–6]*).

The GAL4/UAS system and its extensions allow in vivo experimental dissection of a wide range of biological questions. The major, although not comprehensive, applications in dissecting the Hedgehog (Hh) signaling pathway are as follows: (i) to define in which cells the function of a gene is required, (ii) to determine the hierarchical relationships between the different components of the pathway and thus establish the genetic networks acting in vivo, (iii) to characterize the role, activator or repressor, of wild-type or mutant proteins in the signal transduction cascade, (iv) to test the effects on signal transduction of addressing effectors to specific subcellular compartments, such as the plasma membrane or vesicles, and (v) to define autonomous vs nonautonomous effects of the different components when combined with clonal analysis methods.

In **Section 2**, a list of GAL4 lines commonly used in the study of the Hh pathway as well as a selection of UAS constructs for the major genes involved in the pathway are presented. Also, schematic maps of GAL4 and UAS vectors available for constructs and mutagenesis are given. Finally, some of the web sites for access to useful databases are given.

Section 3 is subdivided into three parts. In the first and second parts, the basic GAL4/UAS system and its extensions are presented, respectively, with emphasis on problems and complications discussed in **Section 4**. In the third part, a subset of the types of biological questions which can be addressed is presented, with examples taken from the literature which analyze the Hh pathway.

2. Materials

2.1. GAL4 Lines Useful for the Study of Hh Signaling

The GAL4 lines (*see* **Table 1**) result either from the insertion of an enhancer-trap GAL4 P-element vector or from the trangenic constructs in which the *gal4* sequence is placed under the control of known regulatory sequences (*see* **Fig. 1**).

2.2. UAS Lines for the Study of the Genes Involved in Hh Signaling

Most of the UAS responder lines correspond to transgenic constructs in which the gene of interest is placed under the control of UAS sequences. A small number correspond to PUAS insertions from an EP mutagenesis. UAS constructs made in PUAST vectors (*see* **Fig. 1**) are only expressed in somatic cells, whereas those made in PUASP vectors (*see* **Fig. 1**) are expressed in both the germline and the somatic cells. UAS constructs for genes involved in the Hh

Table 1
GAL4 Drivers for Tissue-Specific Expression in *Drosophila*

GAL4 strain	Expression domain	Other expression domains	Type of transgene	Chromosome	Stock	Refs.
Ubiquitous						
da-GAL4			C	1) 3 (w+),	1) 8641,	*(34)*
1) P{da-GAL4.w-}3,				2) 3 (w−)	2) 5460	
2) P{GAL4-da.G32}UH1						
act5c-GAL4			C	1) 2,	1) 4414,	*(35)*
1) P{Act5C-GAL4}25FO1,				2) 3	2) 3954	
2) P{Act5C-GAL4} 17bFO1						
tub-GAL4			C	3	5138	*(14)*
P{tubP-GAL4}LL7						
hsp70-GAL4	Heat shock inducible		C	1) 2,	1) 2077,	*(8)*
1) P{GAL4-Hsp70.PB}2,				2) 3	2) 1799	
2) P{GAL4-Hsp70.PB} 89-2-1						
arm-GAL4			C	1) 2,	1) 1560,	*(36)*
1) P{GAL4-arm.S}11,				2) 3	2) 1561	
2) P{GAL4-arm.S}4a,				(2 copies)		
P{GAL4-arm.S}4b						
Flp-out clones						
act5c>y+>GAL4			C	1) 2,	1) 3953,	*(35)*
1) P{AyGAL4}25,				2) 3	2) 4413	
2) P{AyGAL4}17b						
act5c>CD2>GAL4			C	1) 1,	1) 4779,	*(37)*
1) P{GAL4-Act5C (FRT.CD2).P}D,				2) 3	2) 4780	
2) P{GAL4-Act 5C (FRT.CD2).P}S						

(Continued)

Table 1 (*Continued*)

GAL4 strain	Expression domain	Other expression domains	Type of transgene	Chromosome	Stock	Refs.
tub>y+>GAL4 tub84B(FRT,y+GFP) GAL4.B			C		Fbtp 0014139	*(38)*
tubα1 >flu-GFP;y+> GAL4 tub84B(FRT:flu-GFP;y+) GAL4.Z			C		Fbtp 0015860	*(39)*
tubα1>CD2>GAL4 GAL4-aTub84B (FRT:CD2).P			C		Fbtp 0010169 Fbrf 0105382	*(40)*
tubα1>CD2, y+>GAL4:VP16 GAL4-atub84B (FRT:CD2).VP16F442A			C		Fbtp 0012215	*(41)*
Wing disc						
24B-GAL4 P{GawB}24B	All discs	Early embryonic mesoderm, later muscles	E	3	1767	*(1)*
30A-GAL4 P{GawB}30A	Wing hinge	Salivary glands	E	2	1795	*(1)*
32-GAL4 P{GawB}32B	Wing except notum	Embryonic ectoderm in stripes, eye disc (posterior to furrow)	E	3	1782	*(1)*
34B-GAL4 P{GawB}34B	Wing (details undocumented)	Salivary glands, posterior midgut unconfirmed	E	2	1967	*(1)*

69B-GAL4 P{GawB}69B	All wing	Embryonic ectoderm in stripes, all other discs	E	3	1774	*(1)*
71B-GAL4 P{GawB}71B	Dorsal and ventral posterior wing pouch (excluded from margin)	Salivary glands	E	3	1747	*(1)*
C765-GAL4 P{GawB}C-765	Wing pouch (low level)		E		Fbtr 0001340	*(1)*
MS1096-GAL4 P{GawB}Bx[MS1096]	Dorsal wing pouch (high), ventral wing pouch (low), Bx veination phenotype		E	1	8860	*(42,43,29)*
ap-GAL4 P{GawB}ap[md544]	Dorsal wing	Embryonic ventral nerve chord	E	2	3041	*(44,45)*
bs-GAL4 P{GAL4}bs[1348]	Most intervein cells of wing		E	2	6354	
dpp-GAL4 P{GAL4-dpp.blk1}40C.6	Wing disc, expressed in a stripe of cells along the anterior–posterior border	Eye disc morphogenetic furrow	C	3	1553	*(46,47)*
en-GAL4 P{en2.4-GAL4}e16E	en expression pattern (posterior compartment)	Embryonic en expression pattern (posterior compartment of each segment), terminal filament in the ovary	E	2	Fbti 0003572	*(48–50)*

(Continued)

Table 1 (*Continued*)

GAL4 strain	Expression domain	Other expression domains	Type of transgene	Chromosome	Stock	Refs.
en-GAL4-33/en-GAL4-54 P{GAL4}enGAL4-33 P{GAL4}enGAL4-54	Wing posterior compartment		E	2	Fbrf 0173137	*(51)*
e22c-GAL4 P{en2.4-GAL4}e22c	Wing ubiquitous	Embryo ubiquitous, ovarian somatic stem cells	E	2	1973	*(52)*
gug-GAL4 P{GAL4}Gug[AGiR]	Wing peripodial membrane		E	3	6773	*(53)*
hh-GAL4	Wing posterior compartment	Embryonic posterior compartment of each segment, terminal filament in ovaries	E	3	none	*(54,50)*
omb-GAL4 P{GawB}bi[md653]	Center of wing blade (from LV1-2 to LV4-5)	Brain (from embryo onwards)	E	1	3045	*(44,55)*
ptc-GAL4 P{w[+mW.hs]=GawB} ptc[559.1])	Wing disc, expressed in a stripe of cells along the anterior–posterior border	Embryonic brain, region surrounding foregut, adult mushroom body and posterior of thoracic, gangilon anterior somatic cells of the germarium in the ovary	E	2	2017	*(56,50)*
rn-GAL4	Central region and the notum of the	Ring in the leg disc, in antennal portion of	E	3	7405	*(57)*

	the eye-antennal disc and in the central region of the halter disc wing disc				
sd-GAL4 (P{GawB}sd[SG29.1])	Wing pouch	E	1	8609	*(29,44, 1)*
vgBE-GAL4 P{GAL4-vg.M}2	Wing pouch at the dorso-ventral boundary	C	2	6819	*(58)*
vgMQ-GAL4 1) P{vgMQ-GAL4.Exel}1, 2) P{vgMQ-GAL4.Exel}2, 3) P{vgMQ-GAL4.Exel}3	Wing pouch	C	1) 1 2) 2 3) 3	1) 8231 2) 8230 3) 8229	
Embryo					
h-GAL4 P{GawB}h[1J3]	h pair-rule pattern	E	3	1734	*(1)*
prd-GAL4 P{GAL4-prd.F}RG1	Odd pair-rule pattern	C	3	1947	*(59)*
wg-GAL4 P{GAL4-wg.M}MA1	wg pattern in embryo (anterior compartment of each segment) No expression in discs	C	3	4918	
en-GAL4,hh-GAL4, ptc-GAL4 (see Wing) **Eye**					
ey-GAL4 P{GAL4 ey.H}4–8	Strongly expressed in ey pattern in eye disc, other insertions exist, some are less strong	C	2	5535	*(60)*

(Continued)

Table 1 (*Continued*)

GAL4 strain	Expression domain	Other expression domains	Type of transgene	Chromosome	Stock	Refs.
elav-GAL4 P{GawB}elav[C155]	All postmitotic neurons of the PNS and CNS throughout development and in the adult		E	1	458	(*61*)
shortGMR-GAL4 1) P{GAL4-nina E.GMR}12 nina E=rh1, 2) P{GMR-GAL4.w[−]}2	Eye disc-strong expression in all cells behind the morphogenetic furrow (nonspecific eye defects)		C	2	1) 1104 2) 9146	(*30, 62*)
longGMR-GAL4 1) P{longGMR-GAL4}2, 2) P{longGMR-GAL4}3	Eye disc (less dramatic eye phenotypes than short version)		C		1) 8605 2) 8121	(*63*)
lz-GAL4 P{GawB}lz[GAL4]	Eye disc	Hemocytes		1	6313	(*64*)
Trachea						
btl-GAL4	Embryonic/larval tracheal system		C	2	8807	(*65*)
Salivary glands						
sgs3-GAL4 P{Sgs3-GAL4.PD}TP1	Salivary glands		C	3	6870	(*66*)
55B-GAL4 P{GawB}55B	Salivary glands	Ovarian follicle cells	E	3	1803	(*67*)

Maternal

tub-GAL4

Name	Expression	C/E	Chr.	Stock	Ref.
1) P{matalpha4-GAL-VP16}V2H,		C	1) 2,	1) 7062,	(27)
2) P{matalpha4-GAL-VP16}V37			2) 3	2) 7063	
nos-GAL4 P{GAL4::VP16-nos.UTR} MVD1	Germ cells after stage 9 of embryogenesis through adult oogenesis	C	3	4937	(25,26)

Ovary somatic

Name	Expression	C/E	Chr.	Stock	Ref.
T155-GAL4 P{GawB}T155	Follicle cells	E	3	5076	(48)
slbo-GAL4 P{GAL4-slbo.2.6}1206	Border cells, centripetal cells	E		6458	(68)
e22c-GAL4 P{en2.4-GAL4}e22c	Embryo ubiquitous, ovary somatic stem cells Wing ubiquitous	E	2	1973	
neur-GAL4 P{GawB}neur[GAL4-A101]	Ovarian somatic cells All sensory organs and their precursors	E	3	6393	(69)
en-GAL4, hh-GAL4, ptc-GAL4 (see Wing)					

C, construction; E, enhancer trap.
All stock numbers are from the Bloomington Collection and when no stock is available the Flybase identification is given.

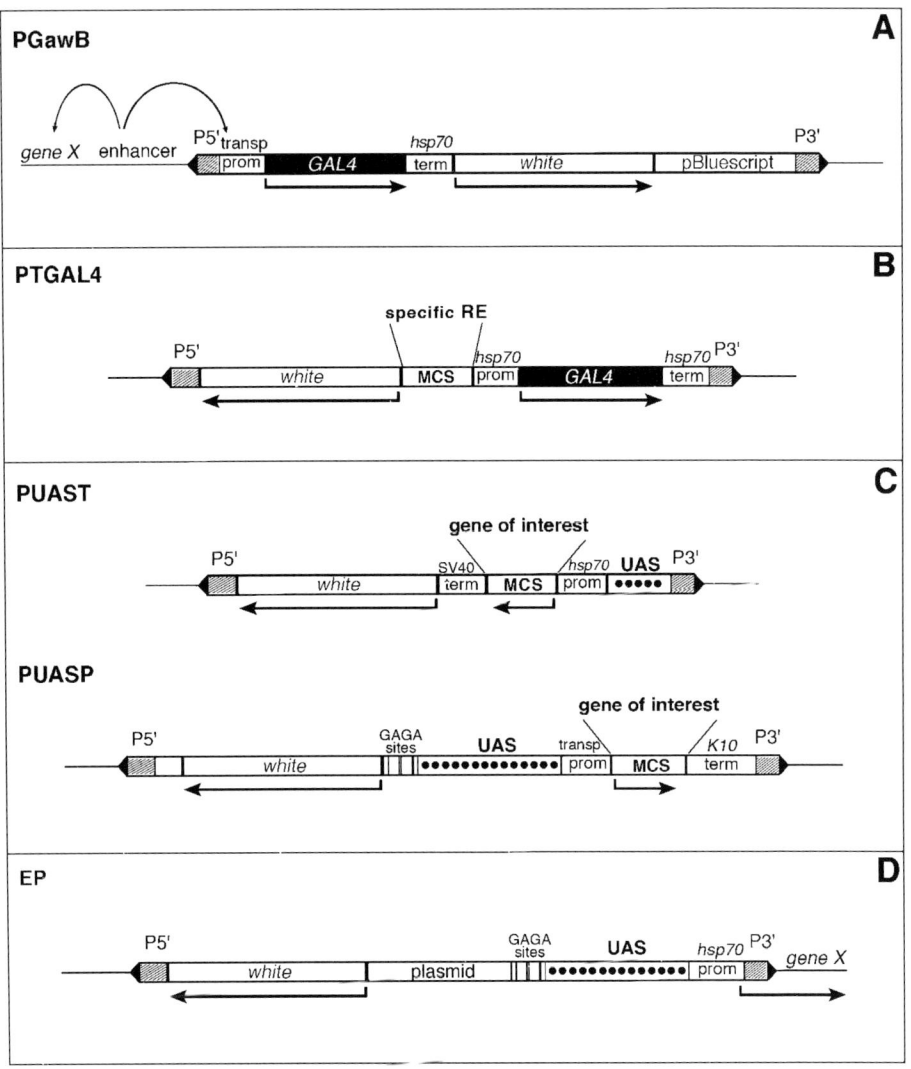

Fig. 1. Schematic of some common P-element transformation vectors used in the UAS/GAL4 system. P5′ and P3′ represent the inverted terminal repeats of the P element; the thin line on both sides of the P element represents genomic sequences. (**A**) PGawB (approx. 11.2 kb) is the first enhancer-trap vector constructed to generate driver lines; neighboring trapped genomic enhancers (enhancer from *gene X*) should drive *GAL4* expression. The transposase promoter (transp prom) is present at the P5′ end. Unique restriction sites (not shown) on both sides of the pBluescript element allow plasmid rescue of upstream and downstream genomic sequences. (**B**) PTGAL4 (approx. 8.7 kb) contains the *GAL4*-coding sequence driven by a minimal promoter (*hsp70* prom) downstream of a multicloning site (MCS) for insertion of specific regulatory elements (RE).

pathway are presented in **Table 2**. A list of available UAS-*lacZ* and UAS-*GFP* lines is given in **Table 3**.

2.3. Vectors Available

The structure of the main types of vectors available to perform GAL4 and EP enhancer-trap mutagenesis or to generate GAL4 and PUAST or PUASP constructs is presented (*see* **Fig. 1**).

2.4. Useful Databases

A nonexhaustive list is as follows:

> BDGP (Berkeley Drosophila Genome Project):
>   http://www.fruitfly.org/p_disrupt/index.html
> FlyBase: http://flybase.bio.indiana.edu/
> Bloomington: http://flystocks.bio.indiana.edu/
> Exelixis: http://drosophila.med.harvard.edu/
> Szeged: http://expbio.bio.u-szeged.hu/fly/index.php

2.4.1. GAL4 Lines

> Flybase: http://flystocks.bio.indiana.edu/Browse/misc-browse/gal4.htm
> Univ. Muenster GAL4 lines: http://flyview.uni-muenster.de/html/overview.html
> GETDB (Gal4EnhancerTrapInsertionDataBase): http://flymap.lab.nig.ac.jp/

Fig. 1. (*Continued*) Unique restriction sites at the MCS include *StuI, XbaI, SstI, NotI, BglII, EcoRI, KpnI,* and *BamHI*. (**C**) PUAST and PUASP are cloning vectors designed to construct UAS responder lines for genes of interest. PUAST (approx. 6.3 kb) contains five tandemly arrayed optimized GAL4-binding sites (UAS), upstream of the *hsp70* promoter (*hsp 70* prom), a polylinker (MCS) with unique restriction sites for *EcoRI, BglII, NotI, XhoI, KpnI,* and *XbaI*, and the SV40 small t intron and polyadenylation site (SV40 term). PUASP (approx. 8 kb) has been modified to allow transcription in the female germline. The enhancer (from the EP vector) with 14 UAS sites and 2 adjacent GAGA sites (to prevent position effect) is placed upstream of the germline competent P-transposase promoter (transp prom, with or without the first intron), followed by a MCS with unique restriction sites for *KpnI, NotI, BamHI,* and *XbaI*. Downstream of the cloning sites are 3'-UTR sequences and terminator from the *K10* gene (*K10* term), which allow transcript stabilization in the germline. (**D**) EP (approx. 7.6 kb) is one of the vector used to perform over- or misexpression screens. The enhancer, which includes 14 UAS sites and 2 GAGA sites, is placed upstream of the *hsp70* promoter (*hsp70* prom), near the P3' end of the transposon. When EP lines are crossed with a specific GAL4 line, the GAL4 activator should bind to UAS sites within the EP vector and activate adjacent endogenous genes (*gene X*). The plasmid fragment allows plasmid rescue of adjacent genomic sequences. References: PGawB (*1*), PTGAL4 (*33*), PUAST (*1*), PUASP (*25*), and EP (*7*).

Table 2
UAS Constructs for Main Genes Involved in *Drosophila* Hh Signaling

Gene Constructs	References	Characteristics
en		
wild-type forms		
-UAS-en	*(58,59,70,71)*	*en* cDNA in pUAST vector
-UAS>y+>HA-en	*(72)*	Allows to obtain overexpressing *en* clones, HA tag
fusion proteins		
-UAS-en.VP16	*(73)*	VP16 activator domain replaces En repressor domain
hh		
wild-type forms		
-UAS-hh	*(42,74,75)*	*hh* cDNA in pUAST vector
-UAS-shh	*(74)*	Zebrafish *shh* cDNA in pUAST vector
-UAS-hh.HA	*(76)*	3× HA between aa 254 and 255
-UAS-hh.GFP	*(77)*	GFP between aa 254 and 255
truncated forms		
-UAS-hh-N	*(78)*	N-term active (aa 1–257) moiety, no cholesterol added
-UAS-hh-N.HA	*(76)*	3× HA between aa 254 and 255 in Hh-N
mutated or modified forms		
-UAS-hh-N.CD2	*(13)*	Membrane-tethered version of Hh-N (transmb. rat CD2)
-UAS-hh-N.GPI	*(76)*	Membrane-tethered version of HH-N (GPI anchor)
-UAS-hh.C85S	*(75,79,80)*	First residue of mature Hh mutated, no palmitoylation
-UAS-hh-N.C85S	*(80)*	First residue of mature Hh-N mutated, no palmitoylation
ptc		
wild-type forms		
-UAS-ptc	*(49)*	*ptc* cDNA in pUAST vector
-UAS>CD2, y+>ptc	*(81)*	Allows to obtain overexpressing *ptc* clones
-UAS-ptc.Myc	*(82)*	Myc tag (N-terminal)
-UAS-ptc.GFP	*(77)*	GFP tag (C-terminal)
truncated forms		
-UAS-ptc.N	*(83)*	First moiety of Ptc (aa 1–676)
-UAS-ptc.C	*(83)*	Second moiety of Ptc (Δ aa 9–676)

(Continued)

Table 2 (*Continued*)

Gene Constructs	References	Characteristics
-UAS-ptc.1130X	*(83)*	Δ C-tail (Δ aa 1131–1286)
-UAS>ptc.Δloop2	*(84)*	Δ extracellular loop2 (aa 738–939)
mutated forms		
-UAS-ptc.D584N	*(85,86)*	Mutation in SSD domain (cf. *ptc*^{S2} mutant)
-UAS-ptc.Y442C	*(86)*	Mutation in SSD domain
-UAS-ptc.E1185K	*(85)*	Mutation in C-tail (cf. *ptc*^{13} mutant)
-UAS-ptc^{mut}.Myc	*(82)*	Two mutations (R111W, G276D) in loop 1
-UAS-ptc^{S2}.GFP	*(77)*	D584N, GFP tag
-UAS-ptc^{14}.GFP	*(77)*	L83Q (First transmb. domain), GFP tag
smo		
wild-type forms		
-UAS-smo	*(87)*	*smo* cDNA in pUAST vector
-UAS-smo.HA	*(88,89)*	1× or 3× HA tag at C-terminal end
-UAS-smo.Flag	*(89,90)*	Flag tag at N-terminal or C-terminal end
-UAS-smo.Myc	*(91,92)*	Myc tag at N-terminal end (aa 35)
-UAS-smo.TAP	*(87)*	TAP tag at C-terminal end
-UAS-smo.GFP	*(91,93)*	GFP tag at N-terminal or C-terminal end
truncated forms		
-UAS-smo.ΔC$^{(0–4)}$.Flag	*(93)*	A series of deletions from aa 562, 638, 742, 837, 939, Flag N-term
-UAS-smo.ΔC $^{(T,730,818)}$.GFP	*(91)*	A series of deletions from aa 556, 731, 819, GFP N-terminal
-UAS-smo.ΔCT.Myc	*(91)*	Deletion from aa 556, Myc tag N-terminal
-UAS-smo.N(T1).Myc	*(92)*	Deletion from aa 256 (289), Myc tag N-terminal
-UAS-smo.ΔN.Flag	*(93)*	Deletion of the N-terminal extracellular domain, Flag tag
-UAS-smo.CT.Flag (Myr)	*(91)*	C-tail (aa 556–1035), Flag or FlagMyr tag N-terminal
-UAS-smo.C.Myc (MycMyr)	*(92)*	C-tail (aa 554–1035), Myc or MycMyr tag N-terminal
mutated or modified forms		
-UAS-smo.PKA.HA	*(88)*	Four PKA phosphorylation sites (667, 687, 740, and 741) mutated

(Continued)

Table 2 (*Continued*)

Gene Constructs	References	Characteristics
-UAS-smo.PKA [3,23,123].Flag	*(90)*	One, two, or three PKA sites (740, 687, and 667) mutated
-UAS-smo.CK1.HA	*(88)*	Three CK1 sites (670, 690, and 743) mutated
-UAS-smo.CK1.Flag	*(90)*	Three clusters of CK1 sites (670–677, 690–693, and 743–746) mutated
-UAS-smo.GSK3.HA	*(88)*	One GSK3 site (683) mutated
-UAS-smo.SD [1,12,123].Flag	*(90)*	One, two, or three clusters mutated, Smo active constitutively
-UAS-smoA479Y.EGFP	*(93)*	One site (sixth TM) mutated, Smo active constitutively
-UAS-smoK580Q.Flag	*(93)*	One site (C-tail) mutated, Smo active constitutively
-UAS-smo.W553L.Flag	*(93)*	One site (seventh TM) mutated, Smo active constitutively
-UAS-smo.M1.GFP	*(23)*	K580Q, Smo active constitutively, GFP tag N-terminal
-UAS-smo.M2.GFP	*(23)*	W553L, Smo active constitutively, GFP tag N-terminal
-UAS-smo.M1. KKDE.GFP	*(23)*	KKDE signal to ER delivery added to M1 construct
-UAS-smo.M2. KKDE.GFP	*(23)*	KKDE signal to ER delivery added to M2 construct
-UAS-smo.GAP43.GFP	*(23)*	Smo addressed to plasma membrane, GAP43 at C-terminus
-UAS-smo.GPI.GFP	*(23)*	Smo addressed to plasma membrane, GPI at C-terminus
fusion proteins		
-UAS-smo.CT.sev	*(91)*	Sev TM domain fused to Smo C-tail (aa 556–1035)
-UAS-fz=smo.SSF (and others)	*(92)*	All combinations between Smo and Fz domains
cos2		
wild-type forms		
-UAS-cos2	*(94,95)*	*cos2* cDNA in pUAST vector
-UAS-cos2.GFP	*(95)*	GFP tag at C-terminal end
-UAS-cos2.HA	*(94)*	HA tag at N-terminal end
-UAS-cos2.Flag	*(94)*	Flag tag at N-terminal end

(*Continued*)

Table 2 (*Continued*)

Gene Constructs	References	Characteristics
truncated forms		
-UAS-cos2.ΔC.HA	*(91)*	Deletion aa 994–1201(Smo binding domain), HA N-terminal
UAS-cos2.MB.HA	*(91)*	aa 1–389 (motor domain), HA tag N-terminal
-UAS-cos2.ΔN1.HA	*(91)*	aa 389–1201, deletion of the motor domain, HA N-terminal
-UAS-cos2.ΔN2.HA	*(91)*	aa 642–1201, keeps Smo binding domain, HA N-terminal
-UAS-cos2.CT1.HA	*(91)*	aa 906–1201, keeps Smo binding domain, HA N-terminal.
-UAS-cos2.CT2.HA	*(91)*	aa 991–1201, keeps Smo binding domain, HA N-terminal.
-UAS-cos2.CC.HA	*(91)*	aa 642–993, HA tag N-terminal
-UAS-cos2.ΔNeck.GFP	*(95)*	Deletion of the neck domain, GFP tag C-terminal
-UAS-cos2.ΔMotor.GFP	*(95)*	Deletion aa 1–313, GFP tag C-terminal
-UAS-cos2.ΔC.GFP	*(95)*	Deletion aa 1058–1201, GFP tag C-terminal
mutated forms		
-UAS-cos2.S182N	*(95)*	Mutant in the P-loop (motor domain), Dom.Neg. effects
-UAS-cos2.S182N.GFP	*(95)*	Same as above, GFP tag C-terminal
-UAS-cos2.S182T	*(95)*	Mutant in the P-loop (motor domain), behaves as *cos2*[+]
-UAS-cos2.S182T.GFP	*(95)*	Same as above, GFP tag C-terminal
fu		
wild-type forms		
-UAS-fu	*(20)*	*fu* cDNA in pUAST vector
-UAS-fu.GFP		GFP tag at N-terminal end (S. Claret and A. Plessis, pers. com.)
-UAS-fu.RFP		RFP tag at N-terminal end (S. Claret and A. Plessis, pers. com.)
-UASp-fu		*fu* cDNA in pUASP vector (F. Besse and A.M. Pret, pers. com.)

(*Continued*)

Table 2 (*Continued*)

Gene Constructs	References	Characteristics
mutated or modified forms		
UAS-fu.GAP.CFP		Fu addressed to plasma membrane, GAP, CFP at N-terminal (S. Claret and A. Plessis, pers. com.)
Su(fu)		
wild-type forms		
-UAS-Su(fu)	*(21)*	*Su(fu)* cDNA in pUAST vector
-UAS-Su(fu).HA		3× HA tag at C-terminus (F. Dussillol and D. Busson, pers. com.)
truncated forms		
-UAS-Su(fu).ΔPEST.HA		PEST region deleted (aa 309–326), HA tag C-terminal
-UAS-Su(fu). Nterm.HA		PEST and C-term regions deleted (aa 309–484), HA tag (F. Dussillol and D. Busson, pers. com.)
ci		
wild-type forms		
-UAS-ci	*(96–98)*	*ci* cDNA in pUAST vector
-UAS-ci.His	*(99)*	6× His tag at the N-terminus
-UAS-ci.HA	*(43,98,100,101)*	HA tag at the N-terminus
-UAS-ci.TBP	*(98)*	TBP tag at the N-terminus
-UAS-ci.GFP	*(99)*	GFP tag at the N-terminus
truncated forms		
-UAS-ci76	*(99)*	N-terminal half (aa 1–703), corresponds to repressor form
-UAS-ci.N/Zn.HA	*(97)*	N-terminal. half (aa 1–684), includes DBD, HA tag at N-terminal.
-UAS-ci.N/Z	*(102)*	N-terminal half (aa 1–616), includes DBD and NLS
-UAS-ci.N/ZΔNLS	*(102)*	N-terminal half (aa 1–609), includes DBD, no NLS
-UAS-ci.Δ3′	*(96)*	N-terminal (aa 1–969), deletion activator domain (aa 970–1235)
-UAS-ci.Δ5′	*(96)*	Deletion N-terminal (aa 1–313)
-UAS-ci.Zn/C	*(97)*	Deletion N-terminal. (aa 1–440), extends from DBD to C-terminus

(Continued)

Table 2 (Continued)

Gene Constructs	References	Characteristics
-UAS-ci.ΔN1	(94)	Deletion N-terminal (aa 1–345), includes DBD to C-terminus
-UAS-ci.ΔN2	(94)	Deletion N-terminal (aa 1–439), includes DBD to C-terminus
-UAS-ci.U.HA	(22)	Deletion cleavage domain (aa 611–760), Ci uncleavable
-UAS-ci.ΔC1	(102)	Deletion cytosolic retaining domain (aa 685–836)
-UAS-ci.ΔNC1	(94)	Deletion N-terminal (aa 1–349) and C-terminal (aa 1161–1397)
-UAS-ci.cyt.GFP	(99)	Cytosolic-retaining domain (aa 675–860), GFP tag
-UAS-ci.C.GFP	(99)	C-terminal end (aa 1066–1396), contains CBP -binding domain
-UAS-ci.Zn	(97)	Ci DNA-binding domain, Myc tag C-terminal
mutated forms		
-UAS-ci.Ce2.HA	(22)	Equal to ci^{cell2} mutant (C-terminal truncation from aa 975)
-UAS-ci.PKA1 or 4	(103)	One or four PKA phosphorylation sites mutated
-UAS-ci.3m (5 m, 7 m)	(98)	Three, five, or seven PKA phosphorylation sites mutated
-UAS-ci.3P.HA	(101)	Three PKA phosphorylation sites mutated
-UAS-ci.(m1-4).HA	(100)	Four PKA phosphorylation sites mutated
-UAS-ci.m1-3.HA	(104)	Three PKA phosphorylation sites mutated
-UAS-ci.U.3P.HA	(101)	Three PKA phosphorylation sites mutated on Ci uncleavable form
-UAS-ci.m1 (m2, m3)	(105)	One, two, or three GSK3 phosphorylation sites mutated
-UAS-ci.Nm.HA	(106)	Two GSK3 phosphorylation sites mutated
-UAS-ci.Cm.HA	(106)	Three CK1 phosphorylation sites mutated
-UAS-ci.NCm.HA	(106)	Two GSK3 + three CK1 phosphorylation sites mutated

(Continued)

Table 2 (*Continued*)

Gene Constructs	References	Characteristics
fusion proteins		
-UAS-ci.Δ3'-En	*(96)*	Ci (aa 1–970) fused to En repressor domain
-UAS-ZFci-VP16	*(96)*	Ci DBD (aa 314–609) fused to VP16 activator domain
-UAS-ci.Zn/EnRD	*(97)*	Ci DBD fused to En repressor domain, Myc tag C-terminal
-UAS-ciZn/Gal4AD	*(97)*	Ci DBD fused to GAL4 activator domain, Myc tag
pka-C1		
wild-type forms		
-UAS-pka-C1	*(107)*	*pka* cDNA in pUAST vector
-UAS-pka-C1.Flag	*(107)*	Flag tag at C-terminus
mutated forms		
-UAS-pka-C1.W224R	*(107)*	Invariant aa 224 mutated
-UAS-pka-C1.K75A	*(108)*	Invariant aa 75 mutated, gives a dominant-negative form
pka-R1		
wild-type forms		
-UASP-pka-R1	*(109)*	EST clone for RA isoform in pUASP vector
mutated forms		
-UAS-pka-R1.BDK	*(110)*	Mutated PKA-R, constitutive inhibitor of PKA-C
-UAS-pka-R1.BDK.HA	*(108)*	Mutated PKA-R, HA tag at N-terminus
-UAS-pka-R1.Δ.HA	*(108)*	Deletion of N-terminal dimerization domain
-UAS-pka-R1.GG.HA	*(108)*	Mutations R91G and R92G
ck1		
RNAi constructs		
-UAS-ck1α.RNAi	*(90)*	*ck1* genomic and cDNA (aa 153–337) in reverse orientation
-UAS-ck1α.dsRNA	*(88)*	Part of genomic DNA in reverse orientation

Table 3
Some Useful UAS-lacZ and UAS-GFP Reporter Lines

Constructs	References	Chrom. insert.	B.S.C.[*] no.		Characteristics
UAS-lacZ					
-UAS-lacZ.B	*(67)*	chr.2	1776	→	Construct in pUAST vector,
		chr.3	1777		cytoplasmic β-galactosidase
-UAS-lacZ.NZ		chr.2	3955	→	pUAST vector, SV40 NLS
		chr.3	3956		fused to lacZ, nuclear
					β-galactosidase
-UAS-lacZ. Excel		chr.2	8529	→	Exelixis donor
		chr.3	8530		
-UAS-tau. lacZ.B	*(111)*	chr.2	5829	→	tau-lacZ fusion in pUAST
		chr.3	7467		vector, microtubule targeted
					lacZ, entire cell visualized
-UAS-lacZ. btau.YES	*(14)*	chr.2	5148	→	tau-lacZ fusion in pYES,
					vector, microtubule targeted
					lacZ, entire cell visualized
-UAS-kinesin -lacZ	*(32)*		-	→	Microtubule targeted lacZ,
					excluded from nucleus,
					reveals cell shape
-UASp-lacZ	*(25)*		-	→	Construct in pUASP vector,
					expression in germinal and
					somatic cells
UAS-GFP					
-UAS-GFP. S65T		chr.2	1521	→	Construct in pUAST vector,
		chr.3	1522		vital marker
-UAS-EGFP	*(112)*	chr.X	5428	→	pUAST vector, GFP with
		chr.2	5431		two enhancing mutations
		chr.3	5430		
-UAS-2× EGFP	*(113)*	chr.X	6873	→	pUAST vector, two EGFP
		chr.2	6874		copies separated by IRES
		chr.3	6658		sequences, brighter than
					1× EGFP
-UAS-2× EYFP	*(113)*	chr.X	6661	→	Yellow fluorescent variant of
		chr.2	6659		EGFP, useful for FRET
		chr.3	6660		experiments
-UAS-GFP.nls	*(65)*	chr.2	4775	→	pUAST vector, NLS fused
		chr.3	4776		N-terminal to GFP, nuclear
					GFP (low in 2N cells)
-UAS-GFP-lac Z.nls	*(65)*	chr.2	6451	→	pUAST vector, GFP fused to
		chr.3	6452		lacZ, nuclear GFP detectable
					in 2N cells

(Continued)

Table 3 (*Continued*)

Constructs	References	Chrom. insert.	B.S.C.[*] no.		Characteristics
-UAS-tau-GFP	*(32)*		-		Microtubule targeted GFP, excluded from the nucleus, reveals cell shape
-UAS-eGFP-DLG	*(114)*		-		pUAST vector, EGFP fused to DLG-coding region, cell membranes visualized
-UAS-GFP. dsRNA.R	*(115)*	chr.2	9331	→	pUAST v., EGFP cDNA repeated head-to-head
		chr.3	9330		specific inactivation of GFP-tagged trans-genes
-UAS-GFP. S65T. DC5	*(116)*	X,FM7c	5193	→	GFP-tagged balancers bearing Kr-GAL4 and UAS-GFP
DC7		2,CyO	5194		transgenes, useful for
DC10		3,TM3	5195		recovery of homozygous mutant embryos
-UAS-GFP.Y		2,CyO	5702	→	GFP-tagged balancers bearing hsp70-Gal4 and UAS-GFP
		3,TM3	5704		transgenes
-UASp-Act 5C.T:GFP		chr.1	7309	→	Construct in pUASP vector, expression in germinal and
		chr.2	7310		somatic cells, GFP-tagged
		chr.3	7311		actin5C, allows observation of cytosqueletal events

[*]B.S.C. Bloomington Stock Center.

Kyoto: http://shigen.lab.nig.ac.jp/fly/nigfly/npListAction.do;jsessionid=DC 2B017F49A1967758FBE3BD3F9F4B60?browseOrSearch=browse

2.4.2. Gal80 Lines

Bloomington: http://flystocks.bio.indiana.edu/Browse/misc-browse/gal80.htm

2.4.3. UAS Misexpression Lines

Bloomington:
http://flystocks.bio.indiana.edu/Browse/insertions/misexpression-top.htm
Pscreen database: http://flypush.imgen.bcm.tmc.edu/pscreen/
Rorth EP lines: http://expbio.bio.u-szeged.hu/fly/modules.php?name=Other_Stocks &op=OtherStocksList&stock_gr=4

2.4.4. RNAi Lines

Bloomington: http://flystocks.bio.indiana.edu/Browse/misc-browse/RNAi.htm

Kyoto:
> http://shigen.lab.nig.ac.jp/fly/nigfly/rnaiListAction.do?browseOrSearch=browse

2.4.5. Vectors

FlyBase: http://www.flybase.bio.indiana.edu/staticpages/lists/vectors.html
DrosophilaGenomicsResourceCenter:
> http://dgrc.cgb.indiana.edu/vectors/store/vectors.html

Gateway vectors: http://www.ciwemb.edu/labs/murphy/Gateway%20vectors.html

3. Methods

3.1. The Basic GAL4/UAS System

The GAL4/UAS system as it was adapted for use in *Drosophila* by Brand and Perrimon (1993) provides a method for in vivo targeting of gene expression in a spatially controlled fashion. The yeast GAL4 transcriptional activator is placed under the control of specific regulatory elements providing stage, tissue, or cell-specific expression and, in *trans*, GAL4 activates the specific expression of a target gene placed under the control of a basal promoter associated with UAS sites optimized for GAL4 binding (**Fig. 2A**). This bipartite system is particularly advantageous because each part of the system is maintained in separate parental lines which are viable since no, or very little, activation is possible when the system is uncoupled (*see* **Note 1**). Therefore, with only one cross, GAL4 driver line crossed to a UAS-*geneX* responder line, the system is set under way immediately in the embryos produced (*see* **Note 2**).

Control of UAS-*geneX* expression depends on the spatio-temporal characteristics of the promoter used to drive *gal4* transcription. Two types of GAL4 drivers have been generated. The enhancer trap strategy has been widely used to screen for PGAL4 insertions with spatially restricted expression patterns (**Fig. 1**). When specific regulatory sequences have been characterized, it is also possible to generate transgenic constructs in which those sequences are associated with the GAL4-coding sequences (**Fig. 1**). The UAS-*geneX* responder lines also come in two varieties, in vitro constructs for the study of a specific gene, and enhancer traps for screens *(7)*. One of the major advantages of the GAL4/UAS system is thus the number of GAL4 driver and UAS responder lines available, allowing numerous combinatorial possibilities (*see* Databases in **Section 2.4.**).

3.1.1. Gain-of-Function Analysis Using the GAL4/UAS System

The GAL4/UAS system was originally designed for gain-of-function genetics to be carried out for the study of gene function. This system is particularly interesting when loss-of-function analysis provides little information due primarily to functional redundancy between genes. The type of misexpression to be induced depends on the type of question to be addressed and the corresponding choice of appropriate GAL4 drivers. Ectopic expression of a gene (ubiquitous or targeted), which provides the possibility of creating new dominant phenotypes, indicates whether a particular gene's function is sufficient in a given process. It is also possible to express a constitutively active form of the protein of interest, rendering it independent of specific activation and allowing for epistatic analysis to be carried out. Finally, determination of the cells where a gene function is specifically needed is possible by expressing the gene in a restricted population of cells and assaying for rescue of the mutant phenotype associated with that gene.

3.1.2. Loss-of-Function Analysis Using the GAL4/UAS System

The GAL4/UAS system can also be used to generate at least partial loss-of-function states by expression of dominant-negative forms of the proteins studied or of double-stranded hairpin RNAs for RNA interference (RNAi)-based inactivation of specific genes. These approaches are particularly useful when there is no available mutant allele for the gene of interest or for tissue-specific inactivation. In addition, collections of UAS-*RNAi* lines directed against the totality of predicted genes in *Drosophila* are becoming available for use in systematic screening (*see* **Section 2.4.**). However, for both approaches, the expression of dominant-negative forms or RNAi, only hypomorpic states can be attained and a strict correlation between the phenotypes observed and inactivation of a specific gene product is not easy to establish. Thus, it is always necessary to generate mutant alleles (preferably null alleles) of the gene of interest to confirm results obtained using the GAL4/UAS system.

3.1.3. Modulating Expression Levels of the GAL4/UAS System

Varying the expression levels using the GAL4/UAS system can be of interest in several ways to show dosage effects, to generate physiological or, on the contrary, overexpresssion levels, or to circumvent lethality effects. The expression level of a gene of interest using the GAL4/UAS system depends on several parameters, which can be manipulated if different expression levels are required.

- The level of expression induced by independent GAL4 drivers in the same tissue can vary significantly (*see* **Note 3**).
- The same GAL4 construct can be modified to generate weaker and stronger versions.
- The higher the number of UAS sites used in vectors for constructing UAS-*geneX* transgenes, the higher the expression levels.
- The same UAS-*geneX* construct inserted at different sites in the genome can be subject to regulation by nearby *cis*-acting sequences (position effect) (*see* **Note 4**).
- The level of expression in this system can also be increased by increasing the number of copies of both the GAL4 driver and the UAS responder transgenes.
- Finally, temperature has a significant effect on the activity of the GAL4 protein (*see* **Note 5**).

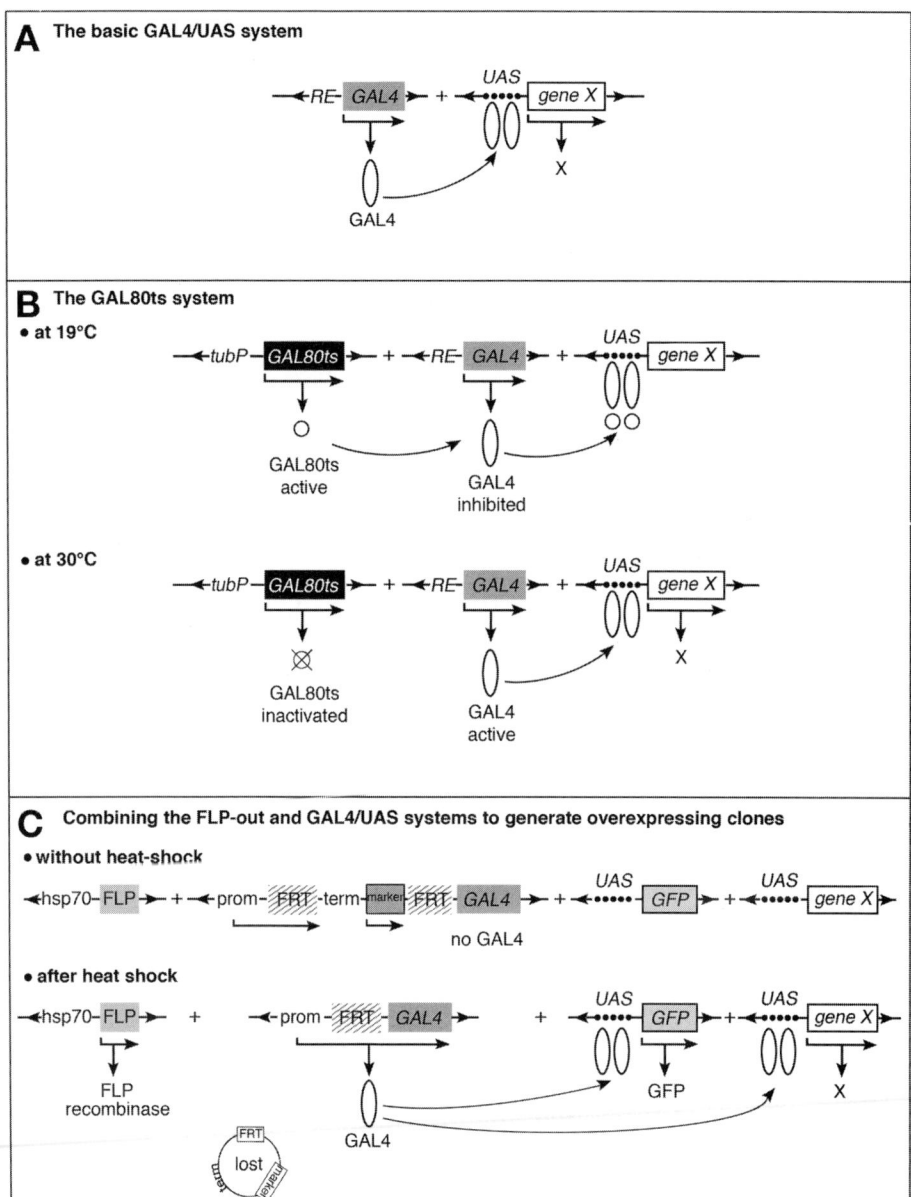

Fig. 2. Genotypes constructed and conditions necessary to make the GAL4/UAS system operational. (**A**) The basic GAL4/UAS system. The two elements are put together in the same flies. The yeast transcriptional activator *GAL4* under the control of a defined regulatory element (RE) is expressed in a specific spatial pattern. Transcription of *gene X* cloned downstream of the UAS sites is activated on binding of the GAL4 protein to

3.1.4. Tissue-Specificity of the GAL4/UAS System

Although an indication of specific expression patterns of GAL4 driver lines can be obtained from Flybase and the literature, it is important to note that most of the time not all stages of development and tissues have been examined. Indeed, in many cases, the expression of a given GAL4 driver is multi-stage and multi-tissue (**Table 1**). In all cases, it is important to verify the expression pattern of a GAL4 driver using one of the UAS-reporter gene transgenic constructs available (*see* **Note 6** and **Table 3**). Also, many GAL4 drivers are expressed during embryogenesis and therefore lethal effects may hinder observation of later stages. In some cases, temperature shifts can be used to circumvent this problem (*see* **Note 5**). If, on the other hand, the expression is specifically required during early embryogenesis, it is important to note that for the original GAL4 constructs, earliest expression is observed only 3—4 h after egg laying *(1)*. More recently, modified versions of the GAL4 transgenic constructs have been made and these have been associated with maternal promoters allowing somewhat earlier embryonic expression (*see* **Note 7**). Although these GAL4 drivers thus provide germline expression of GAL4 protein, it is not possible to activate UAS-*geneX* constructs in the germline during oogenesis with the original PUAST vectors (**Fig. 1**). This is the only tissue for which this problem is encountered and it presents a major drawback to the system. Rorth (1998) generated an alternative UAS responder construct (PUASP) which allows both somatic and germline expressions (*see* **Note 8** and **Fig. 1**). However, most available UAS-*geneX* lines do not use this vector and the systematic cloning system Gateway (*see* **Section 2.4.**) has not been generated with PUASP either.

---

the UAS sequences; *gene X* has the same spatio-temporal expression as the *GAL4* gene. (**B**) The GAL80ts system (TARGET). The three transgenes are carried in the same flies. At permissive temperature (19°C), the GAL80 protein is active, binds and inhibits the GAL4 protein; *gene X* is not transcribed. At the nonpermissive temperature (30°C), GAL80 is inactivated, GAL4 is active and activates *gene X* transcription. Returning to the permissive temperature brings back the noninduced state. (**C**) Genotype constructed for combining the FLP-out method with the GAL4/UAS system. In the *prom*-FRT-*marker*-FRT-*GAL4* transgene, transcriptional terminators are present between the two FRT sequences (here upstream of the marker). Without heat shock, the *GAL4* gene is not transcribed and the UAS transgenes are not expressed. Heat shock at 37°C at specific developmental stages leads to production of FLP recombinase which induces mitotic recombination between the two FRT sequences; the progeny of the cell in which mitotic recombination takes place form a clone of cells in which the *GAL4* gene is transcribed; they express the *gene X* of interest; in the example presented, these cells are recognized both negatively as they do not express the marker and positively as they express the vital GFP marker *(4–6)*.

Finally, with the original Brand and Perrimon GAL4 enhancer trap construct (**Fig. 1**) and its close derivatives *(8)*, salivary gland expression is almost always observed indicating the fortuitous presence of a salivary gland enhancer in the construct (*see* **Note 9**).

3.1.5. Undesired Phenotypic Effects of the GAL4/UAS System

Several components of the GAL4/UAS system can produce phenotypes independently of the misexpression of the gene of interest, thereby interfering with proper interpretation of the experimental results. For example, the PGAL4 and PUAS transposons (enhancer traps) may disrupt genes at the site of insertion, thereby creating new alleles that can be associated with phenotypes including, in some cases, lethality of homozygotes (*see* **Note 10**). In the case of the Hh pathway, several useful GAL4 drivers have been obtained within genes of the pathway itself allowing for very specific expression patterns at the level of tissue compartments and compartment boundaries (both in the embryo and in imaginal discs) (*see* **Note 11**). However, special care must be taken in the use of these GAL4 drivers for the study of the Hh pathway since even in the heterozygous state genetic interactions are possible. It is, therefore, imperative to obtain similar results with several drivers including drivers independent of the Hh pathway.

There are also other potential sources of extraneous phenotypes. For example, the expression of the GAL4 protein itself may give specific phenotypes (*see* **Note 12**). Also, in some cases, the UAS-*geneX* transgenes may have leaky expression, in which case expression, though it may be low, becomes ubiquitous and independent of the presence of GAL4. Expression of tagged versions of proteins can also induce phenotypes due to the presence of the tag (*see* **Note 13**). For all these reasons, it is important that the GAL4 driver and UAS responder lines be tested independently. These tests should be carried out in parallel with those on individuals carrying both the GAL4 driver and the UAS responder, since environmental conditions, such as temperature, crowding, batch of fly food, humidity, and other parameters can influence many aspects of growth and development. Finally, depending on the phenotypes assayed, it is important to remember that accumulation of modifiers in the genetic background of either GAL4 driver or UAS responder lines can occur and therefore periodic outcrossing is sometimes necessary.

3.2. Extensions and Refinements of the GAL4/UAS System

The basic GAL4/UAS system by itself allows neither precise temporal control nor reversible expression of the UAS transgene. Several modifications to render it inducible are presented below.

3.2.1. The TARGET System (Temporal and Regional Gene Expression Targeting)

The TARGET system was developed to allow inducible, temporally controlled UAS transgene expression. This system is based on the use of a temperature-sensitive GAL80 protein *(9)*. In yeast, in absence of galactose the GAL80 protein inhibits GAL4 activity. If galactose is present, this inhibition is relieved and GAL4 is able to activate the transcription of its target genes. A GAL80ts version of the protein was made and fly lines carrying the *gal80ts* gene under the control of the ubiquitous *tubulin-1α* promoter were constructed. Use of this system requires flies bearing the combination of GAL80ts, GAL4, and UAS transgenes (*see* **Fig. 2B**). At the permissive temperature (19°C), the GAL80 protein is active and the UAS trangene is not expressed. In flies treated with heat-shock exposure at 30°C for several hours (6 h), the GAL80ts protein is inhibited thus allowing GAL4 activity and targeted expression of the UAS transgene. Shifting experiments from 19°C to 30°C and vice versa allows determination of the period necessary for UAS-*gene X* to rescue the corresponding mutant phenotype. It has been shown that this system is active at all stages of development without deleterious effects of the GAL80 protein. This inducible system is especially useful if early dominant effects of mis- or overexpression lead to lethality and thus preclude analysis at later stages or in adult animals. One limitation is the slow kinetics, with induction taking several hours (6–24 h) and the return to uninduced levels even longer.

The *tubP*-GAL80ts stocks available at the Bloomington Stock Center are given in **Section 2.4**.

3.2.2. The GeneSwitch System

This system is also inducible. It is based on the construction of a fusion protein between the GAL4 activator and the Progesterone Receptor (GAL4-PR). In absence of hormone (RU486 = mifepristone, Sigma), the GAL4 protein is inactive, whereas upon feeding (or bathing) larvae with hormone, the GAL4 protein is activated *(10,11)*. The levels of activation can be controlled by the dose of RU486 given in the food, with high drug dose at 20 µg/ml and low drug dose at 1 µg/ml *(11)*.

3.2.3. The GAL4/UAS and FLP/FRT Connection

A method combining the GAL4/UAS and FLP/FRT systems has been developed which allows the induction of cellular clones with targeted UAS-*geneX* expression *(12)*. In this system, the GAL4 driver is silent since a transcription termination signal is present between the promoter and the GAL4-coding sequence. However, upon heat-shock-induced expression of a *hs-flp* transgene, the transcriptional terminator can be removed by recombination in *cis* between the two FRT sequences flanking the terminator (**Fig. 2C**).

This system presents two major strengths. Firstly, temporal control of UAS-*geneX* expression can be achieved. This is useful for staging the effect of the expression of *geneX*. Also, the possibility for temporal control is an advantage when expression of *geneX* may be associated with deleterious effects on growth and/or development (*see* **Note 14**). Secondly, this system is ideal for mosaic analysis purposes such as defining the autonomous vs nonautonomous phenotype of a given gene or to demonstrate the gradient effect of a signaling molecule (cf. *hh*⁺ clones and expression of *en*, *ptc*, and *dpp* *[13]*). However, the limitation to this system is the fact that it is not reversible, therefore, control of expression by the experimenter is lost after induction. Examples of the use of this system to study the Hh signaling pathway is given in **Section 3.3.**

3.2.4. The MARCM and Positively Marked Mosaic Lineage (PMML) Methods

The MARCM (Mosaic Analysis with a Repressible Cell Marker) system is designed to create clones of positively marked homozygous mutant cells *(14)*. It combines the properties of the GAL80 protein (which inhibits GAL4 activity), and the GAL4/UAS and Flp/FRT systems. In flies of the following genotype: *hs-flp/+; FRT, P-tubP-GAL80, m⁺/ FRT, m⁻; P-RE-GAL4/UAS-GFP* (or other cell marker or gene), GAL80 inhibits GAL4, thus preventing the expression of the UAS-cell marker or UAS-gene. Following heat shock, homozygous *m⁻/m⁻* clones are produced which lack the P-GAL80 transgene and thus express the UAS-cell marker or UAS-gene. With this system, the lineage of a single cell can be visualized and it is possible to determine the temporal function of a gene. The limitation relies on the perdurance of the GAL80 protein (*see* **Note 15**). The PMML system is a labeling technique for lineage tracing and lineage-specific gene over expression *(15)*. It uses the FLP recombinase to reconstruct a functional *actin5C-GAL4* gene from two complementary inactive alleles located on homologous chromosomes, *actin5C*-FRT52B, and FRT52B-*GAL4*. Flies of *hs-flp/+;UAS-GFP/+;actin5C-FRT52B/ FRT52B-GAL4* genotype do not express GFP. Following heat shock, the cell affected by a mitotic recombination event produces one daughter cell with an active *actin5C-GAL4* gene allowing GFP expression in this cell and its progeny. This method can also activate or knockdown gene function by using convenient UAS constructs. *See* **Section 2.4.** for available GAL80 fly lines.

3.3. Some Uses of the GAL4/UAS System for the Study of the Hh Pathway

In this section, we wish to show how the use of the GAL4/UAS system has helped to elucidate specific questions in the study of the Hh-signaling pathway in *Drosophila*. Examples presented concern (i) the role of two effectors, the

Fused (Fu) serine–threonine kinase and the SUPPRESSOR OF FUSED (Su[fu]) protein, known as activator and inhibitor of Hh signal transduction, respectively, (ii) the role of different domains of the transcription factor Cubitus interruptus (Ci) in regulating Hh target gene expression, and (iii) the relationship between the subcellular localization of the transmembrane protein Smoothened (Smo) and activation of the pathway (reviewed in Refs. *[16–18]*).

3.3.1. Targeted Expression to Determine Where the Function of a Gene is Required

The *fused* (*fu*) gene is transcribed ubiquitously in all tissues (ovary, embryo, and imaginal discs); however, the Fu protein shows some specific accumulation in cells receiving the Hh signal, both in the embryo *(19)* and in the imaginal discs *(20)*. To identify precisely the cells in which Fu is required, the GAL4/UAS system is used to express Fu in restricted regions, looking for the rescue of the *fu* mutant phenotype.

In the embryo, two different GAL4 drivers are used: the *wg-GAL4* and *en-GAL4* drivers (*see* **Table 1**). Rescue is assayed by looking at the cuticular phenotype and at the expression of two Hh targets, *wg* and *ptc* (*see* **Fig. 3**). Embryos of *fu*^{A}, *UAS-fu$^+$/fu^A; wg-Gal4/+* genotype are fully rescued both for cuticular phenotype (**Fig. 3A–C**) and for *wg* (**Fig. 3D–F**) and *ptc* (**Fig. 3G–I**) expression, whereas embryos of *fu*^{A}, *UAS-fu$^+$/fu^A; en-GAL4/+* genotype are not rescued. These results show that Fu is only necessary in cells expressing *wg*. Late *ptc* expression in two narrow stripes flanking the En/Hh-expressing cells depends on Hh and is broadened if *hh* is overexpressed as seen in *fu$^+$; en-GAL4/+; UAS-hh$^+$/+* embryos (**Fig. 3J**); in *fu^A; en-GAL4/+; UAS-hh$^+$/+* embryos, the *ptc* anterior stripe decays while the posterior *ptc* stripe is maintained (**Fig. 3K**). This shows that Hh can signal independently of Fu in cells which express *ptc* posteriorly to the *en/hh* domain.

A similar study was performed in the wing imaginal disc, with the *ptc-GAL4* and *dpp-GAL4* drivers (*see* **Table 1**) which showed, respectively, total and partial rescue of the *fu* wing phenotype. This result suggests that Fu activity is necessary in cells expressing high levels of *ptc*, which are also the cells that receive the highest levels of Hh.

3.3.2. Overexpression for Establishing Epistatic Relationships Between Members of the Pathway

Combining dominant phenotypes created by the use of the GAL4/UAS system for a particular gene with phenotypes associated with the loss-of-function of other genes, has allowed the hierarchical relationships between members of the same pathway to be determined. For instance, *hh* overexpression in the posterior compartment of wing imaginal discs produces a dominant ectopic wing phenotype

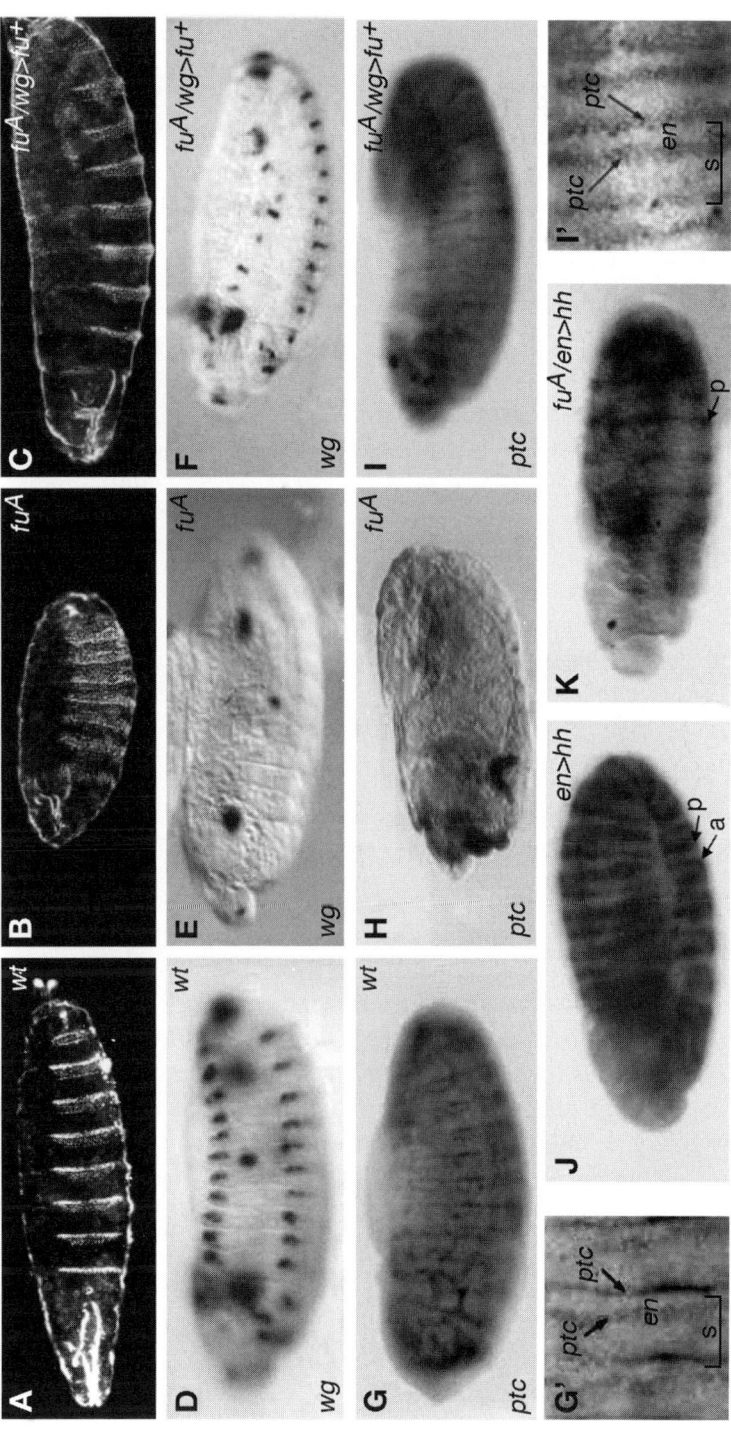

Fig. 3. Fu activity is required in *wg*-expressing cells in the embryo for ventral patterning, *wg* and late *ptc* expression in cells anterior to *en/hh* expressing cells; it is not required for *ptc* expression in cells posterior to *en/hh* expressing cells. (**A–C**) Cuticle patterns of wild-type embryos (**A**), *fu^A* embryos derived from *fu^−* mothers (**B**), *fu^A*, *UAS-fu/+*; *wg-GAL4/+* embryos derived from homozygous *fu^A*, *UAS-fu* females mated with *fu^A*; *wg-GAL4/+* males at 25°C (**C**). (**D–F**) *wg* expression in stage 13 embryos, wild-type (**D**), *fu^A* (**E**) and *fu^A*, *UAS-fu*; *wg-GAL4/+* embryos (**F**). (**G–I**) *ptc* expression in stage 13 wild-type embryos (**G**, enlargement in **G′**), stage 12

189

which is completely suppressed by *fu* mutations in *fu¹; UAS-hh⁺/+; en-Gal4/+* flies. On the other hand, a partial rescue of the *fu* wing phenotype is obtained by *ci* overexpression in *fu¹; UAS-ci⁺/+; ptc-Gal4/+* flies. These results place Fu downstream of Hh and upstream of Ci in a pathway in which Hh signal transduction via Fu leads to the activation of Ci *(20)*.

3.3.3. Overexpression to Reveal New Phenotypes

The Su(fu) protein is known as a negative effector of the Hh-signaling pathway. It behaves as a Fu antagonist as *Su(fu)⁻* mutations fully suppress all the effects of *fu* mutations. Nevertheless, amorphic *Su(fu)* mutations do not present an obvious mutant phenotype. In order to get more information on its function, *Su(fu)* was overexpressed in *fu⁺, UAS-Su(fu)/+; da-Gal4/+* animals *(21)*. In this context, anterior wing duplications and ectopic *dpp* and *ptc* expression in the anterior compartment of wing imaginal discs, indicative of overactivation of Hh signaling are observed. This suggests an unexpected activator role of Su(fu) in those anterior cells which normally do not receive the Hh signal. In agreement with Fu and Su(fu) being antagonists, the effects of *Su(fu)* overexpression are stronger in a mutant *fu* context (*fuᴬ (or fu¹), UAS-Su(fu)/+; da-Gal4/+* flies). This, in turn, suggests an unexpected negative role of Fu in anterior wing disc cells.

3.3.4. Overexpressing Clones for Distinguishing Autonomous vs Non Autonomous Effects

To test whether the ectopic anterior effects due to overexpression of *Su(fu)* in the wing imaginal disc (*see* **Section 3.3.3.**) were independent from its overexpression at the anterior–posterior compartment boundary where Hh-signaling occurs, *Su(fu)* overexpression clones were induced in female larvae of the genotype, *hs-flp/w, UAS-Sufu; dpp-lacZ/+; act5c>CD2>GAL4, UAS-GFP/+* (*see* **Fig. 2C**) *(21)*. Ectopic *dpp* (but no *ptc*) expression was observed in anterior clones located far from the anterior–posterior boundary. This result shows that Hh signal transduction is autonomously activated in anterior clones that are

---

fuᴬ embryos (**H**), and stage 12 *fuᴬ, UAS-fu; wg-GAL4/+* embryos (**I**, enlargement in **I′**). Note the rescue of the *fu* mutant phenotype in **C, F, I**, when compared with **B, E, H**, respectively. In **G′** and **I′**, S represents the limits of a segment. (**J,K**) Increasing the Hh signal in stage 11 *en-GAL4/+; UAS-hh/+* embryos (**J**) results in a broader *ptc* expression in cells posterior (p) to En/Hh expressing cells when compared with *ptc* expression in cells anterior (a) to En/Hh-expressing cells; in *fuᴬ; en-GAL4/+; UAS-hh/+* embryos (**K**) at the same stage, the anterior *ptc* expression is completely absent but the posterior *ptc* expression is present showing that the former depends on Fu activity but the latter does not. All embryos with anterior to the left and dorsal up. From Ref. *(19)*.

overexpressing *Su(fu)*. However, as *dpp*, but not *ptc*, is expressed, Hh pathway activation is not at its highest.

3.3.5. Targeted Expression of Truncated and Mutated Proteins

Many GAL4/UAS experiments are performed with UAS constructs bearing truncated or mutated versions of the proteins of interest (*see* **Table 2**). The aim of these experiments is to evaluate, in vivo, the role of functional domains or post-translational modifications, otherwise identified from molecular analysis or from sequence data. In *Drosophila*, the Gli-family transcription factor Ci is thought to regulate most (if not all) Hh targets. It is found as two major iso-forms, a full-length 155 kDa form (Ci155) and a C-terminally truncated 75 kDa form (Ci75), which play, respectively, activator and repressor roles on different Hh target genes. In their study, Méthot and Basler *(22)* took advantage of the GAL4/UAS method to set up an in vivo assay system in the wing imaginal disc posterior (P) compartment. They used the C765-GAL4 to drive moderate levels of UAS-*ci* construct expression in the entire wing pouch (*see* **Table 1**). As P–compartment cells do not normally express *ci*, the effects of ectopic UAS-*ci* construct expression in this compartment could be analyzed without any ambiguity. In particular, since P–compartment cells express *hh*, inducing *smo⁻* clones in this compartment allows the effects of UAS-*ci* to be followed under two conditions, cells responding (*smo⁺*) vs those not responding (*smo⁻*) to Hh. The reporter strain *hh-lacZ* was used to assay for the repressor effect of Ci, and *ptc-lacZ* for the activator effect of Ci. Finally, the effects of modified forms of *ci*, in particular, UAS-*ci^{cell}* and UAS-*ci^U* which encode, respectively, truncated Ci and uncleavable Ci (*see* **Table 2**), were also tested. The following results were obtained:

(i) in heat-shock treated *y, w, hs-flp; smo³, FRT39/hs-CD2, FRT39; UAS-ci/C765-GAL4, hh-lacZ* larvae, P-compartment *smo⁺* wing disc cells express *hh-lacZ*, whereas *smo⁻* cells do not. By contrast, heat-shock treated *y, w, hs-flp; act5c>CD2>GAL4/UAS-ci^{cell}; hh-lacZ/+* larvae, *smo⁺* cells that express *ci^{cell}* do not express *hh-lacZ*. These results indicate that Ci can function as a potent repressor of *hh* in its cleaved form and that the presence of Hh blocks repressor function. Hh inhibits the cleavage of Ci to its repressor form, but Ci^{cell} (aa 1–975), which is truncated and lacks the C-terminal activation domain, behaves as a constitutive repressor.

(ii) in heat-shock treated *y, w, hs-flp; smo³, FRT40, ptc-lacZ/hs-GFP, FRT40; UAS-ci/C765-GAL4* larvae, P-compartment *smo⁺* wing disc cells ectopically express *ptc-lacZ*, whereas *smo⁻* cells do not. This result indicates that Ci is unable to induce *ptc-lacZ* expression in the absence of Hh signal transduction. Thus, the activator activity of Ci is not constitutive and depends on transduction of the Hh signal.

(iii) in heat-shock treated *y, w, hs-flp; smo³, FRT39/hs-CD2, FRT39; UAS-ci^U/C765-GAL4, hh-lacZ* larvae, both *smo⁺* and *smo⁻* P-compartment cells express *hh-lacZ*. However, in heat-shock treated *y, w, hs-flp; smo³, FRT40, ptc-lacZ/hs-GFP, FRT40;*

UAS-ci^U/C765-GAL4 larvae, P-compartment *smo$^+$* cells express *ptc-lacZ* whereas *smo$^-$* cells do not. These results indicate that Ci^U is unable to provide repressor function even in the absence of the Hh signal; thus, the cleavage of Ci^{155} to Ci^{75} is a necessary step in the formation of Ci repressor. On the other hand, Ci^U can function as an activator but only in Hh-receiving cells.

3.3.6. Targeted Expression of Tagged Proteins to Follow Their Subcellular Localization and Trafficking

In their study, Zhu et al. *(23)* generated GFP-tagged versions of Smo fused to protein domains that address Smo to specific subcellular compartments (*see* **Table 2**). The UAS-*smo* constructs were expressed in salivary gland cells using the 71B-GAL4 driver (*see* **Table 1**) and the subcellular localization of GFP-tagged Smo in the presence and absence of Hh followed by confocal microscopy. Activation of the pathway was assayed by following the expression of a *ptc-lacZ* reporter. The authors showed that:

(i) in *UAS-smo.GFP/+;71B-GAL4/+* salivary gland cells in culture, without exogenously-added Hh, Smo is present at a low level in a network of punctate cytoplasmic structures. Upon addition of Hh, Smo undergoes an obvious shift to the cell surface; this was accompanied by a 20-fold increase in the activity of the *ptc-lacZ* reporter. Thus, Hh induces a relocalization of Smo which correlates with activation of the pathway.

(ii) in *UAS-smo.GFP;UAS-hhN/UAS-ptc;71B-GAL4/+* salivary gland cells, over-production of Ptc blocks most of the movement of Smo to the cell surface despite the presence of Hh. These results suggest that Ptc is necessary to confine Smo to internal cellular locations.

(iii) in *UAS-smo.M1 (or M2).GFP/+;71B-GAL4/+* salivary gland cells, the constitutively active mutant forms of Smo (M1 and M2) accumulate at the cell surface, even in the absence of Hh, and *ptc-lacZ* reporter activity is 7–13-fold higher than when unmodified Smo is expressed. In *UAS-smo.GAP43 (or GPI).GFP/+;71B-GAL4/+* cells, Smo is addressed to the cell surface which also correlates with significantly increased levels of *ptc-lacZ* activity. In contrast, when Smo is retained in the endoplasmic reticulum with a KKDE signal in *UAS-smo.M1 (or M2).KKDE. GFP/+;71B-GAL4/+* cells, Smo is not present at the cell surface and *ptc-lacZ* activity is down to control levels.

Thus, the control of Smo subcellular localization appears to be a crucial step in Smo activation and Hh signal transduction in *Drosophila*.

4. Notes

1. The bipartite nature of the system presents several advantages. For cell ablation studies, lines carrying UAS transgenes of genes encoding toxic products can be maintained and only activated once crossed to a GAL4 driver line. Also, the

function of genes whose overexpression is associated with lethality at early stages of development can be studied during later stages using appropriate GAL4 drivers.

2. A particular orientation to the cross is needed only when GAL4 maternal contribution is desired for earliest embryonic expression. In this case, females from the GAL4 driver line must be crossed to males from the UAS-*geneX* responder line.

3. Although the relative expression levels between GAL4 drivers can be roughly estimated using UAS-*lacZ* responder constructs, more accurate quantification has recently been reported using UAS-*geneX: GFP* fusion responder constructs and laser-scanning confocal microscopy *(24)*.

4. It is in general advisable when generating UAS-*geneX* transgenic lines to isolate several independent insertions *(5,6)* and to test their relative levels of expression. These independent lines can be maintained for use as a phenotypic series.

5. The GAL4 protein presents basal activity at 18°C, significantly higher activity at 25°C and even greater activity at 29°C. It is not advisable to use temperatures lower than 18°C or higher than 29°C, as problems with viability and male fertility arise.

6. The most commonly used UAS-reporter constructs to determine or confirm tissue-specific expression of a particular GAL4 driver are UAS-*lacZ* and UAS-*GFP* which allow nuclear expression of β-galactosidase and GFP, respectively (**Table 3**). However, depending on the cell types to be identified, it can be useful to use modified versions of these UAS-reporter constructs in which β-galactosidase and GFP-coding sequences are fused to other protein-coding sequences allowing the fusion proteins to be addressed to specific subcellular compartments (e.g., the plasma membrane, microtubules, axons, apical and basal compartments; *see* **Table 3** for examples).

7. To generate GAL4 drivers allowing early embryonic expression, several modifications have been made to these constructs (*see* **Table 1**, Maternal GAL4 drivers). The GAL4 activation domain has been replaced by the HERPES Virus VP16 activation domain and the *hsp70* 5′UTR and 3′UTR (which can lead to mRNA degradation in the absence of heat shock) have been replaced with corresponding sequences from other transcripts such as *nanos*. In addition, the promoters associated with these modified forms of the GAL4-transcribed sequences are strongly expressed in the germline (*arm, otu, nos, alpha-tub [25–27]*) to provide a maternal contribution for optimum expression in the early embryo. Indeed with these drivers, *gal4* mRNA can be detected by cellular blastoderm.

8. The modifications made to generate the UASP construct to allow germline expression include an increased number of UAS sites (X14), GAGA sites, the P-transposase promoter and first intron, and the *K10* 3′UTR (**Fig. 1**). The UASP-*geneX* construct can be efficiently expressed in the germline when drivers carrying GAL4:VP16 fusions and non-*hsp70* 3′UTR are used (*see* **Note 7**). It is important to note that the UASP-*geneX* constructs can also be expressed in somatic cells and that the expression tends to be more leaky than with the UAST version. Indeed, in our laboratory, two copies of a UASP-*fused* transgene rescues *fused* embryonic mutant phenotypes without the presence of a GAL4 driver (F. Besse and A.M. Pret, unpublished results).

9. Gerlitz et al. (2002) *(28)* report that PGAWB and PGALW possess a salivary gland enhancer present in the *hsp70*-derived 5′UTR sequences, but unfortunately variants of PGALW without this sequence do not function as well as enhancer traps.

10. The GAL4 drivers obtained via enhancer trap screens generate, in some cases, new alleles (sometimes lethal) of the genes at the sites of insertion. For example, the MS1096 GAL4 driver for wing expression is an allele of the *Beadex* gene and is associated with a veination phenotype *(29)*.

11. Several useful drivers have been obtained which are PGAL4 insertions into genes implicated in Hh signaling (*ptc-gal4*, *dpp-gal4*, *en-gal4*, *hh-gal4*) (*see* **Table 1**).

12. The expression of GAL4 in the developing eye under the control of the glass multiple reporter (GMR) promoter element induces eye phenotypes that can be attributed to accumulation of the GAL4 protein *(30,31)*. In particular, homozygotes have a highly disorganized ommatidial array and high levels of apoptosis in eye imaginal discs of third instar larva.

13. UAS-*khc-lacZ* expressed in neuronal cells and UAS-*tau-lacZ* expressed in imaginal disc cells can lead to embryonic and pupal lethality, respectively *(32)*.

14. The *hsp70*-GAL4 driver also affords temporal control, but the *hsp70* promoter presents basal expression without heat shock. Therefore, the use of the Flp-out method increases the precision of the temporal control.

15. A study of the perdurance of the GAL80 protein in wing imaginal discs shows total perdurance 24 h after heat shock, partial perdurance 36 h after heat shock, and almost no perdurance 48 h after heat shock *(14)*.

Acknowledgments

The authors are grateful to Myriam Barre for helping to prepare **Figs. 1–3**. **Fig. 3** was adapted from Ref. *(19)* and reproduced with permission. This work was supported by grants from the Centre National de la Recherche Scientifique.

References

1. Brand, A. H. and Perrimon, N. (1993) Targeted gene expression as a means of altering cell fates and generating dominant phenotypes. *Development* **118**, 401–415.

2. Giniger, E., Varnum, S. M., and Ptashne, M. (1985) Specific DNA binding of GAL4, a positive regulatory protein of yeast. *Cell* **40**, 767–774.

3. Ptashne, M. (1988) How eukaryotic transcriptional activators work. *Nature* **335**, 683–689.

4. Duffy, J. B. (2002) GAL4 system in Drosophila: a fly geneticist's Swiss army knife. *Genesis* **34**, 1–15.

5. Blair, S. S. (2003) Genetic mosaic techniques for studying Drosophila development. *Development* **130**, 5065–5072.

6. McGuire, S. E., Roman, G., and Davis, R. L. (2004) Gene expression systems in Drosophila: a synthesis of time and space. *Trends Genet.* **20**, 384–391.

7. Rorth, P. (1996) A modular misexpression screen in Drosophila detecting tissue-specific phenotypes. *Proc. Natl Acad. Sci. USA* **93**, 12,418–12,422.

8. Brand, A. H., Manoukian, A. S., and Perrimon, N. (1994) Ectopic expression in Drosophila. *Methods Cell. Biol.* **44,** 635–654.

9. McGuire, S. E., Le, P. T., Osborn, A. J., Matsumoto, K., and Davis, R. L. (2003) Spatiotemporal rescue of memory dysfunction in Drosophila. *Science* **302,** 1765–1768.

10. Osterwalder, T., Yoon, K. S., White, B. H., and Keshishian, H. (2001) A conditional tissue-specific transgene expression system using inducible GAL4. *Proc. Natl Acad. Sci. U S A* **98,** 12,596–12,601.

11. Rogulja, D. and Irvine, K. D. (2005) Regulation of cell proliferation by a morphogen gradient. *Cell* **123,** 449–461.

12. Struhl, G. and Basler, K. (1993) Organizing activity of wingless protein in Drosophila. *Cell* **72,** 527–540.

13. Strigini, M. and Cohen, S. M. (1997) A Hedgehog activity gradient contributes to AP axial patterning of the Drosophila wing. *Development* **124,** 4697–4705.

14. Lee, T. and Luo, L. (1999) Mosaic analysis with a repressible cell marker for studies of gene function in neuronal morphogenesis. *Neuron* **22,** 451–461.

15. Kirilly, D., Spana, E. P., Perrimon, N., Padgett, R. W., and Xie, T. (2005) BMP signaling is required for controlling somatic stem cell self-renewal in the Drosophila ovary. *Dev. Cell.* **9,** 651–662.

16. Lum, L. and Beachy, P. A. (2004) The Hedgehog response network: sensors, switches, and routers. *Science* **304,** 1755–1759.

17. Ogden, S. K., Ascano, M. Jr., Stegman, M. A., and Robbins, D. J. (2004) Regulation of Hedgehog signaling: a complex story. *Biochem. Pharmacol.* **67,** 805–814.

18. Hooper, J. E. and Scott, M. P. (2005) Communicating with Hedgehogs. *Nat. Rev. Mol. Cell Biol.* **6,** 306–317.

19. Therond, P. P., Limbourg Bouchon, B., Gallet, A., et al. (1999) Differential requirements of the fused kinase for hedgehog signaling in the Drosophila embryo. *Development* **126,** 4039–4051.

20. Alves, G., Limbourg-Bouchon, B., Tricoire, H., Brissard-Zahraoui, J., Lamour-Isnard, C., and Busson, D. (1998) Modulation of Hedgehog target gene expression by the Fused serine-threonine kinase in wing imaginal discs. *Mech. Dev.* **78,** 17–31.

21. Dussillol-Godar, F., Brissard-Zahraoui, J., Limbourg-Bouchon, B., et al. (2006) Modulation of the Suppressor of fused protein regulates the Hedgehog signaling pathway in Drosophila embryo and imaginal discs. *Dev. Biol.* **291,** 53–66.

22. Methot, N. and Basler, K. (1999) Hedgehog controls limb development by regulating the activities of distinct transcriptional activator and repressor forms of Cubitus interruptus. *Cell* **96,** 819–831.

23. Zhu, A. J., Zheng, L., Suyama, K., and Scott, M. P. (2003) Altered localization of Drosophila Smoothened protein activates Hedgehog signal transduction. *Genes Dev.* **17,** 1240–1252.

24. Goentoro, L. A., Yakoby, N., Goodhouse, J., Schupbach, T., and Shvartsman, S. Y. (2006) Quantitative analysis of the GAL4/UAS system in Drosophila oogenesis. *Genesis* **44,** 66–74.

25. Rorth, P. (1998) Gal4 in the Drosophila female germline. *Mech. Dev.* **78**, 113–118.
26. Van Doren, M., Williamson, A. L., and Lehmann, R. (1998) Regulation of zygotic gene expression in Drosophila primordial germ cells. *Curr. Biol.* **8**, 243–246.
27. Hacker, U. and Perrimon, N. (1998) DRhoGEF2 encodes a member of the Dbl family of oncogenes and controls cell shape changes during gastrulation in Drosophila. *Genes Dev.* **12**, 274–284.
28. Gerlitz, O., Nellen, D., Ottiger, M., and Basler, K. (2002) A screen for genes expressed in Drosophila imaginal discs. *Int. J. Dev. Biol.* **46**, 173–176.
29. Milan, M., Diaz-Benjumea, F. J., and Cohen, S. M. (1998) Beadex encodes an LMO protein that regulates Apterous LIM-homeodomain activity in Drosophila wing development: a model for LMO oncogene function. *Genes Dev.* **12**, 2912–2920.
30. Freeman, M. (1996) Reiterative use of the EGF receptor triggers differentiation of all cell types in the Drosophila eye. *Cell* **87**, 651–660.
31. Kramer, J. M. and Staveley, B. E. (2003) GAL4 causes developmental defects and apoptosis when expressed in the developing eye of Drosophila melanogaster. *Genet. Mol. Res.* **2**, 43–47.
32. Phelps, C. B. and Brand, A. H. (1998) Ectopic gene expression in Drosophila using GAL4 system. *Methods* **14**, 367–379.
33. Sharma, Y., Cheung, U., Larsen, E. W., and Eberl, D. F. (2002) PPTGAL, a convenient Gal4 P-element vector for testing expression of enhancer fragments in Drosophila. *Genesis* **34**, 115–118.
34. Wodarz, A., Hinz, U., Engelbert, M., and Knust, E. (1995) Expression of crumbs confers apical character on plasma membrane domains of ectodermal epithelia of Drosophila. *Cell* **82**, 67–76.
35. Ito, K., Awano, W., Suzuki, K., Hiromi, Y., and Yamamoto, D. (1997) The Drosophila mushroom body is a quadruple structure of clonal units each of which contains a virtually identical set of neurones and glial cells. *Development* **124**, 761–771.
36. Sanson, B., White, P., and Vincent, J. P. (1996) Uncoupling cadherin-based adhesion from wingless signaling in Drosophila. *Nature* **383**, 627–630.
37. Pignoni, F. and Zipursky, S. L. (1997) Induction of Drosophila eye development by decapentaplegic. *Development* **124**, 271–278.
38. Monge, I., Krishnamurthy, R., Sims, D., et al. (2001) Drosophila transcription factor AP-2 in proboscis, leg and brain central complex development. *Development* **128**, 1239–1252.
39. Zecca, M. and Struhl, G. (2002) Subdivision of the Drosophila wing imaginal disc by EGFR-mediated signaling. *Development* **129**, 1357–1368.
40. Huang, Z. and Kunes, S. (1998) Signals transmitted along retinal axons in Drosophila: Hedgehog signal reception and the cell circuitry of lamina cartridge assembly. *Development* **125**, 3753–3764.
41. Struhl, G. and Adachi, A. (1998) Nuclear access and action of notch in vivo. *Cell* **93**, 649–660.

42. Capdevila, J. and Guerrero, I. (1994) Targeted expression of the signaling molecule decapentaplegic induces pattern duplications and growth alterations in Drosophila wings. *EMBO J.* **13,** 4459–4468.

43. Dominguez, M., Brunner, M., Hafen, E., and Basler, K. (1996) Sending and receiving the Hedgehog signal: control by the Drosophila Gli protein Cubitus interruptus. *Science* **272,** 1621–1625.

44. Calleja, M., Moreno, E., Pelaz, S., and Morata, G. (1996) Visualization of gene expression in living adult Drosophila. *Science* **274,** 252–255.

45. O'Keefe, D. D., Thor, S., and Thomas, J. B. (1998) Function and specificity of LIM domains in Drosophila nervous system and wing development. *Development* **125,** 3915–3923.

46. Staehling-Hampton, K., Jackson, P. D., Clark, M. J., Brand, A. H., and Hoffmann, F. M. (1994) Specificity of bone morphogenetic protein-related factors: cell fate and gene expression changes in Drosophila embryos induced by decapentaplegic but not 60A. *Cell Growth Differ.* **5,** 585–593.

47. Chanut, F., Woo, K., Pereira, S., et al. (2002) Rough eye is a gain-of-function allele of amos that disrupts regulation of the proneural gene atonal during Drosophila retinal differentiation. *Genetics* **160,** 623–635.

48. Harrison, D. A., Binari, R., Nahreini, T. S., Gilman, M., and Perrimon, N. (1995) Activation of a Drosophila Janus Kinase (JAK) causes hematopoietic neoplasia and developmental defects. *Embo J.* **14,** 2857–2865.

49. Johnson, R. L., Grenier, J. K., and Scott, M. P. (1995) patched overexpression alters wing disc size and pattern: transcriptional and post-transcriptional effects on Hedgehog targets. *Development* **121,** 4161–4170.

50. Forbes, A. J., Spradling, A. C., Ingham, P. W., and Lin, H. (1996) The role of segment polarity genes during early oogenesis in Drosophila. *Development* **122,** 3283–3294.

51. Rintelen, F., Hafen, E., and Nairz, K. (2003) The Drosophila dual-specificity ERK phosphatase DMKP3 cooperates with the ERK tyrosine phosphatase PTP-ER. *Development* **130,** 3479–3490.

52. Duffy, J. B., Harrison, D. A., and Perrimon, N. (1998) Identifying loci required for follicular patterning using directed mosaics. *Development* **125,** 2263–2271.

53. Gibson, M. C., Lehman, D. A., and Schubiger, G. (2002) Lumenal transmission of decapentaplegic in Drosophila imaginal discs. *Dev. Cell* **3,** 451–460.

54. Tanimoto, H., Itoh, S., ten Dijke, P., and Tabata, T. (2000) Hedgehog creates a gradient of DPP activity in Drosophila wing imaginal discs. *Mol. Cell* **5,** 59–71.

55. Lecuit, T., Brook, W. J., Ng, M., Calleja, M., Sun, H., and Cohen, S. M. (1996) Two distinct mechanisms for long-range patterning by Decapentaplegic in the Drosophila wing. *Nature* **381,** 387–393.

56. Hinz, U., Giebel, B., and Campos-Ortega, J. A. (1994) The basic-helix-loop-helix domain of Drosophila lethal of scute protein is sufficient for proneural function and activates neurogenic genes. *Cell* **76,** 77–87.

57. St Pierre, S. E., Galindo, M. I., Couso, J. P., and Thor, S. (2002) Control of Drosophila imaginal disc development by rotund and roughened eye: differentially expressed transcripts of the same gene encoding functionally distinct zinc finger proteins. *Development* **129,** 1273–1281.

58. Simmonds, A. J., Brook, W. J., Cohen, S. M., and Bell, J. B. (1995) Distinguishable functions for engrailed and invected in anterior-posterior patterning in the Drosophila wing. *Nature* **376,** 424–427.

59. Yoffe, K. B., Manoukian, A. S., Wilder, E. L., Brand, A. H., and Perrimon, N. (1995) Evidence for engrailed-independent wingless autoregulation in Drosophila. *Dev. Biol.* **170,** 636–650.

60. Hazelett, D. J., Bourouis, M., Walldorf, U., and Treisman, J. E. (1998) Decapentaplegic and wingless are regulated by eyes absent and eyegone and interact to direct the pattern of retinal differentiation in the eye disc. *Development* **125,** 3741–3751.

61. Lin, D. M. and Goodman, C. S. (1994) Ectopic and increased expression of Fasciclin II alters motoneuron growth cone guidance. *Neuron* **13,** 507–523.

62. Perrin, L., Bloyer, S., Ferraz, C., Agrawal, N., Sinha, P., and Dura, J. M. (2003) The leucine zipper motif of the Drosophila AF10 homologue can inhibit PRE-mediated repression: implications for leukemogenic activity of human MLL-AF10 fusions. *Mol. Cell Biol.* **23,** 119–130.

63. Wernet, M. F., Labhart, T., Baumann, F., Mazzoni, E. O., Pichaud, F., and Desplan, C. (2003) Homothorax switches function of Drosophila photoreceptors from color to polarized light sensors. *Cell* **115,** 267–279.

64. Crew, J. R., Batterham, P., and Pollock, J. A. (1997) Developing compound eye in lozenge mutants of Drosophila: lozenge expression in the R7 equivalence group. *Dev. Genes Evol.* **206,** 481–493.

65. Shiga, Y., Tanaka–Matakatsu, M., and Hayashi, S. (1996) A nuclear/β-galatosidase fusion protein as a marker for morphogenesis in living Drosophila. *Dev. Growth Differ.* **38,** 99–106.

66. Cherbas, L., Hu, X., Zhimulev, I., Belyaeva, E., and Cherbas, P. (2003) EcR isoforms in Drosophila: testing tissue-specific requirements by targeted blockade and rescue. *Development* **130,** 271–284.

67. Brand, A. H. and Perrimon, N. (1994) Raf acts downstream of the EGF receptor to determine dorsoventral polarity during Drosophila oogenesis. *Genes Dev.* **8,** 629–639.

68. Rorth, P., Szabo, K., Bailey, A., et al. (1998) Systematic gain-of-function genetics in Drosophila. *Development* **125,** 1049–1057.

69. Jhaveri, D., Sen, A., Reddy, G. V., and Rodrigues, V. (2000) Sense organ identity in the Drosophila antenna is specified by the expression of the proneural gene atonal. *Mech. Dev.* **99,** 101–111.

70. Tabata, T., Schwartz, C., Gustavson, E., Ali, Z., and Kornberg, T. B. (1995) Creating a Drosophila wing de novo, the role of engrailed, and the compartment border hypothesis. *Development* **121,** 3359–3369.

71. Guillen, I., Mullor, J. L., Capdevila, J., Sanchez-Herrero, E., Morata, G., and Guerrero, I. (1995) The function of engrailed and the specification of Drosophila wing pattern. *Development* **121,** 3447–3456.

72. Lawrence, P. A., Casal, J., and Struhl, G. (1999) Hedgehog and engrailed: pattern formation and polarity in the Drosophila abdomen. *Development* **126,** 2431–2439.

73. Alexandre, C. and Vincent, J. P. (2003) Requirements for transcriptional repression and activation by Engrailed in Drosophila embryos. *Development* **130,** 729–739.

74. Ingham, P. W. and Fietz, M. J. (1995) Quantitative effects of Hedgehog and decapentaplegic activity on the patterning of the Drosophila wing. *Curr. Biol.* **5,** 432–440.

75. Lee, J. D., Kraus, P., Gaiano, N., et al. (2001) An acylatable residue of Hedgehog is differentially required in Drosophila and mouse limb development. *Dev. Biol.* **233,** 122–136.

76. Burke, R., Nellen, D., Bellotto, M., et al. (1999) Dispatched, a novel sterol-sensing domain protein dedicated to the release of cholesterol-modified Hedgehog from signaling cells. *Cell* **99,** 803–815.

77. Torroja, C., Gorfinkiel, N., and Guerrero, I. (2004) Patched controls the Hedgehog gradient by endocytosis in a dynamin-dependent manner, but this internalization does not play a major role in signal transduction. *Development* **131,** 2395–2408.

78. Porter, J. A., von Kessler, D. P., Ekker, S. C., et al. (1995) The product of Hedgehog autoproteolytic cleavage active in local and long-range signaling. *Nature* **374,** 363–366.

79. Chamoun, Z., Mann, R. K., Nellen, D., et al. (2001) Skinny hedgehog, an acyltransferase required for palmitoylation and activity of the Hedgehog signal. *Science* **293,** 2080–2084.

80. Gallet, A., Rodriguez, R., Ruel, L., and Therond, P. P. (2003) Cholesterol modification of hedgehog is required for trafficking and movement, revealing an asymmetric cellular response to Hedgehog. *Dev. Cell* **4,** 191–204.

81. Chen, Y. and Struhl, G. (1996) Dual roles for patched in sequestering and transducing Hedgehog. *Cell* **87,** 553–563.

82. Vegh, M. and Basler, K. (2003) A genetic screen for Hedgehog targets involved in the maintenance of the Drosophila anteroposterior compartment boundary. *Genetics* **163,** 1427–1438.

83. Johnson, R. L., Milenkovic, L., and Scott, M. P. (2000) In vivo functions of the patched protein: requirement of the C terminus for target gene inactivation but not Hedgehog sequestration. *Mol. Cell* **6,** 467–478.

84. Briscoe, J., Chen, Y., Jessell, T. M., and Struhl, G. (2001) A Hedgehog-insensitive form of patched provides evidence for direct long-range morphogen activity of sonic Hedgehog in the neural tube. *Mol. Cell* **7,** 1279–1291.

85. Strutt, H., Thomas, C., Nakano, Y., et al. (2001) Mutations in the sterol-sensing domain of Patched suggest a role for vesicular trafficking in Smoothened regulation. *Curr. Biol.* **11,** 608–613.

86. Johnson, R. L., Zhou, L., and Bailey, E. C. (2002) Distinct consequences of sterol sensor mutations in Drosophila and mouse patched homologs. *Dev. Biol.* **242,** 224–235.

87. Denef, N., Neubuser, D., Perez, L., and Cohen, S. M. (2000) Hedgehog induces opposite changes in turnover and subcellular localization of Patched and Smoothened. *Cell* **102,** 521–531.

88. Apionishev, S., Katanayeva, N. M., Marks, S. A., Kalderon, D., and Tomlinson, A. (2005) Drosophila Smoothened phosphorylation sites essential for Hedgehog signal transduction. *Nat. Cell Biol.* **7,** 86–92.

89. Ingham, P. W., Nystedt, S., Nakano, Y., et al. (2000) Patched represses the Hedgehog signaling pathway by promoting modification of the Smoothened protein. *Curr. Biol.* **10,** 1315–1318.

90. Jia, J., Tong, C., Wang, B., Luo, L., and Jiang, J. (2004) Hedgehog signaling activity of Smoothened requires phosphorylation by protein kinase A and casein kinase I. *Nature* **432,** 1045–1050.

91. Jia, J., Tong, C., and Jiang, J. (2003) Smoothened transduces Hedgehog signal by physically interacting with Costal2/Fused complex through its C-terminal tail. *Genes Dev.* **17,** 2709–2720.

92. Hooper, J. E. (2003) Smoothened translates Hedgehog levels into distinct responses. *Development* **130,** 3951–3963.

93. Nakano, Y., Nystedt, S., Shivdasani, A. A., Strutt, H., Thomas, C., and Ingham, P. W. (2004) Functional domains and sub-cellular distribution of the Hedgehog transducing protein Smoothened in Drosophila. *Mech. Dev.* **121,** 507–518.

94. Wang, G., Amanai, K., Wang, B., and Jiang, J. (2000) Interactions with Costal2 and Suppressor of fused regulate nuclear translocation and activity of cubitus interruptus. *Genes Dev.* **14,** 2893–2905.

95. Ho, K. S., Suyama, K., Fish, M., and Scott, M. P. (2005) Differential regulation of Hedgehog target gene transcription by Costal2 and Suppressor of Fused. *Development* **132,** 1401–1412.

96. Alexandre, C., Jacinto, A., and Ingham, P. W. (1996) Transcriptional activation of hedgehog target genes in Drosophila is mediated directly by the cubitus interruptus protein, a member of the GLI family of zinc finger DNA-binding proteins. *Genes Dev.* **10,** 2003–2013.

97. Hepker, J., Wang, Q. T., Motzny, C. K., Holmgren, R., and Orenic, T. V. (1997) Drosophila cubitus interruptus forms a negative feedback loop with patched and regulates expression of Hedgehog target genes. *Development* **124,** 549–558.

98. Price, M. A. and Kalderon, D. (1999) Proteolysis of cubitus interruptus in Drosophila requires phosphorylation by protein kinase A. *Development* **126,** 4331–4339.

99. Aza-Blanc, P., Ramirez-Weber, F. A., Laget, M. P., Schwartz, C., and Kornberg, T. B. (1997) Proteolysis that is inhibited by hedgehog targets Cubitus interruptus protein to the nucleus and converts it to a repressor. *Cell* **89,** 1043–1053.

100. Chen, Y., Cardinaux, J. R., Goodman, R. H., and Smolik, S. M. (1999) Mutants of cubitus interruptus that are independent of PKA regulation are independent of Hedgehog signaling. *Development* **126,** 3607–3616.

101. Wang, G., Wang, B., and Jiang, J. (1999) Protein kinase A antagonizes Hedgehog signaling by regulating both the activator and repressor forms of Cubitus interruptus. *Genes Dev.* **13**, 2828–2837.

102. Wang, Q. T. and Holmgren, R. A. (1999) The subcellular localization and activity of Drosophila cubitus interruptus are regulated at multiple levels. *Development* **126**, 5097–5106.

103. Methot, N. and Basler, K. (2000) Suppressor of fused opposes Hedgehog signal transduction by impeding nuclear accumulation of the activator form of Cubitus interruptus. *Development* **127**, 4001–4010.

104. Chen, Y., Goodman, R. H., and Smolik, S. M. (2000) Cubitus interruptus requires Drosophila CREB-binding protein to activate wingless expression in the Drosophila embryo. *Mol. Cell Biol.* **20**, 1616–1625.

105. Jia, J., Amanai, K., Wang, G., Tang, J., Wang, B., and Jiang, J. (2002) Shaggy/ GSK3 antagonizes Hedgehog signaling by regulating Cubitus interruptus. *Nature* **416**, 548–552.

106. Price, M. A. and Kalderon, D. (2002) Proteolysis of the Hedgehog signaling effector Cubitus interruptus requires phosphorylation by Glycogen Synthase Kinase 3 and Casein Kinase 1. *Cell* **108**, 823–835.

107. Kiger, J. A. Jr., Eklund, J. L., Younger, S. H., and O'Kane, C. J. (1999) Transgenic inhibitors identify two roles for protein kinase A in Drosophila development. *Genetics* **152**, 281–290.

108. Kiger, J. A. Jr. and O'Shea, C. (2001) Genetic evidence for a protein kinase A/cubitus interruptus complex that facilitates processing of cubitus interruptus in Drosophila. *Genetics* **158**, 1157–1166.

109. Yoshida, S., Muller, H. A., Wodarz, A., and Ephrussi, A. (2004) PKA-R1 spatially restricts Oskar expression for Drosophila embryonic patterning. *Development* **131**, 1401–1410.

110. Li, W., Ohlmeyer, J. T., Lane, M. E., and Kalderon, D. (1995) Function of protein kinase A in hedgehog signal transduction and Drosophila imaginal disc development. *Cell* **80**, 553–562.

111. Hidalgo, A., Urban, J., and Brand, A. H. (1995) Targeted ablation of glia disrupts axon tract formation in the Drosophila CNS. *Development* **121**, 3703–3712.

112. Cormack, B. P., Valdivia, R. H., and Falkow, S. (1996) FACS-optimized mutants of the green fluorescent protein (GFP). *Gene* **173**, 33–38.

113. Halfon, M. S., Gisselbrecht, S., Lu, J., Estrada, B., Keshishian, H., and Michelson, A. M. (2002) New fluorescent protein reporters for use with the Drosophila Gal4 expression system and for vital detection of balancer chromosomes. *Genesis* **34**, 135–138.

114. Koh, Y. H., Popova, E., Thomas, U., Griffith, L. C., and Budnik, V. (1999) Regulation of DLG localization at synapses by CaMKII-dependent phosphorylation. *Cell* **98**, 353–363.

115. Roignant, J. Y., Carre, C., Mugat, B., Szymczak, D., Lepesant, J. A., and Antoniewski, C. (2003) Absence of transitive and systemic pathways allows cell-specific and isoform-specific RNAi in Drosophila. *Rna* **9**, 299–308.

116. Casso, D., Ramirez-Weber, F., and Kornberg, T. B. (2000) GFP-tagged balancer chromosomes for Drosophila melanogaster. *Mech. Dev.* **91**, 451–454.

14

Biochemical Fractionation of Drosophila Cells

Melanie Stegman and David Robbins

Abstract

This chapter describes how to perform basic biochemical fractionations of Drosophila cells, and how to begin to characterize the proteins in the resulting fractions. The protocols include maintenance and transfection of Drosophila cell lines (**Section 3.1.**), hypotonic lysis (**Section 3.2.**), and separation of cellular lysates into cytosolic and membrane enriched fractions (**Section 3.3.**). Cytosolic proteins and those extracted from the membrane enriched fraction can be characterized by size exclusion liquid chromatography (**Section 3.4.**), while the membrane enriched fraction can be subjected to equilibrium density centrifugation to separate different types of cellular membranes from dense, nonmembranous cellular components (**Section 3.5.**). The resulting fractions can be used to examine the subcellular localization of a given protein, or the activity of a given protein in various subcellular localizations. When the protein of interest is involved in a signaling pathway, its subcellular localization can provide insight into its mechanism of action in the pathway.

Key Words: Hedgehog; subcellular fractionation; Drosophila; Kinesin; signaling; methods, Costal2.

1. Introduction

Hh was one of several genes reported in 1980 by Nüsslein-Volhard and Wieschaus to be required for embryonic development in *Drosophila melanogaster (1)*. Since then thier labs and many others have genetically described a group of genes involved in Hh signal transduction *(2)*. Because several of these genes appear to function at the same level of the pathway, some geneticists concluded that one or more protein complexes must play a role in Hh signaling (*see* Ref. *[3]*). Our lab, and several others, have biochemically determined the subcellular localizations and interactions among several proteins involved in Hh signaling. By taking advantage of the ability to do biochemistry and genetics in the same system, we have begun to characterize the roles these protein interactions

From: *Methods in Molecular Biology: Hedgehog Signaling Protocols*
Edited by: J. Horabin © Humana Press Inc., Totowa, NJ

play in fly development. In particular, we have used the protocols in this chapter to determine that a large multi-protein complex binds vesicular membranes in a Hh sensitive manner. This complex consists of the Kinesin related protein Costal2 (Cos2), a serine-threonine protein kinase Fused (Fu), and a Zn^{2-} finger transcription factor cubitus interruptus (Ci). Additionally, Smoothened (Smo) a seven trans-membrane spanning protein, and Suppressor of fused [Su(fu)] a protein with no known sequence homology are, under certain circumstances, members of this complex. The importance of this complex in Hh signaling is demonstrated by *fu* mutants that show a Hh loss of function phenotype, and produce Fu proteins that do not interact with Cos2 *(3-10)*. The biochemical activities of these proteins are still uncharacterized, so determining their sub-cellular localization under conditions of pathway activation and inactivation may allow a directed search for the active fraction of these proteins. Additionally, knowing the binding partners and localizations of Cos2 and Fu, in addition to the Hh phenotypes that result in their absence, will aid in the search for vertebrate homologs (*see* for example *[11]*). This chapter describes how to perform basic biochemical fractionations of Drosophila cells. The protocols below include maintenance and transfection of Drosophila cell lines (**Section 3.1.**), hypotonic lysis (**Section 3.2.**); separation of Drosophila lysates into cytosolic and membrane enriched fractions (**Section 3.3.**), and further separation of these fractions by either size exclusion liquid chromatography (**Section 3.4.**) or equilibrium density centrifugation (**Section 3.5.**). The resulting fractions can be used for activity assays, immunoprecipitation, or any number of assays, which are not described here but have been well described in other textbooks *(12,13)*. Therefore, the methods described in this chapter should aid in the biochemical analysis of proteins determined genetically to be required for cellular processes in *D. melanogaster*.

2. Materials

2.1. Transfection of Drosophila Cells

1. Tissue culture hood.
2. 27°C incubator.
3. Schneider 2 (S2) cells, (ATCC).
4. Cl8 cells (Martin Milner at The University of St Andrews, UK).
5. Fetal bovine serum (FBS; Invitrogen).
6. 5 mL penicillin/streptomycin (Pen/Strep; Invitrogen).
7. S2 media: Schneider's media (Invitrogen), each 500 mL supplemented with 50 mL FBS and 5 mL Pen/Strep.
8. 100× Insulin (*see* **Note 1**).
9. Drosophila extract, "Fly Blood" (*see* **Note 1**).
10. Cl8 media: Shield's and Sang media (Sigma), each 500 mL supplemented with 10 mL FBS, 5 mL Pen/Strep, 5 mL of 100× insulin, and 12.5 mL Fly Blood and then filter sterilized.

11. Cellfectin (Invitrogen Carlsbad, CA).
12. Phosphate buffered saline (PBS), sterile (Invitrogen).

2.2. Hypotonic Lysis of Drosophila Cells

1. 7-mL glass dounce for cell homogenization.
2. Rubber or silicon cell scraper.
3. Hypotonic lysis buffer: GNE is 50 mM β-glycerophosphate, 10 mM NaF, 1.5 mM EGTA, and pH 7.6 with HCl. Immediately before using, add protease inhibitor (PI) cocktail at 1 part PI to 250 parts GNE and add dithiothreitol (DTT) to a final concentration of 1 mM. PI consists of 1 mM benzamidine, 1 mg/mL aprotinin, 1 mg/mL leupeptin, and 1 mg/mL pepstatin A in 100% ethanol. Store PI at −20°C. DTT is a 0.5 M stock in H$_2$O, stored at −80°C and not repeatedly thawed.
4. PBS, sterile.

2.3. Separation of Cellular Lysates into Cytosolic and Membrane Enriched Fractions

1. 7-mL glass dounce.
2. Ultracentrifuge and rotor, such as a Beckman benchtop with rotor TLA-100.3.
3. Ultracentrifuge tubes, must be appropriate for rotor.
4. Low-speed centrifuge, such as Beckman J6.
5. Lysis and extraction buffers: 1× GNE (*see* **Section 2.2.3.**) with 0.15, 0.5, 0.75, or 1 M NaCl and 1× GNE (*see* **Section 2.2.3.**) with 1 M NaCl and 1% NP-40.

2.4. Size Exclusion Liquid Chromatography

1. Liquid chromatography system (Äkta FPLC, GE Healthcare Piscataway, NJ USA).
2. Superose 6 gel filtration column (10/300 GL, GE Healthcare).
3. Fraction collector (Frac920, GE Healthcare).
4. Standard proteins, of known Mr, to calibrate the column (GE Healthcare). Ultracentrifuge.
6. 10% NP-40 in H$_2$O (Sigma St. Louis, MO).
7. 150 mM NaCl GNE (*see* **Section 2.3.5.**) supplemented to 0.001% NP-40.

2.5. Separating Total Membranes into Plasma and Vesicular Membrane Enriched Fractions

1. Ultracentrifuge such as a Beckman Optima L-XP with rotor SW60.Ti or equivalent.
2. Centrifuge tubes appropriate for the rotor.
3. Glass dounce.
4. TNE buffer: 100 mM Tris, 150 mM NaCl, 0.2 mM EDTA, pH 7.5 with HCl.
5. 2 M sucrose in TNE.
6. 1.22 M sucrose in TNE.
7. 0.1 M sucrose in TNE.
8. 4 mM Pefabloc SC (Sigma) in H$_2$O, store stock at −20°C.
9. PI (*see* **Section 2.2.3.**).

3. Methods

3.1. Transfection of Drosophila Cells

For general comments and maintenance of Drosophila cell lines (*see* **Note 2**).

1. Three days before transfection, split S2 or Cl8 cells to $4 \times 10^5/\text{cm}^2$ (*see* **Note 2**).
2. Eighteen hours before transfection, plate cells at $1.6 \times 10^5/\text{cm}^2$.
3. Prepare Cellfectin and DNA mixture according to product instructions, using 1 µg DNA/4 µL Cellfectin per million cells in the appropriate media with no supplements.
4. Wash cells twice with 10 mL unsupplemented media. Add the transfection mixture and incubate at 27°C for 4–6 h, do not exceed 6 h.
5. Remove the transfection mixture and replace with 10 mL complete media (*see* **Notes 2** and **3**).
6. Lyse cells at the time appropriate for greatest expression or activity of your protein or pathway of interest.

3.2. Hypotonic Lysis of Drosophila Cells

This protocol is for lysis of either cells that were transfected 40 h before lysis, as in **Section 3.1.**, or for cells that are plated 20 h before lysis as described in **step 1** below (*see* **Notes 3–6**). The volumes of wash and lysis buffer given are for 100-mm plates; if different plate sizes are used, maintain the buffer volume to centimeter squared ratios. Drosophila cells are smaller, and thus do not lyse as readily in hypotonic buffer as mammalian cells. For this reason, it is necessary to wash the cells with hypotonic lysis buffer prior to homogenizing. The hypotonic wash must be done quickly, as cells will begin to lyse, which necessitates the cells being adherent. Therefore, the numbers of cells plated per centimeter squared and the time between plating and lysis are important factors as Drosophila cells (S2 especially) are less adherent as their numbers increase. Additionally, if the ratio of cell number to buffer is varied from experiment to experiment, lysis and extraction efficiencies may vary as well (*see* **Notes 8** and **11** for helpful controls).

1. Twenty hours prior to lysis, plate S2 or Cl8 cells at 6.7×10^5 cells/cm^2 on six 100-mm plates.
2. Prechill buffers (except PBS), all tubes and the glass dounce on ice. Prepare the following: Hypo Wash (GNE plus 1 mM DTT) and Lysis buffer (GNE plus 1 mM DTT and 1:250 PI).
3. Remove media from the first plate of cells by aspiration, wash gently with 10 mL PBS, and remove PBS by aspiration. To avoid washing the cells off of the plate, apply all solutions by touching tip of pipette to the side of the dish, not dripping directly onto cells. Rock the plate gently, so that the PBS moves over cells in a straight front; do not swirl.
4. Gently, and quickly, pipette 10 mL ice-cold Hypo Wash onto plate, rock plate back, and forth twice, then quickly dump the liquid off the plate, aspirate off the rest, and immediately place plate on ice (*see* **Note 6**).

5. Quickly apply 0.333 mL lysis buffer and rotate plate on the ice so the lysis buffer washes over all cells. Scrape cells off plate using a rubber or silicon cell scraper.
6. Transfer the cells to a 15-mL conical tube on ice.
7. Repeat steps 3–6 for the remainder of the plates, combining the cells into one 15-mL tube. The final volume should be about twice the volume of lysis buffer added.
8. Let cells sit on ice for 20 min after the last plate is scraped.
9. Use the glass dounce to homogenize cells (*see* **Note 7**).
10. This lysate is the total lysate (TOT). Keep 150 µL aside for later analysis, and immediately proceed to **Section 3.3**. (If the entire fractionation cannot be completed on the same day as lysis, proceed through the low speed centrifugation, and snap freeze the LSS and the resuspended LSP.)
11. A protein assay (Bradford Assay from Pierce, for example) should reveal that the TOT is 1–4 mg/mL total protein (*see* **Note 5**).
12. Store all fractions by snap freezing them in liquid nitrogen, or in a dry ice and ethanol bath, and then storing at −80°C.

3.3. Separation of Cellular Lysates into Cytosolic and Membrane Enriched Fractions

For general comments (*see* **Notes 4–6** and refer to **Fig. 1**).

1. Prechill all buffers, tubes, and the glass dounce on ice and add PI (1:250) and DTT (1 m*M*) to all buffers.
2. Transfer 3 mL TOT (prepared in **Section 3.2.**) to a centrifuge tube, and centrifuge 20 min at 2000*g*, at 4°C.
3. Promptly remove the supernatant (LSS) and keep it on ice.
4. Resuspend the pellet (LSP) in GNE, to a final volume of 3 mL, thus normalizing the LSP and LSS by volume.
5. Keep a 200-µL aliquot of the LSS and the resuspended LSP for later analysis.
6. Centrifuge 2.75-mL LSS for 60 min at 100,000*g*, at 4°C.
7. Promptly remove the supernatant (HSS1) and keep it on ice.
8. Resuspend the pellet (HSP1) in 150 m*M* NaCl GNE to a final volume of 2.75 mL (*see* **Note 9**). Save 200 µL for later analysis.
9. Centrifuge 2.5 mL of the resuspended HSP1 for 60 min at 100,000*g*, at 4°C.
10. Promptly remove the supernatant (HSS2) and keep it on ice.
11. The pellet (HSP2) contains total washed membranes, and is called the total membrane enriched fraction. (The HSP2 may be separated into vesicular membranes and plasma membranes; to do so, proceed to **Section 3.5.**) To extract the total membranes, resuspend the HSP2 to 2.5 mL with 500 m*M* NaCl GNE.
12. Centrifuge the resuspended HSP2 for 60 min at 100,000*g*, at 4°C.
13. Remove the supernatant (HSS3) and resuspend the pellet (HSP3) in 750 m*M* NaCl GNE. The HSS3 contains proteins extracted from total membranes, and can be used in **Section 5**.
14. Centrifuge the resuspended HSP3 for 60 min at 100,000*g*, at 4°C, and resuspend the pellet (HSP4) in 1 *M* NaCl GNE.

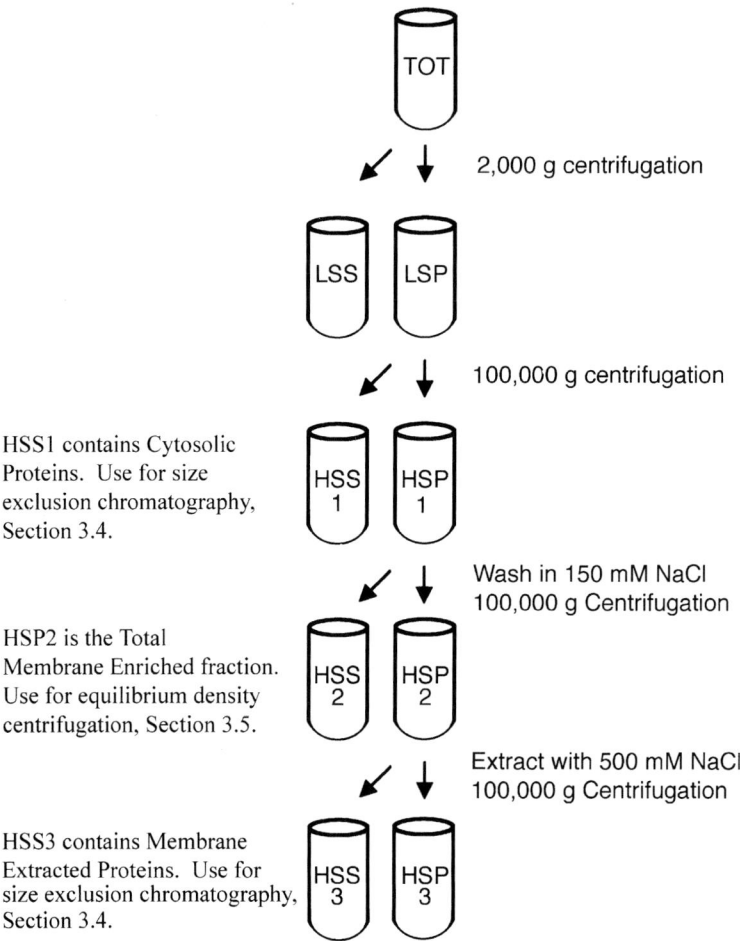

Fig. 1. A schematic of the basic cell fractionation presented here.

15. Centrifuge the resuspended HSP4 for 60 min at 100,000*g*, at 4°C, and resuspend the pellet (HSP5) in 1 *M* NaCl GNE with 1% NP-40.
16. Aliquots of all the fractions, including the TOT, LSS, LSP, and the various HSS and HSP, should be analyzed by immunoblotting. All fractions should be normalized to NaCl, detergent concentration, and normalized to each other by volume prior to loading on an SDS-polyacrylamide gel (*see* **Note 10**). Additionally, perform a protein assay on all fractions, to know how much protein is loaded in each lane (*see* **Note 11**).
17. Store all fractions by snap freezing them in liquid nitrogen, or in a dry ice and ethanol bath, and then storing at −80°C.

3.4. Size Exclusion Liquid Chromatography

For general comments (*see* **Note 12**).

1. Thaw the HSS1 on ice (from **Section 3.3.**).
2. Bring the HSS1 to 150 m*M* NaCl and 0.001% NP-40.
3. Centrifuge the HSS1 for 30 min at 100,000*g* to pellet any aggregates that may have formed during freezing/thawing, which may clog the column.
4. Inject 240 µL HSS1 onto a Superose 6 column that has been prewashed in the same buffer as the sample (150 m*M* NaCl GNE supplemented to 0.001% NP-40).
5. Elute with 1 column volume of buffer, which is 24 mL for the 10/300 GL, collecting 0.325 mL fractions of the entire elution.
6. Very large molecules and aggregates will elute at fraction 19–20, or about one-third of the column volume. Analyze fractions 18 through 43 by immunoblot. Compare the elution pattern of the protein of interest to the elution patterns of standard proteins on the same column (*see* **Note 12**).

3.5. Separating Total Membranes into Plasma and Vesicular Membrane Enriched Fractions

For general comments (*see* **Note 13**).

1. Prechill buffers, centrifuge tubes, and the dounce on ice. Add PIC (1:250) and Pefabloc (1:250) to buffers and each sucrose solution, and mix well.
2. Obtain a fresh (not frozen) HSP2, as described in **Section 3.3**. Rather than resuspending the pellet in GNE, resuspend the pellet in 2 *M* sucrose TNE, using a glass dounce, slowly as 2 *M* sucrose is very dense. The final concentration of sucrose, given the volume of the pellet plus the volume of 2 *M* sucrose added should be 1.4 *M*.
3. Place 1 mL resupended HSP2 in the bottom of a 3 mL centrifuge tube.
4. Carefully overlay with 1.25 mL of 1.22 *M* sucrose in TNE (a sharp interface between the layers is required. Practicing before hand is recommended.).
5. Carefully overlay with 0.5 mL of 0.1 *M* sucrose in TNE.
6. Centrifuge at 128,000*g* for 18–20 h at 4°C, with centrifuge brakes at zero.
7. Remove 0.3 mL fractions from the top of the centrifuge tube.
8. Resuspend the clear matter (pellet) at the very bottom of the centrifuge tube in 0.3 mL TNE.
9. Analyze fractions by immunoblotting the total, LSS, LSP, HSS1, HSP1, HSS2, HSP2, the sucrose gradient fractions 1–10, and the pellet. Prior to loading on an SDS-polyacrylamide gel, all samples must be normalized to NaCl concentration. Sucrose concentrations can vary.

4. Notes

1. To make Fly Blood, obtain 30–60 g Drosophila that were healthy and frozen alive, weigh them and homogenize them in a standard household blender with 7.5 mL Shields and Sang media per gram of flies. Blend until creamy, transfer to 50 mL conical tubes, and centrifuge for 30 min at 3000*g*. Remove any oily layer, transfer

the supernatant to a glass beaker, and bring mixture to 60°C for 10 min in a preheated water bath; a white precipitate should form. Centrifuge at 3000g for 30 min and store 12.5-mL aliquots of the supernatant at −20°C. To make 100× insulin, dissolve 10 mg insulin (Sigma) in 0.5 mL of 0.1 M HCl, add 20 mL Shields and Sang media, and store for up to 3 months at 4°C.

2. S2 and Cl8 cells should be split every second to third day to 2–4 × 10^5/cm^2. At densities higher than these, Cl8 cells form multiple layers, while S2 cells become less adherent. It is possible to carry these cells at high densities; however, this lowers their transfection efficiencies. Maintaining the cells at 2–4 × 10^5/cm^2 allows us to achieve 30–50% transfection efficiency. S2 cells split into 1.6–6.7 × 10^5/cm^2 will be tightly adherent and appear quite flat, becoming less adherent and more round as their numbers increase on the plate. To split S2 cells, resuspend them in the 2- to 3-d-old media by pipeting. Transfer an appropriate number of cells and about 10% of the old media to a fresh flask. We do not trypsinize our S2 cells. Cl8 cells are trypsinized to split and do not require old media.

3. Cells plated at 1.6 × 10^5/cm^2 and transfected the next day will be confluent 40–44 h after transfection. Cells plated at 6.7 × 10^5/cm^2 will be confluent 18–20 h later. If cells appear granular or if small particles appear to float in media, the cells have become too stressed and will not grow. This is caused by splitting S2 or Cl8 cells from a very high density to a very low density, and also by transfecting cells for longer than 4–6 h. To avoid this, maintain cells as in **Note 2**, and split into 4 × 10^5/cm^2 3 days before plating for transfection.

4. This protocol describes the separation of cytosolic proteins (HSS1) from the total membrane-enriched fraction (HSP2) and a sequential salt extraction of the HSP2 (*see [10]*, and **Fig. 1**). Thus, it begins to characterize the affinity of membrane associated proteins for membranes. It is important to note, however, that the proteins found in the HSP2 may be integral membrane proteins, peripheral membrane proteins, proteins associated with dense proteinaceous structures, such as actin filaments, or nonspecifically aggregated proteins. Therefore, we subjected the HSP2 to equilibrium density centrifugation (**Section 3.5.**) which separates different types of membranes (as well as some protein complexes) based on their ability to float on various sucrose solutions, while nonspecific protein aggregates and dense proteinaceous structures such as actin filaments will pellet through the sucrose *(12)*.

5. To complete this entire protocol, six 100-mm plates of cells are required. If only the HSP2 is needed, three 100-mm plates are required. Always lyse with 0.333 mL lysis buffer per 100-mm plate to maintain cell to buffer ratios. Additionally, note that the minimum effective volume in the 7-mL dounce is 0.5 mL and the protocols in **Section 3.** are optimized for lysates that have 2–4 mg/mL total protein concentrations. Drosophila embryos may be collected, dechrionated, and lysed by dounce homogenization and then fractionated (*see* Sullivan and Hawley *[14]* for detailed protocols).

6. Following lysis, nonspecific proteolysis and posttranslational modification of proteins begins immediately. Minimize these processes by keeping lysates on wet ice at 4°C at all times and freezing and storing lysates as quickly as possible.

7. To homogenize Drosophila cells in a glass dounce, use the tight fitting B-pestle, push the ball of the pestle just below the surface of the cells, ensuring that no air is trapped below the surface. With a swift motion, push the pestle down to bottom of dounce. Slowly raise the pestle so that no air is sucked into the dounce. Repeat this motion 20 times, and return the dounce to ice for 2–3 min. Repeat as needed (*see* **Note 8**).

8. To monitor lysis, add 10 µL lysate to 90 µL PBS, and apply to a hemocytometer. Lysis is complete when less than 1×10^6 cells/ml are visible, usually after three rounds of 20 strokes.

9. To resuspend a high speed pellet, first add a small volume (50 µL) of buffer to the pellet using a 200 µL Pipetteman (or equivalent) and a disposable tip. Create a paste by scraping the pellet off the wall of the tube and mixing it with the buffer, using the 200 µL pipette tip. Measure the volume of this resuspension by adjusting the volume of the Pipetteman so that the entire resuspension can be sucked into the tip, with no air. For these protocols, the pellet in 50 µL buffer should result in a 75–150 µL resuspension. Eject the tip into the resuspension tube, and with a clean tip add buffer to bring the volume to 200 µL. Then using the original tip, transfer the resuspension to the glass dounce. Eject the tip into the tube again, add 500 µL buffer to the tube and using the original 200 µL tip, transfer the 500 µL to the glass dounce, thus washing the tip and the tube and providing a quantitative transfer of the HSP to the glass dounce. Add buffer as needed to the glass dounce to bring the resuspension to the appropriate final volume. Remove all lumps from the resuspension with about 10 slow strokes with the dounce. Resuspending the pellet in a higher salt concentration serves to wash the pellet and extract a certain class of proteins from it (*see* **Note 4**).

10. The TOT and the LSP fraction will contain nuclei, and protein gel loading buffer, i.e. SDS, will cause DNA to be released. Therefore, shear the DNA by repeatedly pulling the sample through the needle of a 50 µL glass syringe (Hamilton, from Fisher) until the sample flows freely. Using 0.5× dilutions of the TOT, LSP, and LSS for immunoblot analysis may also be advised, as their DNA and protein content may be so high that the gel may be overloaded.

11. To determine the percentage of the protein of interest in each fraction, compare the immunoblot signals for the protein in the TOT to the other fractions. If the signal in the HSS fraction is equal to the signal in the 0.5× dilution of the TOT, then 50% of the total protein is present in the HSS fraction. For controls, an unrelated protein should be blotted for, and its fractionation should be consistent with its published records, and should be independent of any signaling event being investigated. For example, Cos2 fractionates in the HSS in the presence of Hh, but Kinesin remains in the HSP in the absence or presence of Hh.

12. An FPLC is not necessary for gel filtration chromatography; columns may be hand packed with various resins that provide resolution of different sized proteins and protein complexes, see, for example, GE Healthcare or Sigma catalogs. We have used the Superose 6 column, with a resolution between >669 and 220 kDa and a Superose 12 column with a resolution between 440 and 40 kDa to characterize various Hh-signaling complexes (*9,10,15*).

13. This protocol is based on a protocol from the Kai Simons' lab *(16)* in which enzymatic assays were used to determine the purity of the fractions, and some vesicular contamination is found in the plasma membrane fraction. The vesicular marker protein kinesin *(17)* (an antibody against Drosophila Kinesin is available from Cytoskeleton, Inc.) is found in the lower interface, about Fraction 7. Fasciclin I (a plasma membrane protein, *see [18]*) is found in the upper interface, in Fractions 2 and 3. Cos2 and Fu are ~20% plasma membrane and ~70% vesicular and 10% pellet *(10)*.

References

1. Nusslein-Volhard, C. and Wieschaus, E. (1980) Mutations affecting segment number and polarity in Drosophila. *Nature* **287,** 795–801.
2. Ingham, P. W. (1998) Transducing Hedgehog: the story so far. *EMBO J.* **17,** 3505–3511.
3. Stegman, M. A., Vallance, J. E., Elangovan, G., Sosinski, J., Cheng, Y., and Robbins, D. J. (2000) Identification of a tetrameric Hedgehog signaling complex. *J. Biol. Chem.* **275,** 21,809–21,812.
4. Ascano, M. Jr., Nybakken, K. E., Sosinski, J., Stegman, M. A., and Robbins, D. J. A. G. (2002) The carboxyl-terminal domain of the protein kinase fused can function as a dominant inhibitor of Hedgehog signaling. *Mol. Cell. Biol.* **22,** 1555–1566.
5. Ascano, M. Jr. and Robbins, D. J. (2004) An intramolecular association between two domains of the protein kinase Fused is necessary for Hedgehog signaling. *Mol. Cell. Biol.* **24,** 10,397–10,405.
6. Ogden, S. K., Ascano, M. Jr., Stegman, M. A., and Robbins, D. J. (2004) Regulation of Hedgehog signaling: a complex story. *Biochem. Pharmacol.* **67,** 805–814.
7. Ogden, S. K., Ascano, M. Jr., Stegman, M. A., Suber, L. M., Hooper, J. E., and Robbins, D. J. (2003) Identification of a functional interaction between the transmembrane protein Smoothened and the kinesin-related protein Costal2. *Curr. Biol.* **13,** 1998–2003.
8. Ogden, S. K., Casso, D. J., Ascano, M. Jr., Yore, M. M., Kornberg, T. B., and Robbins, D. J. (2006) Smoothened regulates activator and repressor functions of Hedgehog signaling via two distinct mechanisms. *J. Biol. Chem.* **281,** 7237–7243.
9. Robbins, D. J., Nybakken, K. E., Kobayashi, R., Sisson, J. C., Bishop, J. M., and Therond, P. P. (1997) Hedgehog elicits signal transduction by means of a large complex containing the kinesin-related protein costal2. *Cell* **90,** 225–234.
10. Stegman, M. A., Goetz, J. A., Ascano, M. Jr., Ogden, S. K., Nybakken, K. E., and Robbins, D. J. (2004) The Kinesin-related protein Costal2 associates with membranes in a Hedgehog-sensitive, Smoothened-independent manner. *J. Biol. Chem.* **279,** 7064–7071.
11. Tay, S. Y., Ingham, P. W., and Roy, S. (2005) A homologue of the Drosophila kinesin-like protein Costal2 regulates Hedgehog signal transduction in the vertebrate embryo. *Development* **132,** 625–634.
12. Deutscher, M. P. (1990) *Guide to Protein Purification,* Vol. 182, Academic Press, San Diego.

13. Harlow, E. and Lane, D. (1999) *Using Antibodies: A Laboratory Manual,* Cold Spring Harbor Laboratory Press, Cold Spring Harbor.
14. Sullivan, W., Ashburner, M., and Hawley, R. S. (2000) *Drosophila Protocols,* Cold Spring Harbor Press, Cold Spring Harbor.
15. Ascano, M. Jr., Nybakken, K. E., Sosinski, J., Stegman, M. A., and Robbins, D. J. (2002) The carboxyl-terminal domain of the protein kinase fused can function as a dominant inhibitor of Hedgehog signaling. *Mol. Cell. Biol.* **22,** 1555–1566.
16. Rietveld, A., Neutz, S., Simons, K., and Eaton, S. (1999) Association of sterol- and glycosylphosphatidylinositol-linked proteins with Drosophila raft lipid microdomains. *J. Biol. Chem.* **274,** 12,049–12,054.
17. Brady, S. T. and Pfister, K. K. (1991) Kinesin interactions with membrane bounded organelles in vivo and in vitro. *J. Cell. Sci. Suppl.* **14,** 103–108.
18. Elkins, T., Hortsch, M., Bieber, A. J., Snow, P. M., and Goodman, C. S. (1990) Drosophila fasciclin I is a novel homophilic adhesion molecule that along with fasciclin III can mediate cell sorting. *J. Cell. Biol.* **110,** 1825–1832.

15

Using Immunoprecipitation to Study Protein–Protein Interactions in the Hedgehog-Signaling Pathway

Chao Tong and Jin Jiang

Abstract

The Hedgehog (Hh)-signaling pathway has been intensively studied in the past decade. Increasing evidence suggests that dynamic formation of protein complexes plays a critical role in the organization and regulation of Hh signaling. Immunoprecipitation (IP) is a powerful tool to study protein–protein interactions and has provided important insights into the regulation of Hh signal transduction. Here, we show how to use IP to study protein–protein interactions in the *Drosophila* Hh-signaling pathway.

Key Words: Hedgehog-signaling pathway; immunoprecipitation; epitope tag; Protein A; Protein G.

1. Introduction

The Hh-signaling pathway directs many aspects of metazoan development *(1)*. Malfunction of this pathway is involved in numerous human birth defects and cancers *(2,3)*. Investigations into the dynamics of complex formation and protein–protein interactions have provided important information about the regulation of transduction of the Hh signal *(4–8)*. However, several critical links are still missing. Further study of protein–protein interactions in the Hh-signaling pathway should help dissect the complex regulatory processes and identify new players.

Immunoprecipitation (IP) is a well-established technique to study protein–protein interactions. By IP, a specific protein together with its interacting proteins can be isolated by binding to a specific antibody attached to a sedimentable matrix. Using IP, one can study the interactions between various known players in the Hh pathway and examine the dynamic changes of the interactions under

From: *Methods in Molecular Biology: Hedgehog Signaling Protocols*
Edited by: J. Horabin © Humana Press Inc., Totowa, NJ

different signaling conditions. One can also identify novel components in the pathway by their co-IP with the known components.

Interactions detected by IP can also be evaluated by other in vivo approaches such as colocalization or fluorescence resonance energy transfer *(9)*, as well as in assays such as the yeast-two-hybrid and glutathione-*S*-transferase (GST) pull-down. Compared with yeast-two-hybrid and GST pull-down assays, the conditions used in IP experiments are closer to physiological. However, IP detects proteins in complexes and does not provide information whether the interaction between two proteins is direct or indirect.

In this chapter, we will discuss how to use IP to study the Hh-signaling pathway in *Drosophila. Drosophila* cultured cells, imaginal discs, and embryos can be used as sample sources. We will introduce the culture and transfection procedures for two commonly used *Drosophila* cell lines: Schneider 2 (S2) and Clone 8 (Cl8) cells. In addition, we will give a brief introduction about how to prepare lysates from *Drosophila* wing imaginal discs for IP. The general procedures for IP, as well as a simple protocol for coupling antibodies to protein A/G beads to prevent their dissociation from the beads when IP products are eluted, are described. In most cases, IP products are analyzed by Western blot which we also describe.

2. Materials

All chemicals are from Sigma–Aldrich, MO except for those specifically mentioned.

2.1. Sample Preparation

2.1.1. S2 Cell Culture, Transfection, and Lysis

1. Special equipment: 25°C incubator, rotator.
2. S2 cells.
3. S2 cell culture medium: Schneider's *Drosophila* medium (Invitrogen, CA) containing 10% heat-inactivated fetal bovine serum (FBS) (Sigma, MO) and 10 mL/L penicillin–streptomycin: 5000 U penicillin and 5000 µg streptomycin/mL (Invitrogen).
4. Hh-conditioned medium: seed 4×10^7 Hh-N producing cells *(5)* in 10 mL fresh S2 cell culture medium and add 10 µL of 0.7 M $CuSO_4$. Incubate cells in 25°C incubator for 24 h and then transfer the cell suspension into a sterile tube before centrifuging at 1000g for 5 min. Transfer the Hh-conditioned medium (supernatant) into a new tube and store at 4°C. The Hh-conditioned medium can be stored for up to 1 wk before it loses activity.
5. Transfection solutions: 2 M $CaCl_2$, 2× HEPES-buffered saline (2× HBS): 50 mM HEPES, 1.5 mM Na_2HPO_4, and 280 mM NaCl (pH 7.1).
6. Sterile tissue culture water (Mediatech, Inc., VA).
7. Constructs: Gal4 expression vectors (e.g., ub-Gal4: ubiquitin promoter driven Gal4); Flag-, 3× HA- or 6× Myc-tagged pUAST expression constructs of proteins of interest (*see* **Note 1** for the vector maps and **Note 2** for the properties of different

epitope tags); carrier DNA (e.g., pcDNA). Other expression systems can also be used (*see* **Note 3**).

8. Phosphate buffered saline (PBS): 140 mM NaCl, 2.7 mM KCl, 10 mM Na$_2$HPO$_4$, and 1.8 mM KH$_2$PO$_4$ (pH 7.4).

9. Lysis buffer: 50 mM Tris–HCl (pH 8.0), 100 mM NaCl, 10 mM NaF, 1 mM Na$_3$VO$_4$, 1% NP40, 10% glycerol, 1.5 mM ethylenediamine tetraacetic acid (EDTA; pH 8.0), protease inhibitor tablets (Roche, IN) (other recipes could also be used; *see* **Note 4**).

2.1.2. Cl8 Cell Culture, Transfection, and Lysis

1. Special equipment: 25°C incubator, rotator, Teflon cell scrapers (Fisher, PA).
2. Cl8 cells.
3. Fly extract: place 200 mL M3 medium and about 30 g frozen flies in a blender and blend for 2–3 min. Centrifuge the mixture, discard the oily layer at the top of the supernatant, and then transfer the supernatant into a new tube. Heat-inactivate the extract in a 60°C water bath for 30 min and then centrifuge at 3000g for 1 h. Sterilize the supernatant through 0.22 µm filter. Store the fly extract as 12.5 mL aliquots in a −20°C freezer.
4. 100× insulin stock: dissolve 10 mg (25 IU) insulin in 0.5 mL of 0.01 N HCl and add M3 medium to 20 mL. Store at −20°C as 500 µL aliquots or make fresh each time.
5. Cl8 culture medium: 2% FBS, 2.5% fly extract, and 0.125 IU/mL insulin in M3 medium (Sigma, MO).
6. Solution of trypsin (0.25%) and EDTA (1 mM) (Invitrogen, CA).
7. Transfection solutions: same as 2.1.1.5.
8. Sterile tissue culture water (Mediatech, Inc., VA).
9. Constructs: same as 2.1.1.7.
10. PBS.
11. Lysis buffer: same as 2.1.1.9.

2.1.3. Drosophila Wing Imaginal Discs Lysis

1. Special equipment: dissection microscope, dissection forceps, and tissue grinders for 500 µL microtubes (VWR, NJ).
2. Late third instar *Drosophila* larvae.
3. PBS.
4. Lysis buffer (same as 2.1.1.9.).

2.2. IP

2.2.1. General IP Procedure

1. Special equipment: bench centrifuge (4°C), rotator.
2. Washing buffer: 50 mM Tris–HCl (pH 8.0), 150 mM NaCl, 10 mM NaF, 1 mM Na$_3$VO$_4$, 1% NP40, 10% glycerol, and 1.5 mM EDTA (pH 8.0), protease inhibitor tablets (Roche, IM).
3. Antibodies: mouse anti-Flag, M2 (Sigma); mouse anti-HA (Santa Cruz Biotechnology, Inc, CA.); mouse anti-c-myc, 9E10 (Santa Cruz Biotechnology, Inc.);

mouse anti-smoothened (Smo) (Developmental Studies Hybridoma Bank (DSHB));
mouse anti-Costal 2 (Cos2) (DSHB, IA); mouse anti-fused (Fu) (DSHB).
4. UltraLink immobilized protein A (Pierce, IL) and protein G agarose (Roche) (*see* **Note 5**).

2.2.2. Coupling Antibody to Protein A/G Beads

1. Special equipment: rotator.
2. PBS.
3. UltraLink immobilized protein A (Pierce) and protein G agarose (Roche).
4. 200 mM sodium borate (pH 9.0).
5. 200 mM ethanolamine (pH 8.0).
6. 0.01% merthiolate (ethylmercurithiosalicyclic acid, sodium salt) in PBS.
7. Dimethyl pimelimidate (DMP).
8. Coomassie blue staining solution: 0.025% Coomassie Brilliant Blue R250, 40% (v/v) methanol, and 7% (v/v) acetic acid.
9. Destaining solution: 40% (v/v) methanol and 7% (v/v) acetic acid.

2.3. Western Blot Analysis

1. Special equipment: Mini-protein III SDS-PAGE system and electrophoretic transfer system (Bio-Rad, CA), X-ray film (Kodak, NY), X-ray exposure box (Kodak), and X-ray processor (Kodak).
2. 10% (w/v) SDS.
3. 10% (w/v) ammonium persulfate (APS), made fresh.
4. TEMED (N,N,N',N'-tetramethylethylethylenediamine).
5. 1.5 M Tris–HCl (pH 8.8).
6. 0.5 M Tris–HCl (pH 6.8).
7. 30% acrylamide bisacrylamide (Acr Bis); Acrylamide: bisacrylamide 29:1.
8. Gel preparation recipe (**Table 1**).
9. 2× SDS loading buffer: 0.1 M Tris–HCl (pH 6.8), 10% (v/v) β-mercaptoethanol, 20% (v/v) glycerol, 4% (w/v) SDS, and 0.02% (w/v) bromophenol blue.
10. Gel running buffer (10× stock): 192 mM glycine, 25 mM Tris base, and 0.1% SDS.
11. Transfer buffer (10× stock): 192 mM glycine and 25 mM Tris base.
12. Transfer buffer working solution: 100 mL of 10× stock solution, 200 mL methanol, add H$_2$O to 1000 mL.
13. Filter paper (Bio-Rad).
14. TBST: 20 mM Tris–HCl, 137 mM NaCl, and 0.1% Tween-20 (pH 7.4).
15. Blocking buffer: TBST containing 5% (w/v) nonfat dry milk (Bio-Rad).
16. Antibody dilution buffer: TBST containing 1% (w/v) nonfat dry milk (Bio-Rad).
17. Blotting membrane: PVDF membrane (Millipore, MA).
18. ECL plus kit (Amersham, NJ).
19. Antibodies: mouse anti-HA, F7 (Santa Cruz Biotechnology, Inc.); rabbit anti-HA (Santa Cruz Biotechnology, Inc.); mouse anti-Flag, M2 (Sigma); mouse anti-c-Myc, 9E10 (Santa Cruz Biotechnology, Inc.); goat anti-c-Myc (Santa Cruz Biotechnology, Inc.); rat anti-Smo (*10*); rabbit anti-Cos2 (*5*); rabbit anti-Fu (*11*); rat anti-Cubitus

Table 1
SDS-PAGE Gel Preparation Recipe

	Separating gel[a]			Stacking gel
	7.5%	10%	12%	4%
30% Acr Bis (mL)	2.5	3.3	4	0.33
1.5 M Tris (pH 8.7) (mL)	2.5	2.5	2.5	–
0.5 M Tris (pH 6.8) (mL)	–	–	–	0.63
10% SDS (µL)	100	100	100	25
MilliQ H_2O (mL)	4.85	4.05	3.35	1.59
APS (10%) (µL)	50	50	50	12.5
TEMED (µL)	5	5	5	2.5

[a]For 10 mL separating gel, the amount (X mL) of 30% Acr Bis could be calculated by the formula: X = gel concentration/0.03.

interruptus (Ci), 2A1 *(12)*; rabbit anti-protein kinase A (PKA)c (Santa Cruz Biotechnology, Inc.); rabbit anti-casein kinase I (CKI) ε *(13)*; rabbit anti-glycogen synthase kinase 3 (GSK3) β (Stressgene, CA); and HRP-conjugated secondary antibodies (Jackson Lab., MA).

3. Methods

In IP experiments, the immunoprecipitated proteins could be exogenously expressed epitope-tagged proteins or endogenous proteins. In the former case, the antibodies against the epitope tags are commercially available, so there is no need to generate specific antibodies against the proteins of interest, which makes the experiments much easier to carry out. Furthermore, using an over-expression system, one could study the interactions between truncated proteins or mutant proteins with specific amino acid substitutions. This type of analysis allows one to define the domains required for interaction. In combination with in vivo functional analysis of various deletion mutants, one can assess the biological relevance of individual protein–protein interaction events. However, there are caveats when studying protein–protein interactions with overexpression systems, as such approaches could pick up weak interactions that may not be physiologically relevant. Therefore, it is advised to test whether the interactions detected in the overexpression systems can also be reproduced by immuno-precipitating the endogenous proteins if suitable antibodies are available. In addition, the interaction detected by IP should be confirmed by independent approaches, such as immunocolocalization, GST pull down, and yeast two hybrid assays.

To study the interactions between exogenously expressed proteins, we provide the culture and transfection protocol for two commonly used *Drosophila* cell lines: S2 cells and Cl8 cells. In addition, we discuss the lysate preparation of *Drosophila* wing imaginal discs. The S2 cell line is derived from a primary culture of late stage (20–24 h) *Drosophila melanogaster* embryos. It is easy to culture and transfect. However, there is no detectable Ci in S2 cells (*5*), so exogenous Ci should be introduced into the cells when studying Ci complexes. Cl8 cells are derived from *Drosophila* wing imaginal discs. They express all known components of the Hh-signaling pathway and respond to Hh properly. The *Drosophila* wing imaginal disc has been one of the best models to study Hh signaling, and provides a more physiologically relevant system to study protein– protein interaction in the Hh-signaling pathway. It is also possible to overexpress epitope-tagged proteins in wing discs by using transgenic approaches. However, it is generally challenging to collect enough material from wing discs for IP experiments.

Appropriate controls are required to exclude nonspecific binding due to immunoglobulin or protein A/G beads. To detect interactions between a particular protein and its binding partners, one should use a nonrelated antibody (or preimmune serum) as a control group to the IP. A specific interaction is indicated when as interaction partner is only found in the experimental but not in the control groups. For example: to detect endogenous Smo/Cos2 interactions, one should separately IP with an anti-Smo antibody and a control antibody (e.g., anti-Flag antibody), and detect whether Cos2 can only be pulled down by the Smo but not the Flag antibody. When an epitope-tagged protein is used for IP, exclusion of the tagged protein or replacement with an irrelevant protein with the same tag can be used as controls.

Following IP, the IP products are usually analyzed by Western blot. Since the antibodies used in IP will fall off from the beads and become part of the sample, there will be two IgG bands on the Western blot membrane—heavy chain (about 55 kDa) and light chain (about 25 kDa)—which could mask the protein bands of interest if they are in the same molecular weight range as the IgG bands. To avoid this problem, the animal source of the antibodies for Western blot should be different from the animal source of the antibodies used for IP. For example, one could IP with mouse anti-HA and blot with rabbit anti-HA. However, secondary antibodies sometime have very strong crossreaction between animal species. A good secondary antibody source is key to reduce the signal of the IgG bands. The heavy chain bands can also be avoided by crosslinking the primary antibodies to protein A/G beads before the IP (3.2.2).

3.1. Sample Preparation

3.1.1. S2 Cell Culture, Transfection, and Lysis

1. S2 cells are suspension growth cells. Routinely S2 cells are grown at $1–5 \times 10^7$ cells/mL and split into fresh medium at the dilution in the ratio of 1:5 every 3 d. It is not necessary to trypsinize the cells before passage.

2. Transfection (*see* **Note 6**): Day 1: seed 1×10^7 cells in a 10-cm plate in 10 mL S2 cell culture medium (1×10^6 cells/mL) and grow cells for 16–24 h in 25°C incubator.

3. Day 2: prepare the following transfection mixtures in two separate Eppendorf tubes: Solution A: mix 20 μg recombinant DNA (usually 4 μg ub-Gal4, 2 μg pUAST expression vector for each protein of interest, and carrier DNA to bring the final DNA amount to 20 μg. Other DNA ratio can also be used, *see* **Note 7**) with 60 μL of 2 M CaCl$_2$. Use sterile tissue culture water to bring the final volume to 500 μL. Solution B: 500 μL 2× HBS.

   Slowly add Solution A to Solution B while mixing thoroughly. Incubate the resulting solute on at room temperature for 30 min (a small precipitate will form). Slowly add the mixture to the cells and swirl the plate to mix the solution and medium well. Incubate in 25°C incubator for 24 h.

4. Day 3: Use a pipet to transfer the cells from the plate to a 15-mL tube. Centrifuge the cells at 1000g for 5 min and remove the supernatant. Resuspend cells in 10 mL fresh S2 cell culture medium and replate the cells in the same plate. (If Hh treatment is necessary, the Hh-conditioned medium should be added on this day. To add Hh-conditioned medium, one should resuspend the centrifuged cells in 6 mL Hh-conditioned medium plus 4 mL fresh S2 culture medium.) Incubate the cells at 25°C for 24 h.

5. Day 4: Use a pipet to transfer the cells from the plate to a 15 mL tube. Centrifuge the cells at 1000g for 5 min and remove the supernatant. Wash the cells twice by resuspending the cells with ice-cold PBS followed by centrifuging the cells at 1000g for 5 min and aspirating the supernatant.

6. Add 400 μL lysis buffer to the cell pellet derived from a 10-cm dish culture (there are around 2–3 × 10^7 cells) and transfer the lysate to a 1.5 mL Eppendorf tube. Rotate the tube on a rotator for 30 min at 4°C.

7. Centrifuge the lysate at 16,000g for 10 min.

8. Transfer the supernatant into a new Eppendorf tube and proceed to IP. Take out 20 μL supernatant and place into another tube. Add SDS loading buffer and use this as the whole cell lysate (WCL). Keep the pellet for trouble shooting analysis (*see* **Note 8**).

3.1.2. Cl8 Cell Culture, Transfection and Lysis

1. Routinely Cl8 cells are grown at $1–5 \times 10^5$/mL and are split into fresh medium at a dilution in the ratio of 1:5 for every 2–3 d. Do not split the cells too sparse or let the cells grow too densely. In either case, they will die quickly. Before passage, aspirate the medium and wash cells twice with PBS and trypsinize the cells.

2. One day before transfection, seed 1×10^6 Cl8 cells in a 10-cm plate with 10 mL Cl8 culture medium and grow cells for 16–24 h in a 25°C incubator. The transfection procedure for Cl8 cells is the same as that for S2 cells except for changing the medium. For Cl8 cells, aspirate the old medium and add fresh medium into the culture dish.

3. Before IP experiments, wash the cells twice with cold PBS and add the lysis buffer (500 μL/dish) into the culture dish. Use a cell scraper to scratch the dish and pipet to collect the lysate into Eppendorf tubes. Centrifuge the lysate at 16,000*g* for 10 min at 4°C and transfer the supernatant into a new tube.
4. The resulting cell lysate can be used to proceed with the IP.

3.1.3. Drosophila Wing Imaginal Discs Lysis

1. Dissect wing imaginal discs (50–300 discs/sample, depending on the expression level of the protein of interest) from *Drosophila* late third instar larvae in PBS.
2. Use a pipet to transfer the discs into an Eppendorf tube. Spin gently (100*g*) to settle the discs to the bottom of the tube.
3. Aspirate and remove the last of the PBS with a pipet.
4. Add 50 μL lysis buffer. Use a tissue grinder to crush the discs for 3 min and add 200 μL more lysis buffer into the tube.
5. Centrifuge the sample at 16,000*g* for 10 min at 4°C and transfer the supernatant into a new tube. The resulting lysate can be used for an IP experiment.

3.2. IP

3.2.1. Coupling Antibody to Protein A/G Beads

1. Combine 1 mg monoclonal antibody or affinity-purified polyclonal antibody with 1 mL protein A/G 50% slurry and 5 mL PBS in a 15-mL tube and shake gently on a rotator at room temperature for 1 h.
2. Centrifuge beads at 3000*g* for 2 min and aspirate the supernatant.
3. Wash the beads twice by resuspending in 10 mL of 200 m*M* sodium borate (pH 9.0) following centrifugation and removing the supernatant.
4. Resuspend beads in 5 mL of 200 m*M* sodium borate (pH 9.0), take out 500 μL slurry and set aside for future analysis. Add the solid DMP to the main slurry and bring the final DMP concentration to 20 m*M*.
5. Rotate gently at room temperature for 30 min and take out 500 μL slurry for further analysis.
6. Stop the reaction by washing the beads once with 200 m*M* ethanolamine (pH 8.0) and then resuspend in 5 mL of 200 m*M* ethanolamine (pH 8.0). Incubate at room temperature for 2 h with gentle shaking on the rotator.
7. Wash the beads with PBS and store in PBS with 0.01% merthiolate. Check the efficiency of coupling by boiling the beads taken before and after coupling, in SDS loading buffer and running out on SDS-PAGE. Stain the gel with Coomassie blue stain solution for 30 min and destain for 1 h. Good coupling is indicated by the heavy chain band (about 55 kDa) being found only in samples before but not after coupling.

3.2.2. General IP Procedure

Keep everything ice cold during the experiment.

1. To preclear the lysate, add 10 μL protein A/G beads to the lysate, and rotate the tube for 30 min at 4°C.

2. Spin down the beads at 16,000*g* for 30 s and transfer the supernatant lysate into a new tube.
3. Add antibody to the lysate (*see* **Note 9** for the dilution of different antibodies) and rotate for 1–2 h at 4°C. If using antibody-conjugated protein A/G beads (3.2.1.), add antibody-conjugated beads (usually 10–20 μL of a 50% slurry) at this step, and rotate the tube for 2 h at 4°C. Skip step 4 and go directly to step 5.
4. Add 15 μL (of a 50% slurry) protein A/G beads to the lysate and rotate for 2 h or overnight at 4°C.
5. Spin down the beads at 5000*g* for 30 s. Take out the supernatant and save it for trouble shooting analysis.
6. Resuspend the beads with 1 mL washing buffer and rotate the tube for 10 min at 4°C. Spin down the beads at 5000*g* for 30 s.
7. Aspirate the supernatant and keep the beads at the bottom of the tube. Repeat four times.
8. Add SDS running buffer to the beads and proceed to Western blot analysis.

3.3. Western Blot Analysis

1. Boil the samples for 4 min at 100°C and put on ice for 1 min. Centrifuge at 16,000*g* for 3 min (in some cases, the sample should not be boiled, *see* **Note 10**).
2. Load the samples on the gel and run SDS-PAGE at 60 V for 20 min and at 120 V for about 2 h until the bromophenol blue runs out of the gel (*see* **Note 11**).
3. Pretreat the PVDF membrane by sequentially soaking it in methanol for 10 s, water for 1 min, and transfer buffer for 10 min.
4. Soak the gel, transfer pads, and filter paper in the transfer buffer for 2 min, and make the transfer "sandwich".
5. Transfer the proteins at 200 mA for 2 h at 4°C.
6. Wash the membrane with TBST once and block with blocking buffer for 1 h at room temperature or overnight at 4°C.
7. After blocking, incubate the membrane with diluted primary antibody (*see* **Note 12** for dilutions) for 1–2 h at room temperature.
8. Wash the membrane with TBST three times at room temperature, 5 min each time.
9. Incubate the membrane with diluted HRP-conjugated secondary antibody (1:10,000) for 45 min at room temperature.
10. Wash the membrane with TBST three times at room temperature, 5 min each time.
11. Incubate the membrane with 2 mL ECL mixture for 2 min, wrap it with plastic wrap, and expose to X-ray film. The exposure time depends on the signal intensity, usually less than 2 min.
12. Develop film in the X-ray processor.

4. Notes

1. pUAST Vector Map (*see* **Fig. 1**).
2. Commonly used epitope tags and their properties are shown in **Table 2**. Choosing a suitable epitope tag could be critical for the success of the IP experiment. Since most IP experiments are carried out under nondenaturing conditions, the

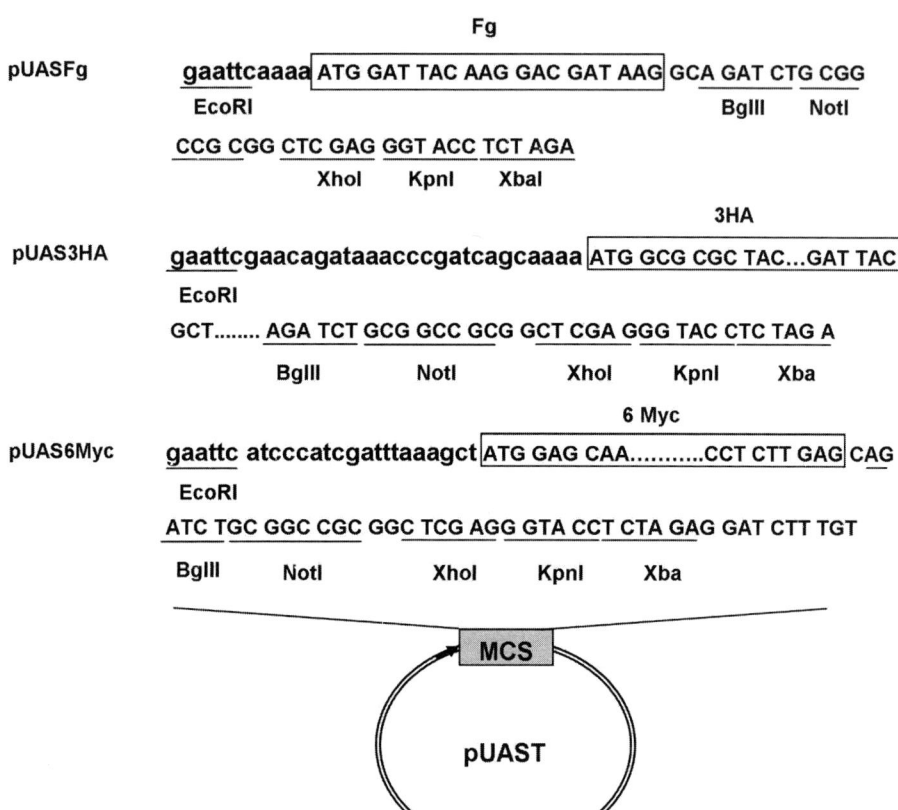

Fig. 1. pUAST vectors with different epitope tags. The region of multiple cloning sites is enlarged and restriction sites are underlined. The boxed nucleotides are the coding sequences for the epitope tags. The reading frames are shown by the capitalized nucleotide triplets.

epitope tag might not be exposed for antibody recognition if it is not in the right position. Usually, tags are fused to the end of a protein (either N- or C-terminus), which is more likely to be exposed. Furthermore, the addition of an epitope tag to a protein should not disturb its normal function. For Smo, the tags can be added to its C-terminus without perturbing activity *(14)*. However, if N-terminal tags are necessary for specific experiments, the epitope tags should be inserted after the Smo signal peptide *(6,15)*.

3. In addition to the Gal4/UAS expression system, the pAc system (Invitrogen) and pMT system (Invitrogen) can also be used for overexpressing proteins in *Drosophila* cell lines.

Table 2
The Properties of Different Epitope Tags

Tag	Number[a]	Position	Comments
HA	2 or 3	N-terminal, C-terminal or internal	Relatively short tag. The antibody is very sensitive for both IP and WB
Flag	1	N-terminal or C-terminal	The M1 antibody can only bind the tag at the N-terminus, M2 antibody can bind the tag at either N- or C-terminus
Myc	3–6	N-terminal, C-terminal or internal	Less than three copies of the Myc tag usually do not give good IP and WB results

[a]Number means how many copies of epitope tags should be fused to the protein of interest in order to get satisfactory IP and WB results.

4. The recipe for the lysis buffer can be modified depending on the experiment. **Table 3** shows the properties of different lysis buffer components. Note that many components of the Hh-signaling pathway are highly phosphorylated. The phosphorylation may affect protein–protein interactions, so it is necessary to add phosphatase inhibitors in the lysis buffer to preserve the phosphorylation states of various signaling components.

5. Proteins A and G are bacteria cell wall proteins that have specific binding sites for the Fc-parts of certain classes of immunoglobulin from different animal sources. Protein A recognizes IgM, IgA, IgD, and most subclasses of IgG. Protein G binds all subclasses of IgG, but not other immunoglobulins. **Table 4.1** shows the affinities of protein A/G for various IgG subclasses and **Table 4.2** for immunoglobulins of different animal species.

6. In addition to calcium phosphate transfection liposome-mediated transfection can also be used for S2 cells. Fugene 6 (Roche) and Effectene (Qiagen, CA) have good transfection efficiencies.

7. Different amounts of a given recombinant DNA could be used in the calcium phosphate transfection. If the protein of interest does not have high-expression levels, more DNA should be used. For example, Smo has very low-expression levels in S2 cells (likely due to its instability), so we usually add 6 μg Smo-recombinant DNA in a single transfection mixture.

8. Trouble shooting for IP is shown in **Table 5**.

9. Final concentrations of the antibodies used in IP experiment are shown in **Table 6**.

10. Smo is a seven trans-membrane protein and tends to aggregate if the loading sample is boiled at high temperature. To reduce Smo aggregation, the loading sample should be heated at 55°C for 5 min rather than being boiled at 100°C.

Table 3
The Properties of the Lysis Buffer Components

Class	Example	Effective concentration	Purpose
Salt	NaCl	50–150 mM	Maintains ionic strength, prevents nonspecific binding
Salt	NaF, Na$_3$VO$_4$	10 mM, 1 mM	Phosphatase inhibitor, prevents protein dephosphorylation
Glycerol		5–10%	Stabilizes protein
Metal chelator	EDTA, EGTA	1 mM	Reduces oxidation damage, prevents protein degradation
Detergent	NP40	0.1–1%	Solubilizes membrane proteins (such as Smo), reduces nonspecific binding

Table 4.1
The Affinities of Protein A/G for Various IgG Subclasses

Antibody	Mouse IgG$_1$	Mouse IgG$_{2a}$	Mouse IgG$_{2b}$	Mouse IgG$_3$	Rat IgG$_1$	Rat IgG$_{2a}$	Rat IgG$_{2b}$	Rat IgG$_{2c}$
Protein A	+	++++	+++	++	–	–	–	+
Protein G	++++	++++	+++	+++	+	++++	++	++

Information from Roche.
The number of "+" indicates the levels of binding affinity. More "+" means higher level of binding affinity. "–" indicates no detectable binding.

Table 4.2
The Affinities of Protein A/G for Immunoglobulins of Different Animal Species

Antibody	Sheep	Goat	Rabbit	Chicken	Hamster	Guinea pig	Rat	Mouse
Protein A	+/–	–	++++	–	+	++++	+/–	++
Protein G	++	++	+++	+	++	++	++	++

Information from Roche.
The number of "+" indicates the levels of binding affinity. More "+" means higher level of binding affinity. "–" indicates no detectable binding.

11. In the Hh-signaling pathway, many proteins including Smo, Fu, Cos2, Su(fu) are hyperphosphorylated. To reveal the electrophoresis mobility shift due to phosphorylation, proper SDS gel concentration should be used. Low voltage (such as 80 V) and extended running times are recommended.

Table 5
Trouble Shooting for IP

Problems	Possible reasons	How to solve
No protein is pulled down	The antibody used is not suitable for IP	Check whether the major protein is still in the supernatant collected in 3.2.2.5. If so, use another antibody
	No protein expression or the expression level is too low	Check the protein expression level in the WCL (3.1.1.5.).
	The epitope tag is not exposed well	Change epitope tag position change to other epitope tags
	The salt or detergent concentration is too high or the pH of the lysis buffer is not suitable for the particular interaction	Reduce salt and detergent concentration and try to use different pH
	The protein is not soluble	Check protein levels in WCL and pellet (3.1.1.5.). If the majority of the protein is in the pellet, try to increase detergent concentration
	The protein is a nuclear protein	The lysis buffer used here barely breaks the nuclear envelope. Try other lysis buffer for a nuclear protein
Background is high	The salt or detergent concentration is too low	Increase salt and detergent concentration
	Washing is not enough	Try to wash longer

12. Final concentrations of the primary antibodies for Western blot are shown in **Table 7**.

Acknowledgments

J.J. is supported by grants from NIH, Leukemia and Lymphoma Society Scholar program, and Welch Foundation (#46303).

Table 6
Final Concentration of the Antibodies Used in IP Experiment

Antibody	Concentration (ng/mL)
mα HA	20
mα Flag	400
mα Myc	20

m, mouse.

Table 7
Final Concentration of the Primary Antibodies for Western Blot

Antibody	Concentration (ng/mL)
mα HA	20
mα Flag	400
mα Myc	20
rα HA	40
gα Myc	40
rα GSK3β	2 µg/mL
rα PKAc	200

m, mouse; r, rabbit; g, goat.

References

1. Ingham, P. W. (2001) Hedgehog signaling: a tale of two lipids. *Science* **294,** 1879–1881.
2. Villavicencio, E. H., Walterhouse, D. O., and Iannaccone, P. M. (2000) The sonic Hedgehog-patched-gli pathway in human development and disease. *Am. J. Hum. Genet.* **67,** 1047–1054.
3. Taipale, J. and Beachy, P. A. (2001) The Hedgehog and Wnt signaling pathways in cancer. *Nature* **411,** 349–354.
4. Zhang, W., Zhao, Y., Tong, C., et al. (2005) Hedgehog-regulated Costal2-kinase complexes control phosphorylation and proteolytic processing of Cubitus interruptus. *Dev. Cell.* **8,** 267–278.
5. Lum, L., Zhang, C., Oh, S., et al. (2003) Hedgehog signal transduction via Smoothened association with a cytoplasmic complex scaffolded by the atypical kinesin, Costal-2. *Mol. Cell.* **12,** 1261–1274.
6. Jia, J., Tong, C., and Jiang, J. (2003) Smoothened transduces Hedgehog signal by physically interacting with Costal2/Fused complex through its C-terminal tail. *Genes Dev.* **17,** 2709–2720.
7. Jia, J., Zhang, L., Zhang, Q., et al. (2005) Phosphorylation by double-time/ CKIepsilon and CKIalpha targets cubitus interruptus for Slimb/beta-TRCP-mediated proteolytic processing. *Dev. Cell* **9,** 819–830.

8. Ruel, L., Rodriguez, R., Gallet, A., Lavenant-Staccini, L., and Therond, P. P. (2003) Stability and association of Smoothened, Costal2 and Fused with Cubitus interruptus are regulated by Hedgehog. *Nat. Cell. Biol.* **5,** 907–913.

9. Miyawaki, A. (2003) Visualization of the spatial and temporal dynamics of intracellular signaling. *Dev. Cell* **4,** 295–305.

10. Denef, N., Neubuser, D., Perez, L., and Cohen, S. M. (2000) Hedgehog induces opposite changes in turnover and subcellular localization of patched and smoothened. *Cell* **102,** 521–531.

11. Therond, P. P., Knight, J. D., Kornberg, T. B., and Bishop, J. M. (1996) Phosphorylation of the fused protein kinase in response to signaling from Hedgehog. *Proc. Natl Acad. Sci. USA* **93,** 4224–4228.

12. Motzny, C. K. and Holmgren, R. (1995) The Drosophila Cubitus interruptus protein and its role in the wingless and Hedgehog signal transduction pathways. *Mech. Dev.* **52,** 137–150.

13. Gao, Z. H., Seeling, J. M., Hill, V., Yochum, A., and Virshup, D. M. (2002) Casein kinase I phosphorylates and destabilizes the beta-catenin degradation complex. *Proc. Natl Acad. Sci. USA* **99,** 1182–1187.

14. Jia, J., Tong, C., Wang, B., Luo, L., and Jiang, J. (2004) Hedgehog signaling activity of Smoothened requires phosphorylation by protein kinase A and casein kinase I. *Nature* **432,** 1045–1050.

15. Zhu, A. J., Zheng, L., Suyama, K., and Scott, M. P. (2003) Altered localization of Drosophila Smoothened protein activates Hedgehog signal transduction. *Genes Dev.* **17,** 1240–1252.

16

Sequence Analyses to Study the Evolutionary History and *Cis*-Regulatory Elements of Hedgehog Genes

Ferenc Müller and Anne-Gaelle Borycki

Abstract

Sequence analysis and comparative genomics are powerful tools to gain knowledge on multiple aspects of gene and protein regulation and function. These have been widely used to understand the evolutionary history and the biochemistry of Hedgehog (Hh) proteins, and the molecular control of *Hedgehog* gene expression. Here, we report on some of the methods available to retrieve protein and genomic sequences. We describe how protein sequence comparison can produce information on the evolutionary history of Hh proteins. Moreover, we describe the use of genomic sequence analysis including phylogenetic footprinting and transcription factor-binding site search tools, techniques that allow for the characterization of *cis*-regulatory elements of developmental genes such as the *Hedgehog* genes.

Key Words: Hedgehog; sequence analysis; evolution; *cis*-regulatory element; bioinformatics.

1. Introduction

The use of bioinformatics to analyze protein and genomic sequences is based on the principle that functional regions in proteins and genomes are less likely to undergo random mutational changes, hence conserved sequences are candidates for important structural or *cis*-regulatory function *(1–7)*. The application of this principle to *Hedgehog* (*hh*) genes and proteins is particularly relevant. Not only *hh* genes are often highly conserved in their protein-coding sequence, but they have also highly conserved expression patterns among distantly related phylogenetic groups *(8–12)*. This implies that homologs can be searched in different taxa based on the conservation of protein domains. A history of the evolution of this protein family can then be deduced from analysis of the number

From: *Methods in Molecular Biology: Hedgehog Signaling Protocols*
Edited by: J. Horabin © Humana Press Inc., Totowa, NJ

of homologs in each taxa, the rate of amino acid substitutions and the evolutionary distance between orthologs. In this chapter, we focus on the use of sequence analysis and comparative genomics for the identification of Hedgehog (Hh) family members in different taxa and the analysis of their evolutionary history.

Cis-regulatory elements (CRMs) of genes play a crucial role in the correct spatial and temporal expression of genes. Mutations in CRMs can cause gene misexpression and disease or expose individuals to higher risk of multifactorial diseases. For example mutations mapping in the vicinity of *sonic hedgehog*-regulatory elements have been suggested to cause preaxial polydactily *(13,14)*. Therefore, identification of CRMs is an important step in understanding the genetic basis of human diseases. We describe here the current methods for the identification of *cis*-acting-regulatory elements of genes. Although no *hh* gene-specific protocols can be established for *cis*-regulatory sequence analysis, this chapter provides examples related to *hh* genes from the published literature. Rather than providing detailed protocols, we aim to give the reader general considerations and advice to apply best, these biocomputing tools to the study of Hh proteins and genes.

2. Materials

All software and algorithms cited in this chapter can be downloaded from the internet. Some of these are commercial packages, but most are free. We have listed their web sites in **Table 1**. Moreover, a selection of useful websites with more software for phylogenetic analyses and tools for analysis of CRMs are also listed in **Table 2**.

3. Methods
3.1. Evolutionary Analysis of Hedgehog Proteins

The phylogenetic relationship and evolution of Hh proteins have been analyzed in considerable detail *(15,16)*. Recently, further members of the Hh gene family have been reported in teleosts with the description of a second *indian hedgehog* and a *desert hedgehog* homologs *(17)*.

3.1.1. Retrieving Protein Sequences for Phylogenetic Analyses

Protein sequences of conserved genes used to be predicted from a cDNA sequence, isolated either by degenerate polymerase chain reaction or by screening of cDNA libraries. Although these methods are still used in the case of nonmodel organisms, protein sequences are now mostly isolated *in silico*. There are numerous possibilities to find the sequences of interest by searching protein databases or genomic databases. Searches can be performed with keywords (i.e., Hh or Shh) and/or using Blast searches. NCBI and EBI have search tools to scan GenBank and Swiss-Prot.

Table 1
Web Sites for Sequence, Phylogenetic and *Cis*-Regulatory Element Analyses

BlastZ	http://www.psc.edu/general/ software/packages/blastz/	Sequence alignment
CisOrtho	http://dev.wormbase.org/cisortho/	Worm TF-binding sites
ClustalW	http://bioweb.pasteur.fr/seqanal/ interfaces/clustalw.html	Sequence alignment
ClustalX	ftp://ftp-igbmc.u-strasbg.fr/ pub/ClustalX/	Sequence alignment
Compare Prospector	http://ai.stanford.edu/~iliu/ CompareProspector/	TF-binding sites
CONREAL	http://conreal.niob.knaw.nl/	TF-binding sites
Consite	http://mordor.cgb.ki.se/cgi-bin/ CONSITE/consite/	TF-binding sites
DBTSS	http://dbtss.hgc.jp	Transcription start sites
DiAlign	http://bibiserv.techfak. unibielefeld.de/dialign/	Sequence alignment
DoOP	http://doop.abc.hu/	Chordate promoters
EBI	http://www.ebi.ac.uk	Genomic tools (searches, alignment)
ECR browser	http://ecrbrowser.dcode.org/	Display alignment
Ensembl	http://www.ensembl.org	Genome sequences
Footprinter	http;//wingless.cs.washington.edu/ htbin-post/unrestricted/FootPrinter Web/FootPrinterInput2.pl	TF-binding sites
Genomatrix	www.genomatix.de	TF-binding sites
JASPAR	http://mordor.cgb.ki.se/cgi-bin/ jaspar2005/jaspar_db.pl	TF-binding sites
LAGAN/ shuffle LAGAN	http://lagan.stanford.edu/ lagan_web/index.shtml	Sequence alignment
MAFFT	http://www.biophys.kyoto-u.ac.jp/ ~katoh/programs/align/mafft/	Sequence alignment
MUSCLE	http://www.ebi.ac.uk/muscle/	Sequence alignment
NCBI	http://www.ncbi.nlm.nih.gov	Genomic tools (searches...)
NJPlot	http://pbil.univ-lyon1.fr/ software/njplot.html	Tree building
PAUP	http://paup.csit.fsu.edu/	Tree building
PHYLIP	http://evolution.genetics. washington.edu/phylip.html	Tree building
PhyloCon	http://ural.wustl.edu/~twang/PhyloCon/	TF-binding sites
PipMaker	http://pipmaker.bx.psu.edu/pipmaker/	Display alignment

(Continued)

Table 1 *(Continued)*

rVISTA	http://rvista.dcode.org/	TF-binding sites
T-Coffee	http://www.ebi.ac.uk/t-coffee/	Sequence alignment
TraFac	http://trafac.cchmc.org/trafac/index.jsp	TF-binding sites
TRANSFAC	http://www.gene-regulation.com/ pub/databases.html	TF-binding sites
TreeView	http://taxonomy.zoology.gla.ac.uk/ rod/treeview.html	Tree drawing
VISTA	http://genome.lbl.gov/vista/index.shtml	Display alignment
Weeder	http://159.149.109.16:8080/weederWeb/	CRM search
WordSpy	http://cic.cs.wustl.edu/wordspy	TF-binding sites

Table 2
Websites with Bioinformatic Tools Mentioned in this Article

http://bioweb.pasteur.fr/ intro-uk.html	Contains several sequence analysis and comparison software
http://evolution.genetics. washington.edu/phylip/ software.html	Has multiple software for phylogenetic analyses
http://taxonomy.zoology. gla.ac.uk/software/index.html	Also multiple software for tree building and analysis
http://tolweb.org/tree/	Tree of life lists all organisms and provides information on their classification
http://www.ucmp.berkeley.edu/ alllife/threedomains.html	Another web site providing information on species and their classification
http://pbil.univ-lyon1.fr/ software/	Multiple phylogenetic tools and software
www.Dcode.org	A series of web-based alignment (zPicture, Mulan, and eShadow) and visualization tools (ECR browser) are provided in this server.

Alternatively, animal model genomes can be searched using Ensembl, which in its newest version (v.37) contains several genomes, although not all complete and annotated (*see* **Table 3**).

3.1.2. Protein Sequence Alignment

Protein sequences must then be aligned. For our purpose, a global alignment method, which performs progressive pairwise alignments should be used. ClustalW *(18)* or Clustal X *(19)* software have been widely used. However, with the recent growth of sequence databases, it has been necessary to develop other algorithms that can align large protein families with speed and accuracy.

Table 3
Genomes Available in Ensembl

Species name	Common name
Anopheles gambiae	Mosquito
Apis mellifera	Honey bee
Bos tauris	Cattle
Caenorhabditis elegans	Nematode
Canis familiaris	Dog
Ciona intestinalis	Sea squirt
Danio rerio	Zebrafish
Drosophila melanogaster	Fruit fly
Gallus gallus	Chick
Homo sapiens	Human
Macaca mulatta	Rhesus monkey
Monodelphis domestica opossum	Short-tailed
Mus musculus	Mouse
Pan troglodytes	Chimpanzee
Rattus Norvegicus	Rat
Saccharomyces cerevisiae	Yeast
Takifugu rubripes	Fugu
Tetraodon nigroviridis	Pufferfish
Xenopus tropicalis	Pipid frog

Thus, new software for multiple sequence alignment have been designed and include: T-Coffee *(20)* which is slower than Clustal but tends to perform better in sequence alignments. MAFFT *(21)* is another program, which performs very well with sequences of different lengths (*see* **Note 1**) and appears to be faster than Clustal. Finally, MUSCLE *(22)* is advertised as faster than T-Coffee or Clustal.

Before proceeding with the inference of a phylogenetic tree, sequence alignments should be checked and edited to realign sequences and eliminate gaps. Jalview provided in Clustal, MUSCLE, and T-Coffee allows you to edit your sequence alignment, whereas the PHYLIP package contains its own sequence editing program. Once the alignment has been performed, the tree file should be saved in the appropriate format (*see* **Note 2**).

3.1.3. Building a Phylogenetic Tree

There are three methods which make up two main classes to infer a phylogenetic tree: Character-based methods, which include maximum parsimony (MP) and maximum likelihood (ML) *(23)*, and distance-based methods, which include

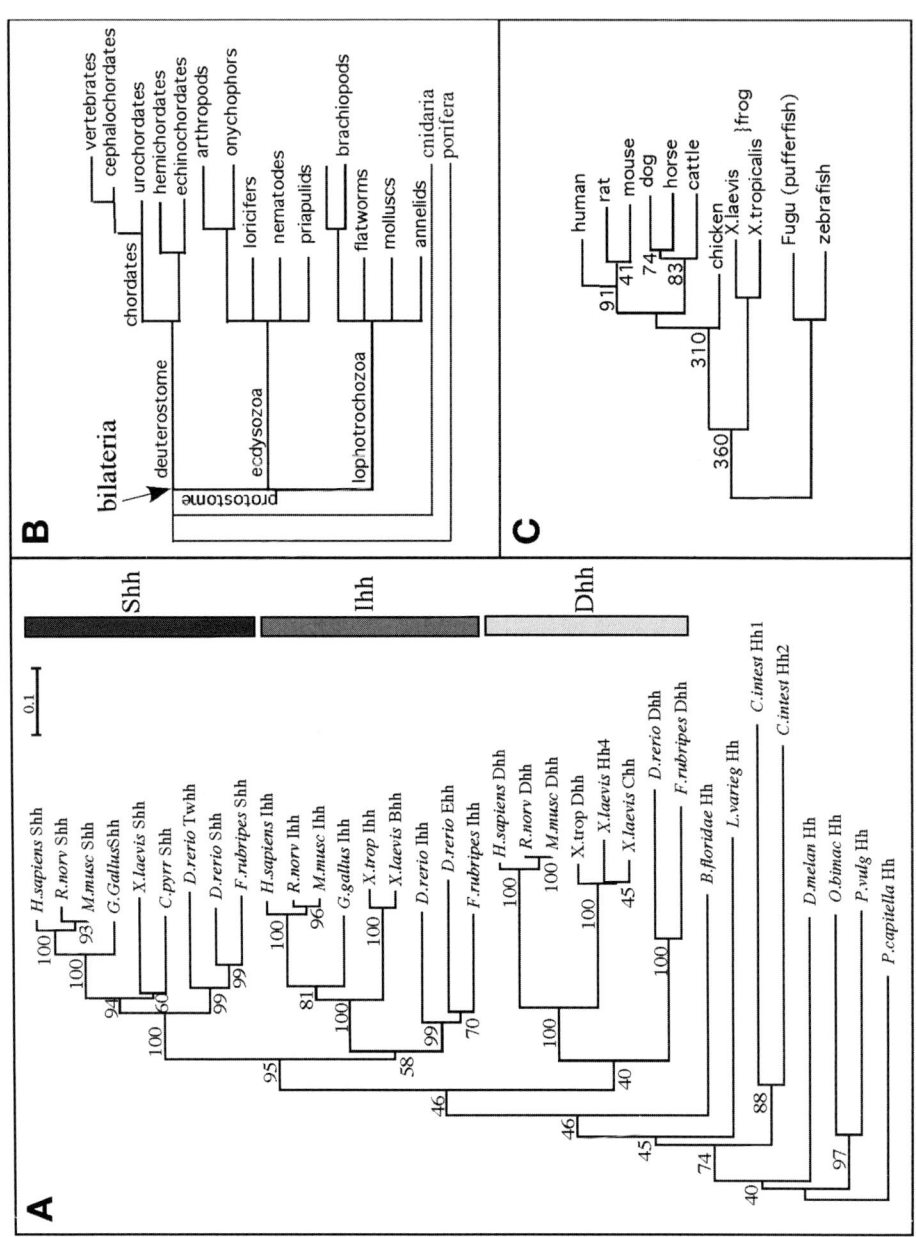

the neighbor-joining (NJ) method *(24)*. The former relies on character states, such as the position of an amino acid at a specific place, whereas with the latter method evolutionary distances are calculated as the number of amino acid replacements between two proteins. None of these methods provide entire satisfaction (i.e., will infer a true tree) because they rely on several assumptions; for instance, a constant rate of divergence of a taxa from an ancestor. NJPlot algorithms will build a tree based on the NJ method, whereas PHYLIP and PAUP allow for the inference of an evolutionary tree using NJ, MP, or ML methods. Because distance-based methods are more amenable to molecular data (such as protein sequences) and several methods including bootstrap analyses have been designed to establish the reliability of an evolutionary tree. NJ methods tend to be more widely used and have been the preferred method for analyses of Hh proteins *(15,25)*. If using NJ Plot open the tree file (.nj) previously saved. If using PHYLIP, a tree can be drawn using DRAWGRAM. Both will draw rooted trees, which allow for evolutionary analyses, in contrast to unrooted trees, which only display the degree of relationship with no mention of the most recent ancestor. TREEVIEW is another software package to draw trees. It supports tree files in pretty much any format and will display bootstrap values.

If the assumption of rate constancy among taxa does not account for the actual rate of divergence, the inferred tree may appear erroneous (i.e., misplace a species or a group of species). These errors can be remedied by choosing an outgroup as a reference (i.e., a species for which we have previous knowledge that it diverged from a common ancestor prior to the other species listed). A new tree is then built based on a new distance matrix established from the reference (**Fig. 1A**).

3.1.4. Phylogenetic Tree Analyses

Tree reliability: One of the advantages of using the NJ method is that it allows for bootstrap analysis, a computational method to apply statistics on a tree topology *(26)*. This technique calculates the level of confidence for each clade

---

Fig. 1. Phylogenetic analysis of Hedgehog (Hh) proteins. (**A**) Inferred phylogenetic tree of Hh proteins: Hh proteins (full length) were aligned using ClustalW. An inferred phylogenetic tree was established with the NJ method after eliminating gaps from the alignment and using the Kimura correction for distances. The annelid *P. capitella* was used as an outgroup. Bootstrap values, indicated at the nodes, were calculated from 1000 pseudosamples within ClustalW. Branch lengths are proportional to the distance. (**B**) Phylogeny of the Metazoa. At the branching between protostome and deuterostome is indicated the position of the bilaterian ancestor. (**C**) Phylogeny of vertebrates. Estimated evolutionary distances between some species are indicated at the node (in my: million year).

of an inferred tree. This is done through a resampling technique where a series of pseudosamples are generated (usually between 500 and 1000, *see* **Note 3**) and the deduced trees are compared with the inferred one. A bootstrap value, expressed as the percentage of trees having the same topology as the inferred tree, is then calculated. It is usually admitted that a bootstrap value of >95 corresponds to a high level of confidence in the clade, whereas values <70 show a low level of confidence. Bootstrap can be run from PHYLIP using Seqboot or Clustal.

Estimating divergence time: An estimation of the evolutionary divergence time can be calculated from a distance-based tree (**Fig. 1A**). This calculation is based on the hypothesis that the rate of amino acid substitutions is constant during evolution. First, the rate of divergence per site per million years, r, is calculated for two species for which the divergence time, T_1, is known from other data (paleontological records, molecular data). Usually, vertebrates are a better choice because there are many records available providing the best approximate divergence time (*see* **Note 4**).

$r = d/2T_1$, where d is the average distance between the two species chosen and the distance is directly proportional to the rate of amino acid substitution. Once r is determined, it can be applied to the equation $T_2 = d_{avg}/2r$, where T_2 is the unknown divergence time between two species/events we are interested in and d_{avg} is the average distance between these two species/event.

Using similar calculations, it was found that the divergence time between Shh and Ihh, and Shh and Dhh was 563 and 662 my, respectively *(15)*, which suggests that the first duplication of the Hh gene to give rise to the Dhh family occurred prior to the emergence of chordates (550 my) *(27,28)*. This is not consistent with the fact that prior to the emergence of vertebrates, a single Hh gene is found in all three phyla, Deuterostomia, Ecdysozoa, and Lophotrochozoa (**Fig. 1A,B**). In particular, the presence of a single Hh gene in the cephalo-chordate amphioxus *Branchiostoma floridae* *(12)* suggests that the duplication event that gave rise to Hh1 and Hh2 in the urochordate *Ciona intestinalis* occurred independently from the duplication events leading to the Dhh, Ihh, and Shh families (**Fig. 1A–C**) *(29)*. An interesting exception to the existence of a single Hh is that of the nematode *Caenorhabditis elegans* for which no true Hh ortholog was found. In contrast, closer sequence comparisons with subdomains of the Hh protein unraveled that several *C. elegans* proteins were homologs to the C-terminal region of Hh and formed a family of proteins, the inteins, with endonuclease activity *(30)*. Because earlier taxa such as the mollusc *Proteus vulgaris* and the Annelid *P. capitella* do have a single Hh gene, this would suggest that nematodes have had Hh proteins but lost them during evolution. Alternatively, there is the possibility that nematodes do not belong to Ecdysozoa and form an earlier taxon *(31)*. There are data consistent with a grouping of Arthropods and vertebrates together (protostome and deuterostome), called Coelomata that leave out the nematodes, which form an earlier phylum, the Pseudocoelomata *(32)*. If this were the case, Hh would have evolved after the emergence of nematodes and before the Coelomata group.

3.2. Detection of Cis-Regulatory Elements of Hedgehog Genes by Sequence Analysis

CRMs do not have stringent directional, positional, and compositional constraints such as coding exons, which makes their automated detection with bioinformatics tools more difficult. One technique often used is phylogenetic footprinting *(33)*, which, is based on the principle that alignment of noncoding sequences from different species reveals evolutionarily conserved segments that are candidates for *cis*-regulatory function *(1,3,5,7,34)*. Bioinformatic tools which utilize phylogenetic footprinting to detect such regions have been reviewed recently *(35–38)*. Phylogenetic footprinting has been used extensively to identify putative CRMs of *sonic hedgehog* orthologs *(36,39–42)*.

3.2.1. Choice of Sequence Alignment and Visualization Tools

Two main strategies can be followed in sequence alignment: The local alignment protocol (e.g., BLASTZ *[43]*) searches for short stretches of similarity between the sequences, which are then extended, whereas global alignment tools (e.g., LAGAN *[44]*) search for best alignment over the entire length of the sequence using local similarities as anchors (*see* **Note 5**). A recent addition to LAGAN also allows for the detection of inversions between the two compared sequences (shuffle-LAGAN *[44]*). Global alignment tools have a higher sensitivity, whereas local tools provide better specificity in detection of shorter conserved blocks *(45)*. Results of sequence alignments are usually displayed through web-based graphical tools, such as PipMaker *(46)*, ECR browser *(47)*, and VISTA *(48,49)*, which indicate conservations above certain threshold levels. Because of their distinct designs, the performance of global and local alignment algorithms differs in the detection of conservation. Notably, the DiAlign tool *(50,51)* allows for both local and global alignment output modes.

3.2.2. Choice of Genomes for Cross Species Comparison

Comparisons of multiple species ("phylogenetic shadowing") *(38)*, using a set of closely related species (e.g., Refs. *[50,52]*), may be applied for the identification of conserved elements. However, the efficiency of finding conserved CRMs by phylogenetic footprinting (both in terms of number and level of conservation) is dependent on the evolutionary distance between the species compared *(38,53)*. Comparisons between mouse and human (approx. 90 million years, **Fig. 1C**) provide close evolutionary distance with high degree of conservation among functionally relevant binding sites placed in conserved blocks *(54–58)*.

However, the slow rate of neutral divergence among vertebrates, may result in the retention of conserved sequences with no regulatory role between species with short evolutionary distance *(59)*. Several vertebrate genomes representing most major classes have recently been sequenced (*see* **Table 3**), providing the raw material for comparative analyses of species with greater evolutionary distances than mammals. A note of caution must be applied though, the greater the evolutionary distance, the more likely regulatory elements will have diverged. Thus, a lower number of regulatory elements will have retained conserved transcriptional activities, reducing the likelihood of identifying conserved CRMs *(60)*. However, it is generally observed that developmentally regulated genes (including *hh* genes) and transcription factors tend to be more conserved in their CRMs than other genes *(40,61)*. This was particularly striking in CRMs of fish and mammals *sonic hedgehog* orthologs that are separated by 450 my and still show remarkable conservation *(36)*.

3.2.3. Variable Divergence of CRMs Within a Locus

CRMs within one gene locus may have different rates of change, as is the case for the *shh* locus itself. For example, four enhancers named *ar-A* to *ar-D*, are involved in *shh* activation in the zebrafish midline tissues. These four CRMs show varying degree of conservation between pufferfish and mouse *(36,62–64)*. Interestingly, *ar-A* and *ar-C* are conserved between fish and mouse, whereas *ar-B* also shows significant sequence similarity when compared with zebrafish and pufferfish (*Tetraodon nigroviridis*), indicating that the phylogenetic footprinting approach can result in the detection of additional functional regulatory elements when the evolutionary distance between the species used in the analysis matches the rate of change in regulatory sequences. The enhancer *ar-C* is significantly conserved in mouse but less than *ar-A*, and is active in the midline in zebrafish and mouse. Strikingly, no function has been assigned to the well-conserved *ar-A* in mouse. This may indicate a conservation due to functional constraints other than CRM (reviewed in Ref. *[65]*). Significant sequence similarity in the 3′ UTR region of *shh* genes has also been observed between fish and mouse. However, no published data is available for a putative function of these conserved sequences.

3.2.4. Identification of Long Distance Regulatory Elements

It is not always trivial to assign a predicted conserved regulatory element to its cognate gene. The distance limit of regulatory elements from their regulated gene is not at all deciphered, and looping of chromatin over 40 Mb to sites of transcriptional activity has been demonstrated *(66)*. Bacterial or phage artificial chromosome vectors provide a technology for analysis of regulatory elements over large distances *(42)*. This approach allowed for the detection of *shh*-regulatory elements that lay several hundred kilobases away from the coding sequence in the mouse. Several of the elements identified in the mouse (SBE 2, 3, and 4) are well conserved among human, chicken, and frog, but not teleost fish sequences *(42)*. Interestingly, the function of these long distance elements is to drive *shh* expression in the ventral diencephalon, an activity covered by the intronic *ar-C* enhancer in the fish. This functional divergence of enhancers may explain the lack of conservation of SBE2-4 and suggests that subfunctionalization mechanisms may be involved in the evolution of *shh* CRMs *(67)*.

A large number of genes are likely to contain CRMs at very long distance from the gene locus *(68)*. An extreme example is the case of the *sonic hedgehog* limb enhancer, which lies 1 Mb away from the *shh* coding sequence in the intron of the *lbmr1* gene *(69)*. This enhancer is highly conserved among vertebrates both in terms of its sequence and its interdigital position in the *lmbr1* gene *(70)*. This example suggests that further regulatory elements placed at a large distance may function in the regulation of *shh*. Indeed, several conserved noncoding elements were found at long distances from *shh* (up to 50 kb in fugu) and when tested in zebrafish embryos, provided enhancer activity *(41)*. It may be possible to identify these elements by limiting the search to chromosomal regions that remain unchanged during evolution. The interdigitation of coding genes with embedded regulatory elements of other neighboring genes also implies an evolutionary constraint on chromosomal rearrangements to avoid breakpoints in such regions. Conserved chromosomal synteny has been suggested to aid in predicting the limits of the regulatory regions of a gene *(71,72)*. Thus, comparisons between multiple species should establish the furthest, long distance CRMs are located from the promoter by analyzing the breakpoints of syntenic fragments. To assist researchers in these analyses, the Ensembl genome server database provides mammalian and chick chromosomal synteny, whereas an independent web server provides fugu and human synteny analysis *(73)*.

3.2.5. Identification of the Transcriptional Start Site and the Core Promoter

Core or basal promoters are positionally defined regulatory regions, which are located about 50–100 base pairs (bp) up- and/or downstream of the transcriptional start site (TSS), and are required for the formation of preinitiation complexes for subsequent transcription initiation *(74)* (*see* **Note 6**). The absence of experimental approaches to characterize TSSs and the diversity of promoter types made it relatively difficult to predict accurately core promoter regions using sequence analysis, despite the large number of programs available on the internet (*see* **Tables 1** and **2** for a selection of tools). Prediction of core promoters has recently improved substantially, due to the accumulation of large-scale data on TSS *(75,76)*. Promoter predictors based on searching for motifs such as the TATA box (reviewed in Ref. *[74]*) failed, as it is now known that only a subset of human genes whose transcription is initiated by the RNA polymerase II contain a TATA box *(77)*. The characterization of motifs involved in transcription initiation of the remaining genes is still in progress *(77,78)*. A TATA box is however present in vertebrate *shh* genes *(79,80)*. Interestingly, transcription factors and brain-specific genes were found to have shorter conserved blocks than other genes *(81)*. The core promoter of vertebrate *shh* genes have been characterized in fish and human *(79,80)* and were shown to contain two TSSs and to be regulated by retinoic acid and Foxa2 (HNF3β).

3.2.6. Transcription Factor-Binding Site Analysis

Information on transcription factor-binding sites are available in either commercial (like TRANSFAC *(82)*, Genomatix) or open access (JASPAR *[83]*) databases. Binding-site clustering is a feature of CRMs *(84)*, which is utilized by several algorithms *(85–91)*. The predictive value of such clustering approaches is enhanced by incorporating sequence conservation criteria (*see* Ref. *(92)* for example). Ahab also detects clusters of weak sites *(93,94)*, and this can be further improved with Stubb, which includes comparative information and allows for the prediction of regulatory modules *(95,96)*. To search entire genomes for coexpressed genes, a software package (CisOrtho *[97]*) was developed which evaluates the co-occurrence of motifs in orthologs regions. CRMs of coregulated genes show "signatures", i.e., transcription factor-binding site combinations with distinct spacing and orientation requirements *(90,98)*, which seem to be retained between species even when the overall sequence similarity is low *(90)*. On the basis of this finding, TraFaC identifies conserved TF-binding sites by scanning regions of conserved sequence similarity to detect co-occurrence of binding sites *(99)*, whereas rVista *(100,101)* and ConSite *(57)* score aligned binding sites in conserved regions. CONREAL *(102)* applies a similar approach and uses binding-site predictions as anchors for sequence alignment, and performs better than other sequence alignment programs when aligning sequences from distant species. As more algorithms for motif detection that take into account phylogenetic conservation (e.g., PhyloCon *[103]*, CompareProspector *[104]*, Footprinter *[105]*) become available, functional-binding sites in *hedgehog* genes and other developmentally regulated genes will be identified.

4. Notes

1. It has been reported that variations in sequence length affect the accuracy of sequence alignments. ClustalW seems to be more sensitive to this issue than

MAFFT. Thus, it is recommended to include sequences covering regions of similar length, although a sufficiently large portion of the protein sequence should be included to make the analysis meaningful. Comparing fragments of Hh protein to other full-length Hh proteins, for instance, can only lead to unmeaningful data.

2. Take care of saving the tree file corresponding to the sequence alignment in the correct format (.nj if you are to use NJPlot to draw the tree or .ph if you are to use PHYLIP).

3. It is common in the literature to see bootstrap samples of 100 or 200. It is recommended to use 500–1000, especially if many species are involved.

4. Listed here are some evolutionary divergence times commonly used (*see* **Fig. 1C**). Rat/mouse, 41 my; mammals/fishes, 450 my; mammals/amphibians, 360 my; mammals/birds, 310 my.

5. A consideration when choosing a particular program is that many algorithms have been optimized for specific-species comparisons (e.g., BlastZ for human-mouse, WABA *(106)* for *C. elegans–C. briggsae*) and may not perform well with other species.

6. A recent larger-scale analysis of mouse and human promoters identified conserved blocks within 500 bp from the start site, thereby defining the likely 5′ limit of proximal promoter regions *(58)*.

References

1. Bejerano, G., Pheasant, M., Makunin, I., et al. (2004) Ultraconserved elements in the human genome. *Science* **304,** 1321–1325.
2. Dermitzakis, E. T. and Clark, A. G. (2002) Evolution of transcription factor binding sites in Mammalian gene regulatory regions: conservation and turnover. *Mol. Biol. Evol.* **19,** 1114–1121.
3. Frazer, K. A., Sheehan, J. B., Stokowski, R. P., et al. (2001) Evolutionarily conserved sequences on human chromosome 21. *Genome Res.* **11,** 1651–1659.
4. Hillier, L. W., Miller, W., Birney, E., (2004) Sequence and comparative analysis of the chicken genome provide unique perspectives on vertebrate evolution. *Nature* **432,** 695–716.
5. Mural, R. J., Adams, M. D., Myers, E. W., et al. (2002) A comparison of whole-genome shotgun-derived mouse chromosome 16 and the human genome. *Science* **296,** 1661–1671.
6. Rubin, G. M., Yandell, M. D., Wortman, J. R., et al. (2000) Comparative genomics of the eukaryotes. *Science* **287,** 2204–2215.
7. Waterston, R. H., Lindblad-Toh, K., Birney, E., Rogers, J., et al. (2002) Initial sequencing and comparative analysis of the mouse genome. *Nature* **420,** 520–562.
8. Echelard, Y., Epstein, D. J., St-Jacques, B., et al. (1993) Sonic Hedgehog, a member of a family of putative signaling molecules, is implicated in the regulation of CNS polarity. *Cell* **75,** 1417–1430.

9. Johnson, R. L., Laufer, E., Riddle, R. D., and Tabin, C. (1994) Ectopic expression of Sonic Hedgehog alters dorsal-ventral patterning of somites. *Cell* **79,** 1165–1173.

10. Krauss, S., Concordet, J. P., and Ingham, P. W. (1993) A functionally conserved homolog of the Drosophila segment polarity gene hh is expressed in tissues with polarizing activity in zebrafish embryos. *Cell* **75,** 1431–1444.

11. Ruiz i Altaba, A., Jessell, T. M., and Roelink, H. (1995) Restrictions to floor plate induction by Hedgehog and winged-helix genes in the neural tube of frog embryos. *Mol. Cell. Neurosci.* **6,** 106–121.

12. Shimeld, S. M. (1999) The evolution of the Hedgehog gene family in chordates: insights from amphioxus hedgehog. *Dev. Genes Evol.* **209,** 40–47.

13. Tsukurov, O., Boehmer, A., Flynn, J., et al. (1994) A complex bilateral polysyndactyly disease locus maps to chromosome 7q36. *Nat. Genet.* **6,** 282–286.

14. Lettice, L. A., Horikoshi, T., Heaney, S. J., et al. (2002). Disruption of a long-range cis-acting regulator for Shh causes preaxial polydactyly. *Proc. Natl Acad. Sci. USA* **99,** 7548–7553.

15. Kumar, S., Balczarek, K. A., and Lai, Z. C. (1996) Evolution of the Hedgehog gene family. *Genetics* **142,** 965–972.

16. Zardoya, R., Abouheif, E., and Meyer, A. (1996) Evolutionary analyses of Hedgehog and Hoxd-10 genes in fish species closely related to the zebrafish. *Proc. Natl Acad. Sci. USA* **93,** 13,036–13,041.

17. Avaron, F., Hoffman, L., Guay, D., and Akimenko, M. A. (2006) Characterization of two new zebrafish members of the Hedgehog family: atypical expression of a zebrafish indian hedgehog gene in skeletal elements of both endochondral and dermal origins. *Dev. Dyn.* **235,** 478–489.

18. Thompson, J. D., Higgins, D. G., and Gibson, T. J. (1994) CLUSTAL W: improving the sensitivity of progressive multiple sequence alignment through sequence weighting, position-specific gap penalties and weight matrix choice. *Nucleic Acids Res.* **22,** 4673–4680.

19. Thompson, J. D., Gibson, T. J., Plewniak, F., Jeanmougin, F., and Higgins, D. G. (1997) The CLUSTAL_X windows interface: flexible strategies for multiple sequence alignment aided by quality analysis tools. *Nucleic Acids Res.* **25,** 4876–4882.

20. Notredame, C., Higgins, D. G., and Heringa, J. (2000) T-Coffee: A novel method for fast and accurate multiple sequence alignment. *J. Mol. Biol.* **302,** 205–217.

21. Katoh, K., Misawa, K., Kuma, K., and Miyata, T. (2002) MAFFT: a novel method for rapid multiple sequence alignment based on fast Fourier transform. *Nucleic Acids Res.* **30,** 3059–3066.

22. Edgar, R. C. (2004) MUSCLE: a multiple sequence alignment method with reduced time and space complexity. *BMC Bioinformatics* **5,** 113.

23. Felsenstein, J. (1988) Phylogenies from molecular sequences: inference and reliability. *Annu. Rev. Genet* **22,** 521–565.

24. Saitou, N. and Nei, M. (1987) The neighbor-joining method: a new method for reconstructing phylogenetic trees. *Mol. Biol. Evol.* **4,** 406–425.
25. Zardoya, R., Abouheif, E., and Meyer, A. (1996) Evolution and orthology of Hedgehog genes. *Trends Genet.* **12,** 496–497.
26. Felsenstein, J. (1985) Confidence limits on phylogenies: An approach using the bootstrap. *Evolution* **39,** 783–791.
27. Conway Morris, S. (1993) The fossil record and the early evolution of the Metazoa. *Nature* **361,** 219–225.
28. Dehal, P., Satou, Y., Campbell, R .K., et al. (2002) The draft genome of *Ciona intestinalis*: insights into chordate and vertebrate origins. *Science* **298,** 2157–2167.
29. Takatori, N., Satou, Y., and Satoh, N. (2002) Expression of Hedgehog genes in *Ciona intestinalis* embryos. *Mech. Dev.* **116,** 235–238.
30. Aspock, G., Kagoshima, H., Niklaus, G., and Burglin, T. R. (1999) *Caenorhabditis elegans* has scores of Hedgehog-related genes: sequence and expression analysis. *Genome Res.* **9,** 909-923.
31. Hedges, S. B. (2002). The origin and evolution of model organisms. *Nat. Rev. Genet.* **3,** 838–849.
32. Blair, J. E., Ikeo, K., Gojobori, T., and Hedges, S. B. (2002) The evolutionary position of nematodes. *BMC Evol. Biol.* **2,** 7.
33. Tagle, D. A., Koop, B. F., Goodman, M., Slightom, J. L., Hess, D. L., and Jones, R. T. (1988) Embryonic epsilon and gamma globin genes of a prosimian primate (Galago crassicaudatus). Nucleotide and amino acid sequences, developmental regulation and phylogenetic footprints. *J. Mol. Biol.* **203,** 439–455.
34. Dermitzakis, E. T., Reymond, A., Lyle, R., Scamuffa, N., et al. (2002) Numerous potentially functional but non-genic conserved sequences on human chromosome 21. *Nature* **420,** 578–582.
35. Wasserman, W. W. and Sandelin, A. (2004) Applied bioinformatics for the identification of regulatory elements. *Nat. Rev. Genet.* **5,** 276–287.
36. Muller, F., Blader, P., and Strahle, U. (2002). Search for enhancers: teleost models in comparative genomic and transgenic analysis of cis regulatory elements. *Bioessays* **24,** 564–572.
37. Nardone, J., Lee, D. U., Ansel, K. M., and Rao, A. (2004) Bioinformatics for the 'bench biologist': how to find regulatory regions in genomic DNA. *Nat. Immunol.* **5,** 768–774.
38. Boffelli, D., Nobrega, M. A., and Rubin, E. M. (2004). Comparative genomics at the vertebrate extremes. *Nat. Rev. Genet.* **5,** 456–465.
39. Lemos, B., Yunes, J. A., Vargas, F. R., Moreira, M. A., Cardoso, A. A., and Seuanez, H. N. (2004) Phylogenetic footprinting reveals extensive conservation of Sonic Hedgehog (SHH) regulatory elements. *Genomics* **84,** 511–523.
40. Woolfe, A., Goodson, M., Goode, D. K., et al. (2004) Highly conserved non-coding sequences are associated with vertebrate development. *PLoS Biol.* **3,** e7.
41. Goode, D. K., Snell, P., Smith, S. F., Cooke, J. E., and Elgar, G. (2005) Highly conserved regulatory elements around the SHH gene may contribute to the

maintenance of conserved synteny across human chromosome 7q36.3. *Genomics* **86,** 172–181.

42. Jeong, Y. and Epstein, D. J. (2003) Distinct regulators of Shh transcription in the floor plate and notochord indicate separate origins for these tissues in the mouse node. *Development* **130,** 3891–3902.

43. Schwartz, S., Kent, W. J., Smit, A., et al. (2003) Human-mouse alignments with BLASTZ. *Genome Res.* **13,** 103–107.

44. Brudno, M., Do, C. B., Cooper, G. M., et al. (2003) LAGAN and Multi-LAGAN: efficient tools for large-scale multiple alignment of genomic DNA. *Genome Res.* **13,** 721–731.

45. Pollard, D. A., Bergman, C. M., Stoye, J., Celniker, S. E., and Eisen, M. B. (2004) Benchmarking tools for the alignment of functional noncoding DNA. *BMC Bioinformatics* **5,** 6.

46. Schwartz, S., Zhang, Z., Frazer, K. A., et al. (2000) PipMaker—a web server for aligning two genomic DNA sequences. *Genome Res.* **10,** 577–586.

47. Ovcharenko, I., Nobrega, M. A., Loots, G. G., and Stubbs, L. (2004) ECR Browser: a tool for visualizing and accessing data from comparisons of multiple vertebrate genomes. *Nucleic Acids Res.* **32**(Web Server issue), W280–W286.

48. Mayor, C., Brudno, M., Schwartz, J. R., et al. (2000) VISTA: visualizing global DNA sequence alignments of arbitrary length. *Bioinformatics* **16,** 1046–1047.

49. Frazer, K. A., Pachter, L., Poliakov, A., Rubin, E. M., and Dubchak, I. (2004) VISTA: computational tools for comparative genomics. *Nucleic Acids Res.* **32**(Web Server issue), W273–W279.

50. Brudno, M., Steinkamp, R., and Morgenstern, B. (2004) The CHAOS/DIALIGN WWW server for multiple alignment of genomic sequences. *Nucleic Acids Res.* **32**(Web Server issue), W41–W44.

51. Morgenstern, B. (1999) DIALIGN 2: improvement of the segment-to-segment approach to multiple sequence alignment. *Bioinformatics* **15,** 211–218.

52. Ovcharenko, I., Boffelli, D., and Loots, G. G. (2004) eShadow: a tool for comparing closely related sequences. *Genome Res.* **14,** 1191–1198.

53. Cooper, G. M. and Sidow, A. (2003) Genomic regulatory regions: insights from comparative sequence analysis. *Curr. Opin. Genet. Dev.* **13,** 604–610.

54. Hardison, R. C. (2000) Conserved noncoding sequences are reliable guides to regulatory elements. *Trends Genet.* **16,** 369–372.

55. Oeltjen, J. C., Malley, T. M., Muzny, D. M., Miller, W., Gibbs, R. A., and Belmont, J. W. (1997) Large-scale comparative sequence analysis of the human and murine Bruton's tyrosine kinase loci reveals conserved regulatory domains. *Genome Res.* **7,** 315–329.

56. Brickner, A. G., Koop, B. F., Aronow, B. J., and Wiginton, D. A. (1999) Genomic sequence comparison of the human and mouse adenosine deaminase gene regions. *Mamm Genome* **10,** 95–101.

57. Lenhard, B., Sandelin, A., Mendoza, L., Engstrom, P., Jareborg, N., and Wasserman, W. W. (2003) Identification of conserved regulatory elements by comparative genome analysis. *J. Biol.* **2,** 13.

58. Suzuki, Y., Yamashita, R., Shirota, M., et al. (2004) Sequence comparison of human and mouse genes reveals a homologous block structure in the promoter regions. *Genome Res.* **14**, 1711–1718.
59. Tautz, D. (2000) Evolution of transcriptional regulation. *Curr. Opin. Genet. Dev.* **10**, 575–579.
60. Thomas, J. W., Touchman, J. W., Blakesley, R. W., et al. (2003) Comparative analyses of multi-species sequences from targeted genomic regions. *Nature* **424**, 788–793.
61. Plessy, C., Dickmeis, T., Chalmel, F., and Strähle, U. (2005) Enhancer sequence conservation between vertebrates is favoured in developmental regulator genes. *Trends Genet.* **21**, 207–210.
62. Muller, F., Chang, B., Albert, S., Fischer, N., Tora, L., and Strahle, U. (1999) Intronic enhancers control expression of zebrafish sonic Hedgehog in floor plate and notochord. *Development* **126**, 2103–2116.
63. Epstein, D. J., McMahon, A. P., and Joyner, A. L. (1999) Regionalization of Sonic Hedgehog transcription along the anteroposterior axis of the mouse central nervous system is regulated by Hnf3-dependent and -independent mechanisms. *Development* **126**, 281–292.
64. Goode, D. K., Snell, P. K., and Elgar, G. K. (2003) Comparative analysis of vertebrate Shh genes identifies novel conserved non-coding sequence. *Mamm Genome* **14**, 192–201.
65. Adams, M. D. (2005) Conserved sequences and the evolution of gene regulatory signals. *Curr. Opin. Genet. Dev.* **15**, 628–633.
66. Osborne, C. S., Chakalova, L., Brown, K. E., et al. (2004) Active genes dynamically colocalize to shared sites of ongoing transcription. *Nat. Genet.*
67. Force, A., Lynch, M., Pickett, F. B., Amores, A., Yan, Y. L., and Postlethwait, J. (1999) Preservation of duplicate genes by complementary, degenerate mutations. *Genetics* **151**, 1531–1545.
68. Vavouri, T., McEwen, G. K., Woolfe, A., Gilks, W. R., and Elgar, G. (2006) Defining a genomic radius for long-range enhancer action: duplicated conserved non-coding elements hold the key. *Trends Genet.* **22**, 5–10.
69. Lettice, L. A., Heaney, S. J., Purdie, L. A., et al. (2003) A long-range Shh enhancer regulates expression in the developing limb and fin and is associated with preaxial polydactyly. *Hum. Mol. Genet.* **12**, 1725–1735.
70. Sagai, T., Hosoya, M., Mizushina, Y., Tamura, M., and Shiroishi, T. (2005) Elimination of a long-range cis-regulatory module causes complete loss of limb-specific Shh expression and truncation of the mouse limb. *Development* **132**, 797–803.
71. Mackenzie, A., Miller, K. A., and Collinson, J. M. (2004) Is there a functional link between gene interdigitation and multi-species conservation of synteny blocks? *Bioessays* **26**, 1217–1224.
72. Flint, J., Tufarelli, C., Peden, J., et al. (2001) Comparative genome analysis delimits a chromosomal domain and identifies key regulatory elements in the alpha globin cluster. *Hum. Mol. Genet.* **10**, 371–382.

73. Halling-Brown, M., Sansom, C., Moss, D. S., Elgar, G., and Edwards, Y. J. (2004) A Fugu-Human Genome Synteny Viewer: web software for graphical display and annotation reports of synteny between Fugu genomic sequence and human genes. *Nucleic Acids Res.* **32**, 2618–2622.

74. Butler, J. E. and Kadonaga, J. T. (2002) The RNA polymerase II core promoter: a key component in the regulation of gene expression. *Genes Dev.* **16**, 2583–2592.

75. Hashimoto, S., Suzuki, Y., Kasai, Y., et al. (2004) 5′-end SAGE for the analysis of transcriptional start sites. *Nat. Biotechnol.* **22**, 1146–1149.

76. Kawaji, H., Kasukawa, T., Fukuda, S., et al. (2006) CAGE Basic/Analysis Databases: the CAGE resource for comprehensive promoter analysis. *Nucleic Acids Res.* **34** (Database issue), D632–D636.

77. FitzGerald, P. C., Shlyakhtenko, A., Mir, A. A., and Vinson, C. (2004) Clustering of DNA sequences in human promoters. *Genome Res.* **14**, 1562–1574.

78. Kadonaga, J. T. (2002) The DPE, a core promoter element for transcription by RNA polymerase II. *Exp. Mol. Med.* **34**, 259–264.

79. Kitazawa, S., Kitazawa, R., Tamada, H., and Maeda, S. (1998) Promoter structure of human sonic Hedgehog gene. *Biochim. Biophys. Acta* **1443**, 358–363.

80. Chang, B. E., Blader, P., Fischer, N., Ingham, P. W., and Strahle, U. (1997) Axial (HNF3beta) and retinoic acid receptors are regulators of the zebrafish sonic Hedgehog promoter. *EMBO J.* **16**, 3955–3964.

81. Suzuki, Y., Yamashita, R., Sugano, S., and Nakai, K. (2004) DBTSS, DataBase of Transcriptional Start Sites: progress report 2004. *Nucleic Acids Res.* **32**(Database issue), D78–D81.

82. Wingender, E., Dietze, P., Karas, H., and Knuppel, R. (1996) TRANSFAC: a database on transcription factors and their DNA binding sites. *Nucleic Acids Res.* **24**, 238–241.

83. Sandelin, A., Alkema, W., Engstrom, P., Wasserman, W. W., and Lenhard, B. (2004) JASPAR: an open-access database for eukaryotic transcription factor binding profiles. *Nucleic Acids Res.* **32**(Database issue), D91–D94.

84. Arnone, M. I. and Davidson, E. H. (1997) The hardwiring of development: organization and function of genomic regulatory systems. *Development* **124**, 1851–1864.

85. Markstein, M., Markstein, P., Markstein, V., and Levine, M. S. (2002) Genomewide analysis of clustered Dorsal binding sites identifies putative target genes in the Drosophila embryo. *Proc. Natl Acad. Sci. USA* **99**, 763–768.

86. Stathopoulos, A., Van Drenth, M., Erives, A., Markstein, M., and Levine, M. (2002) Whole-genome analysis of dorsal-ventral patterning in the Drosophila embryo. *Cell* **111**, 687–701.

87. Rebeiz, M., Reeves, N. L., and Posakony, J. W. (2002) SCORE: a computational approach to the identification of cis-regulatory modules and target genes in whole-genome sequence data. Site clustering over random expectation. *Proc. Natl Acad. Sci. USA* **99**, 9888–9893.

88. Berman, B. P., Nibu, Y., Pfeiffer, B. D., et al. (2002) Exploiting transcription factor binding site clustering to identify cis-regulatory modules involved in

pattern formation in the Drosophila genome. *Proc. Natl Acad. Sci. USA* **99**, 757–762.

89. Halfon, M. S., Grad, Y., Church, G. M., and Michelson, A. M. (2002) Computation-based discovery of related transcriptional regulatory modules and motifs using an experimentally validated combinatorial model. *Genome Res.* **12**, 1019–1028.

90. Erives, A. and Levine, M. (2004). Coordinate enhancers share common organizational features in the Drosophila genome. *Proc. Natl Acad. Sci. USA* **101**, 3851–3856.

91. Markstein, M., Zinzen, R., Markstein, P., et al. (2004) A regulatory code for neurogenic gene expression in the Drosophila embryo. *Development* **131**, 2387–2394.

92. Berman, B. P., Pfeiffer, B. D., Laverty, T. R., et al. (2004) Computational identification of developmental enhancers: conservation and function of transcription factor binding-site clusters in Drosophila melanogaster and Drosophila pseudoobscura. *Genome Biol.* **5**, R61.

93. Rajewsky, N., Vergassola, M., Gaul, U., and Siggia, E. D. (2002) Computational detection of genomic cis-regulatory modules applied to body patterning in the early Drosophila embryo. *BMC Bioinformatics* **3**, 30.

94. Schroeder, M. D., Pearce, M., Fak, J., et al. (2004) Transcriptional control in the segmentation gene network of Drosophila. *PLoS Biol.* **2**, E271.

95. Sinha, S., van Nimwegen, E., and Siggia, E. D. (2003) A probabilistic method to detect regulatory modules. *Bioinformatics* **19**(Suppl 1), i292–i301.

96. Sinha, S., Schroeder, M. D., Unnerstall, U., Gaul, U., and Siggia, E. D. (2004) Cross-species comparison significantly improves genome-wide prediction of cis-regulatory modules in Drosophila. *BMC Bioinformatics* **5**, 129.

97. Bigelow, H. R., Wenick, A. S., Wong, A., and Hobert, O. (2004) CisOrtho: a program pipeline for genome-wide identification of transcription factor target genes using phylogenetic footprinting. *BMC Bioinformatics* **5**, 27.

98. Senger, K., Armstrong, G. W., Rowell, W. J., Kwan, J. M., Markstein, M., and Levine, M. (2004) Immunity regulatory DNAs share common organizational features in Drosophila. *Mol. Cell.* **13**, 19–32.

99. Jegga, A. G., Sherwood, S. P., Carman, J. W., et al. (2002) Detection and visualization of compositionally similar cis-regulatory element clusters in orthologous and coordinately controlled genes. *Genome Res.* **12**, 1408–1417.

100. Loots, G. G. and Ovcharenko, I. (2004) rVISTA 2.0: evolutionary analysis of transcription factor binding sites. *Nucleic Acids Res.* **32**(Web Server issue), W217–W221.

101. Loots, G. G., Ovcharenko, I., Pachter, L., Dubchak, I., and Rubin, E. M. (2002) rVista for comparative sequence-based discovery of functional transcription factor binding sites. *Genome Res.* **12**, 832–839.

102. Berezikov, E., Guryev, V., Plasterk, R. H., and Cuppen, E. (2004) CONREAL: conserved regulatory elements anchored alignment algorithm for identification

of transcription factor binding sites by phylogenetic footprinting. *Genome Res.* **14,** 170–178.

103. Wang, T. and Stormo, G. D. (2003) Combining phylogenetic data with co-regulated genes to identify regulatory motifs. *Bioinformatics* **19,** 2369–2380.

104. Liu, Y., Liu, X. S., Wei, L., Altman, R. B., and Batzoglou, S. (2004) Eukaryotic regulatory element conservation analysis and identification using comparative genomics. *Genome Res.* **14,** 451–458.

105. Blanchette, M. and Tompa, M. (2002) Discovery of regulatory elements by a computational method for phylogenetic footprinting. *Genome Res.* **12,** 739–748.

106. Kent, W. J. and Zahler, A. M. (2000) Conservation, regulation, synteny, and introns in a large-scale *C. briggsae–C. elegans* genomic alignment. *Genome Res.* **10,** 1115–1125.

Index

Printed in the United States of America